Structured Light for Optical Communication

Nanophotonics

Structured Light for Optical Communication

Edited by

Mohammad D. Al-Amri

King Abdulaziz City for Science and Technology (KACST)
Riyadh, Saudi Arabia

David L. Andrews

University of East Anglia
Norwich, United Kingdom

Mohamed Babiker

University of York
York, United Kingdom

Series Editor

Akhlesh Lakhtakia

Elsevier
Radarweg 29, PO Box 211, 1000 AE Amsterdam, Netherlands
The Boulevard, Langford Lane, Kidlington, Oxford OX5 1GB, United Kingdom
50 Hampshire Street, 5th Floor, Cambridge, MA 02139, United States

Copyright © 2021 Elsevier Inc. All rights reserved.

No part of this publication may be reproduced or transmitted in any form or by any means, electronic or mechanical, including photocopying, recording, or any information storage and retrieval system, without permission in writing from the publisher. Details on how to seek permission, further information about the Publisher's permissions policies and our arrangements with organizations such as the Copyright Clearance Center and the Copyright Licensing Agency, can be found at our website: www.elsevier.com/permissions.

This book and the individual contributions contained in it are protected under copyright by the Publisher (other than as may be noted herein).

Notices

Knowledge and best practice in this field are constantly changing. As new research and experience broaden our understanding, changes in research methods, professional practices, or medical treatment may become necessary.

Practitioners and researchers must always rely on their own experience and knowledge in evaluating and using any information, methods, compounds, or experiments described herein. In using such information or methods they should be mindful of their own safety and the safety of others, including parties for whom they have a professional responsibility.

To the fullest extent of the law, neither the Publisher nor the authors, contributors, or editors, assume any liability for any injury and/or damage to persons or property as a matter of products liability, negligence or otherwise, or from any use or operation of any methods, products, instructions, or ideas contained in the material herein.

Library of Congress Cataloging-in-Publication Data
A catalog record for this book is available from the Library of Congress

British Library Cataloguing-in-Publication Data
A catalogue record for this book is available from the British Library

ISBN: 978-0-12-821510-4

For information on all Elsevier publications
visit our website at https://www.elsevier.com/books-and-journals

Publisher: Matthew Deans
Acquisitions Editor: Simon Holt
Editorial Project Manager: Isabella C. Silva
Production Project Manager: Prasanna Kalyanaraman
Designer: Matthew Limbert

Typeset by VTeX

Contents

List of contributors	xi
Preface	xiii
Chapter 1: Basics of quantum communication	**1**
O. Alshehri, Z.-H. Li, and M.D. Al-Amri	
1.1 Introduction	2
1.2 Optical polarization	5
1.3 Dirac's notation	6
1.4 Quantum bits (qubits)	7
1.5 The Bloch sphere	8
1.6 Quantum entanglement and nonlocality	9
1.7 Measurement, decoherence, and irreversibility	11
1.8 Quantum cloning	13
1.9 Quantum communication with single photons	14
1.9.1 Polarization encoding	15
1.9.2 Orbital angular momentum (OAM) encoding	15
1.9.3 Time-bin encoding	16
1.9.4 Path encoding	17
1.9.5 Frequency-bin encoding	19
1.10 Protocols of quantum communications	19
1.10.1 Quantum key distribution (QKD) protocols	20
1.10.2 Quantum teleportation protocol	22
1.10.3 Superdense coding protocol	26
1.11 Ranges of quantum communication	27
1.11.1 Long distance quantum communication	27
1.11.2 Short distance quantum communications	30
1.12 Conclusions	31
Acknowledgments	32
References	32

Contents

Chapter 2: Structured light **37**
    M. Babiker, V.E. Lembessis, Koray Köksal, and J. Yuan

2.1	Introduction	38
2.2	Optical angular momentum	40
2.3	Helmholtz equation and paraxial regime..................................	40
	2.3.1 Linearly polarized 'unstructured' light	40
	2.3.2 Elliptically polarized	41
2.4	Structured light...	42
	2.4.1 Phase-structured light	43
	2.4.2 Laguerre-Gaussian (LG) light beams.............................	44
2.5	Bessel and Bessel-Gaussian vortex beams...............................	46
	2.5.1 Bessel vortex beams	46
	2.5.2 Bessel-Gaussian vortex beams	47
2.6	Non-paraxial LG beams	47
	2.6.1 Extracting the paraxial regime	49
2.7	Paraxial beams with small waists	51
2.8	Chirality and helicity...	52
	2.8.1 Cycle-averaged fields..	52
	2.8.2 Effects of the Gouy and curvature phases	55
2.9	Multiple vortex beams	57
	2.9.1 Linearly polarized LG beams	57
	2.9.2 Axial shift	58
2.10	No axial shift—polarization gradients	61
	2.10.1 Co-propagating LG beams	62
	2.10.2 Counter-propagating beams	65
	2.10.3 Bi-chromatic vortex beams.......................................	67
2.11	Quantization of optical angular momentum	71
	2.11.1 Quantized SAM..	72
	2.11.2 Orbital angular momentum	73
2.12	Conclusions	73
	References	75

Chapter 3: Quantum features of structured light **77**
    David L. Andrews

3.1	Introduction	77
3.2	Basis for the quantization of structured light	79
3.3	Quantum issues in measurement and localization	83
3.4	Quantized angular momentum: light and matter	86
3.5	Entanglement..	88

3.6	Conclusion	89
References		90

Chapter 4: Poincaré beams for optical communications — 95
Enrique J. Galvez, Behzad Khajavi, and Brianna M. Holmes

4.1	Introduction	95
4.2	Vortex Poincaré Gaussian beams	96
4.3	Poincaré–Bessel beams	100
4.4	Asymmetric and monstar patterns	102
4.5	Experimental methods	104
4.6	Discussion	104
Acknowledgments		104
References		105

Chapter 5: Operators in paraxial quantum optics — 107
Gerard Nienhuis

5.1	Introduction		107
5.2	Quantization and conserved quantities of Maxwell field		109
	5.2.1	Discretized plane-wave modes	109
	5.2.2	Continuum of plane-wave modes	110
	5.2.3	Operators for conserved quantities of radiation field	111
5.3	Paraxial quantum fields		114
	5.3.1	Change of variables	114
	5.3.2	Paraxial wave equation	114
	5.3.3	Paraxial limit of quantum field	115
	5.3.4	Discrete transverse modes	116
	5.3.5	Algebra of continuum of bosonic operators	118
5.4	Paraxial modes and harmonic oscillators		119
	5.4.1	Hermite–Gauss modes	120
	5.4.2	Correspondence between paraxial modes and harmonic-oscillator states	121
	5.4.3	Laguerre–Gauss modes	122
5.5	Paraxial energy, momentum and angular momentum		122
5.6	Operator description of Gaussian paraxial modes		124
	5.6.1	Operator description of Hermite–Gauss modes	124
	5.6.2	Operator description of Laguerre–Gauss modes	126
	5.6.3	Elliptical Gaussian modes	128
5.7	Schwinger representation of Laguerre–Gauss modes		130
	5.7.1	The Lie algebra su(2)	130
	5.7.2	The Lie algebra su(1, 1)	132

Contents

5.8	Conclusions	134
References		136

Chapter 6: Quantum cryptography with structured photons ... 139
Alicia Sit, Felix Hufnagel, and Ebrahim Karimi

6.1	Introduction		139
6.2	Generation and detection		143
	6.2.1	Polarization	143
	6.2.2	Holography	145
	6.2.3	Pancharatnam–Berry optical elements	147
6.3	High-dimensional quantum information		149
	6.3.1	Optimal quantum cloning	149
	6.3.2	Protocols	152
	6.3.3	Quantum process tomography	155
6.4	Quantum key distribution implementations		159
	6.4.1	Optical fiber	159
	6.4.2	Free-space	162
	6.4.3	Underwater	167
6.5	Conclusion		173
Acknowledgment			173
References			173

Chapter 7: Spin and orbital angular momentum coupling ... 177
Lorenzo Marrucci

7.1	Introduction	177
7.2	Paraxial spin-orbit coupling: q-plates, meta-surfaces and similar devices	181
7.3	Non-paraxial spin-orbit coupling: spin Hall effect of light and optical fibers	187
7.4	Applications to optical communication	193
7.5	Conclusions	200
References		200

Chapter 8: Quantum communication with structured photons ... 205
Robert Fickler and Shashi Prabhakar

8.1	Introduction		206
8.2	Quantum protocols		208
	8.2.1	Information capacity, dense coding and noise resistance	208
	8.2.2	Quantum key distribution	209
	8.2.3	Quantum coin tossing	210
	8.2.4	Quantum secret sharing	211

	8.2.5	Layered quantum key distribution	212
8.3	Experimental toolbox		215
	8.3.1	Generation and detection methods	215
	8.3.2	Modulation methods	217
8.4	Quantum network		219
	8.4.1	Entanglement sources	219
	8.4.2	Quantum channels	223
	8.4.3	Quantum repeater	228
	8.4.4	Quantum interfaces	231
	8.4.5	Quantum router	231
8.5	Conclusion		233
Acknowledgments			233
References			233

Chapter 9: Optical angular momentum interaction with turbulent and scattering media 237

Mingjian Chen and Martin Lavery

9.1	Atmospheric turbulence variations in real environments	238
9.2	Turbulence-induced phase variations	241
9.3	Turbulence's effect on structured beams	244
9.4	Degradation of beams that carry OAM	246
9.5	Scattering dynamics of beams that carry OAM	254
9.6	Conclusions	257
References		257

Chapter 10: Causes and mitigation of modal crosstalk in OAM multiplexed optical communication links 259

Alan E. Willner, Haoqian Song, Cong Liu, Runzhou Zhang, Kai Pang, Huibin Zhou, Nanzhe Hu, Hao Song, Xinzhou Su, Zhe Zhao, Moshe Tur, Hao Huang, Guodong Xie, and Yongxiong Ren

10.1	Introduction and overview		260
10.2	Causes for channel crosstalk in an OAM multiplexed link		263
	10.2.1	Atmospheric turbulence	263
	10.2.2	Misalignment	264
	10.2.3	Obstruction	265
	Summary		266
10.3	Adaptive optics (AO) for crosstalk (XT) mitigation		266
	10.3.1	AO using wavefront sensor (WFS) and Gaussian probe beam	266
	10.3.2	AO using WFS and Gaussian probe beam in a quantum communication link	269

		10.3.3	AO using camera for beam intensity measurement	271
		10.3.4	Simultaneous demultiplexing and XT mitigation by using multi-plane light converter (MPLC)	273
		Summary		274
	10.4	Spatial modes manipulation for crosstalk mitigation		275
		10.4.1	Turbulence pre-compensation by OAM mode combination	275
		10.4.2	Simultaneous orthogonalizing and shaping of multiple LG beams	277
		10.4.3	Utilizing Bessel-Gaussian (BG) beams with non-zero OAM order	279
		Summary		280
	10.5	Digital signal processing for crosstalk mitigation		280
		10.5.1	MIMO equalization for crosstalk mitigation in laboratory	281
		10.5.2	MIMO equalization for crosstalk mitigation in the link through a flying UAV	282
		Summary		284
	10.6	Summary		284
	Acknowledgment			284
	References			284
Index				291

List of contributors

M.D. Al-Amri NCQOQI, King Abdul-Aziz City for Science and Technology (KACST), Riyadh, Saudi Arabia
O. Alshehri Industrial Engineering Department, King Saud University, Riyadh, Saudi Arabia
David L. Andrews University of East Anglia, Norwich Research Park, Norwich, United Kingdom
M. Babiker Department of Physics, University of York, York, United Kingdom
Mingjian Chen University of Glasgow, Glasgow, United Kingdom
Xidian University, Xi'an, Shaanxi, China
Robert Fickler Tampere University, Physics Unit, Tampere, Finland
Enrique J. Galvez Colgate University, Department of Physics and Astronomy, Hamilton, NY, United States
Brianna M. Holmes Colgate University, Department of Physics and Astronomy, Hamilton, NY, United States
University of Rochester, Institute of Optics, Rochester, NY, United States
Nanzhe Hu University of Southern California, Los Angeles, CA, United States
Hao Huang University of Southern California, Los Angeles, CA, United States
Felix Hufnagel University of Ottawa, Department of Physics, Ottawa, ON, Canada
Ebrahim Karimi University of Ottawa, Department of Physics, Ottawa, ON, Canada
Behzad Khajavi Colgate University, Department of Physics and Astronomy, Hamilton, NY, United States
University of Houston, Department of Biomedical Engineering, Houston, TX, United States
Koray Köksal Physics Department, Bitlis Eren University, Bitlis, Turkey
Martin Lavery University of Glasgow, Glasgow, United Kingdom
V.E. Lembessis Quantum Technology Group, Department of Physics and Astronomy, College of Science, King Saud University, Riyadh, Saudi Arabia
Z.-H. Li Department of Physics, Shanghai University, Shanghai, China
Cong Liu University of Southern California, Los Angeles, CA, United States
Lorenzo Marrucci Dipartimento di Fisica "Ettore Pancini", Università di Napoli Federico II, Napoli, Italy
Gerard Nienhuis Huygens-Kamerlingh Onnes Laboratory, Leiden University, Leiden, the Netherlands
Kai Pang University of Southern California, Los Angeles, CA, United States

List of contributors

Shashi Prabhakar Tampere University, Physics Unit, Tampere, Finland
Yongxiong Ren University of Southern California, Los Angeles, CA, United States
Alicia Sit University of Ottawa, Department of Physics, Ottawa, ON, Canada
Hao Song University of Southern California, Los Angeles, CA, United States
Haoqian Song University of Southern California, Los Angeles, CA, United States
Xinzhou Su University of Southern California, Los Angeles, CA, United States
Moshe Tur Tel Aviv University, Ramat Aviv, Israel
Alan E. Willner University of Southern California, Los Angeles, CA, United States
Guodong Xie University of Southern California, Los Angeles, CA, United States
J. Yuan Department of Physics, University of York, York, United Kingdom
Runzhou Zhang University of Southern California, Los Angeles, CA, United States
Zhe Zhao University of Southern California, Los Angeles, CA, United States
Huibin Zhou University of Southern California, Los Angeles, CA, United States

Preface

Throughout history, the most enduring transformations of human society have often resulted from technological innovation. The massive developments that occurred in the twentieth century, especially in the realm of communications and IT, primarily owe their origin to the birth of electronics. At the turn of the century in 1900, vacuum tube electronics had just arrived, though almost half a century elapsed before semiconductor components fully unleashed the latent potential. That century is already beginning to be seen as the electronics century – 'the era of the electron'.

The vision that the future – our present – might be as much transformed by a field to be called 'photonics' was a mere pipe dream at that point. But the acceleration of optical science and technology was so rapid that this was already becoming a reality before the dawn of the new millennium. With far greater reason, this new era has already been hailed as the 'century of the photon'. Communications and IT are again at the forefront of the revolution, with major advances in optical methods for data handling, transmission and processing.

Apart from obvious advantages such as speed of transmission and lower losses, light quanta offer attributes that go well beyond the capacity of the electron. Two such aspects have prompted an outstanding burst of research activity; the capacity to readily engineer and encode states of a specifically quantum logical nature, and development of the field known as 'structured' or 'complex' light, strongly associated with singular optics, 'twisted light' and optical vortices. In each respect, photonics offers transformational prospects that are already rapidly moving into commercial applications.

Prior to the advent of structured light, developments in optical quantum communication relied primarily on optical spin angular momentum, providing two degrees of freedom for the realization of qubit, quantum logic states. This polarization feature itself proved remarkably successful, as proven for example in the achievement of secure quantum cryptography. However, a quantum state of phase structured light, as in an optical vortex, is further endowed with the degree of freedom represented by an integer topological charge, associated with orbital angular momentum, in addition to spin.

For this reason, structured light has been recognized as affording outstanding prospects for the realization of high bandwidth communication, enhanced tools for more highly secure cryptography, and exciting opportunities for reliable quantum computing. Since the topological

Preface

charge of such a beam can have either positive or negative values, only technical limitations placing any upper bound on its magnitude, this discrete nature in principle suggests using these quantum states as a platform for implementing high-dimensional generalizations of the familiar qubit states, as well as offering an enhanced capacity for optical information transfer.

The main aim of this book is to highlight the wide-ranging principles, new frontiers, and applications in this rapidly evolving field. The first three chapters, contributed by the editors and their collaborators, provide a background on the basics of quantum information, address the fundamental principles of structured light in its most common forms, and then focus on specifically quantum aspects of these optical modes. The next chapter by Galvez et al. introduces another key area of development: modes that are structured in their polarization, representing another dimension of complexity across the beam profile. Then follows a chapter by Nienhuis, exploring an operator formalism for the quantum optical states of paraxial beams. The exploitation of specifically quantum aspects of structured light in secure communications is featured in the chapter on cryptography by Sit et al., and Marrucci's chapter describes the intricacies that arise when spin and orbital angular momentum are no longer separable, their interplay producing many additional and novel effects.

The principles and practicality of communication with structured light are then surveyed in the chapter by Fickler and Prabhakar, which also draw attention to the very different kinds of propagation issues that arise according to the transmission medium. The concluding chapters, by Chen and Lavery, and by Wilner et al., describe the characteristic effects associated with propagation through real, non-ideal media, where complications of turbulence or scattering arise, and cross-talk between modes may become a problem. Practical strategies to optimize the fidelity of data transmission are clearly of key significance for the advantages of using structured light to be fully secured.

It is a great pleasure to thank the many experts who have contributed to this volume, whose authoritative chapters provide an up to date and extensive survey of the subject, richly studded with insights. Their enthusiasm for this project has been immensely encouraging to us, and we are greatly indebted to them for their glad involvement. Our hope is that the compilation before you will serve not only as a resource, but also an inspiration for the many who now and in the future will be joining this exciting field.

M.D. Al-Amri[a], Riyadh
D.L. Andrews, Norwich
M. Babiker, York

December 2020

---

[a] M.D. Al-Amri was involved, along with the co-editors, in editorial and reviewing work only in relation to Chapters 2 to 6.

CHAPTER 1

Basics of quantum communication

O. Alshehri[a,d], Z.-H. Li[b,e], and M.D. Al-Amri[c,f]

[a]Industrial Engineering Department, King Saud University, Riyadh, Saudi Arabia [b]Department of Physics, Shanghai University, Shanghai, China [c]NCQOQI, King Abdul-Aziz City for Science and Technology (KACST), Riyadh, Saudi Arabia

Contents

1.1 Introduction 2
1.2 Optical polarization 5
1.3 Dirac's notation 6
1.4 Quantum bits (qubits) 7
1.5 The Bloch sphere 8
1.6 Quantum entanglement and nonlocality 9
1.7 Measurement, decoherence, and irreversibility 11
1.8 Quantum cloning 13
1.9 Quantum communication with single photons 14
    1.9.1 Polarization encoding 15
    1.9.2 Orbital angular momentum (OAM) encoding 15
    1.9.3 Time-bin encoding 16
    1.9.4 Path encoding 17
    1.9.5 Frequency-bin encoding 19
1.10 Protocols of quantum communications 19
    1.10.1 Quantum key distribution (QKD) protocols 20
        1.10.1.1 *The BB84 protocol* 20
        1.10.1.2 *The E91 protocol* 21
    1.10.2 Quantum teleportation protocol 22
    1.10.3 Superdense coding protocol 26
1.11 Ranges of quantum communication 27
    1.11.1 Long distance quantum communication 27
        1.11.1.1 *Quantum Internet* 29
    1.11.2 Short distance quantum communications 30
1.12 Conclusions 31

---

[d] ORCID: https://orcid.org/0000-0003-1986-0556.
[e] ORCID: https://orcid.org/0000-0003-1870-210X.
[f] ORCID: https://orcid.org/0000-0002-7679-3620.

Structured Light for Optical Communication
https://doi.org/10.1016/B978-0-12-821510-4.00007-8
Copyright © 2021 Elsevier Inc. All rights reserved.

Acknowledgments 32
References 32

1.1 Introduction

With the advent and relentless expansion of the internet as a global information resource, our need to communicate and exchange information has become more than ever dependent on the online-enabled world. The demand for secure information storage and transfer is now involved in all walks of life, from individual's health to commercial and national security data. Information requires not only an increasingly more secure, but also ever faster, means of communication, and such a need will continue to be of paramount importance in the foreseeable future. Recent efforts have been directed at potentially achieving reliable quantum communication, hailed as a promising platform for the fast exchange of information, securely and reliably. The development of this technology demands clear outperformance of current technologies.

This chapter aims to provide an overview of the essentials of quantum communication, so it seems appropriate to begin with a brief note of some important history. The theory of communication is traditionally a mathematical discipline, pioneered within the field of telecommunications by Shannon [1]. Technological challenges became strikingly apparent in what has become well known as Moore's law [2]. This indicates that the complexity of electronic circuits doubles approximately every 18 months, suggesting an inexorable path of size reduction towards atomic scale dimensions, where quantum laws come into play. Feynman [3], followed by Benioff [4] and Manin [5] as first instances highlighted the relationship between computers and quantum physics. Deutsch showed the relevance of quantum physics to the theory of computation [6]. However, the turning point for communication took place when both Bennett and Brassard proposed the protocol, now known as BB84, for implementation of Quantum Key Distribution (QKD) [7]. This work was followed shortly after by Ekert, who proposed another protocol based on entanglement, E91 [8]. These developments fostered the emergence of quantum cryptography—still a growing area of research both theoretically and experimentally, but substantial progress has been made. In many labs around the world we now see experimental effort gradually moving from proof-of-principle demonstrations to in-field implementations and technological prototypes.

There are some fundamental quantum characteristics envisaged to be behind the success of quantum communication in terms of performance over classical communication, and these include: (i) quantum uncertainty, which dictates that measuring one property of a particle inherently weakens the accuracy in measuring its conjugate property at the same time; (ii) superposition, indicating that a particle can be in a superposition of multiple states at the same time; (iii) entanglement, a correlation between particles that does not have a counterpart in the classical world. However, the key difference between classical and quantum communication was most clearly marked in the development of the no-cloning theorem by Wootters and Zurek [9], and separately by Dieks [10].

Classically, the basic unit of information is the bit: a binary unit that allows information to be stored in values either 0 or 1. Quantum mechanically, the basic unit of quantum information is the qubit, an abstract mathematical representation in two-dimensional Hilbert space. There is a broad range of physical systems that serve as candidates to be utilized as a two-level quantum system in order to carry out quantum communication tasks. These include, but are not limited to: (1) photon polarization states; (2) the spin states of an electron; (3) the direction of a vortex in a superconductor; (4) the different energy levels of an atom, ion, molecule or even quantum dot. Fig. 1.1 presents some more examples of those physical representations [11].

The choice of photon polarization for implementation of qubit communication is advocated for several reasons. Some of the most important are: (a) Classical communication is already generally operated using optical pulses of light, sent through either optical fibers or through free space to perform communications tasks. Whereas millions of photons are routinely sent through such channels, quantum information may be carried by single photons. (b) Photons can be transmitted quickly and with low noise from the sender to a distant receiver. (c) Each individual photon has its intrinsic quantum state, which can be controlled by interferometry, for example. (d) The propagation of light through designed channels is commonly associated with very low losses, so the energy cost is low.

This chapter is primarily focused on photon-based qubits, using two-dimensional Hilbert space to illustrate concepts and techniques that should facilitate an understanding of more complex ideas and techniques in the context of a higher-dimensional Hilbert space; the latter is the subject of coverage by contributors in subsequent chapters of this book. The case of three-dimensional space involves "qutrit" states, while in the case of a higher-dimensional Hilbert space we have "qudit" states. Having a high-dimensional Hilbert space may prove advantageous for quantum information as a whole and particularly for quantum communication, as it facilitates a larger information transfer rate and also, surprisingly, noise resilience. One physical system that can represent this expansion of Hilbert space to higher dimensions is provided by structured light in the form of optical vortices as states carrying orbital angular momentum (OAM). This physically offers higher photonic degrees of freedom, capable of spanning high-dimensional quantum states [12–15].

The main body of the chapter in outline is as follows. We first introduce the basics and key concepts starting with a brief introduction of optical polarization and how it can be used to represent the photonic qubit and its graphical representation using the Bloch sphere. This is

Superconducting loops

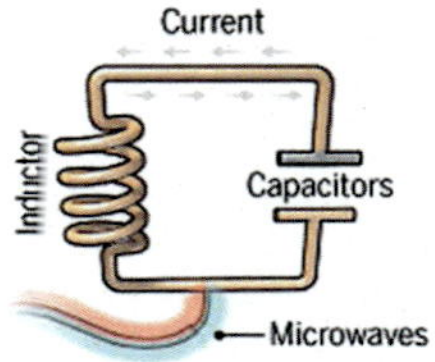

A resistance-free current oscillates back and forth around a circuit loop. An injected microwave signal excites the current into super-position states.

Longevity (seconds) 0.00005
Logic success rate 99.4%
Number entangled 9

Company support
Google, IBM, Quantum Circuits

⊕ **Pros**
Fast working. Build on existing semiconductor industry.

⊖ **Cons**
Collapse easily and must be kept cold.

Trapped ions

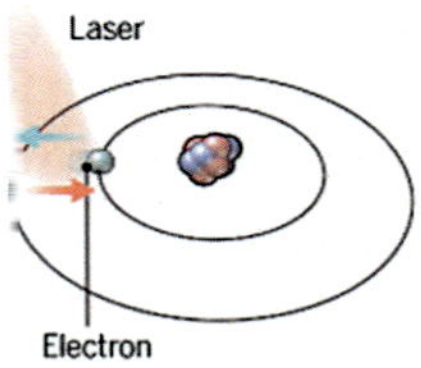

Electrically charged atoms, or ions, have quantum energies that depend on the location of electrons. Tuned lasers cool and trap the ions, and put them in super-position states.

Longevity (seconds) >1000
Logic success rate 99.9%
Number entangled 14

Company support
ionQ

⊕ **Pros**
Very stable. Highest achieved gate fidelities.

⊖ **Cons**
Slow operation. Many lasers are needed.

Silicon quantum dots

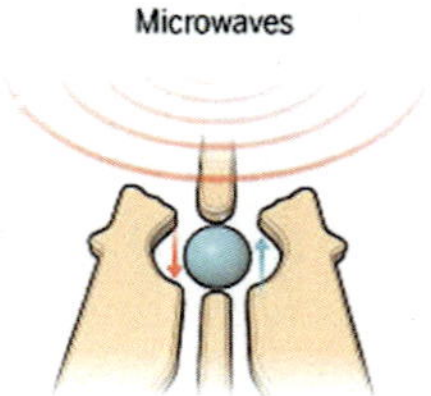

These "artificial atoms" are made by adding an electron to a small piece of pure silicon. Microwaves control the electron's quantum state.

Longevity (seconds) 0.03
Logic success rate ~99%
Number entangled 2

Company support
Intel

⊕ **Pros**
Stable. Build on existing semiconductor industry.

⊖ **Cons**
Only a few entangled. Must be kept cold.

Topological qubits

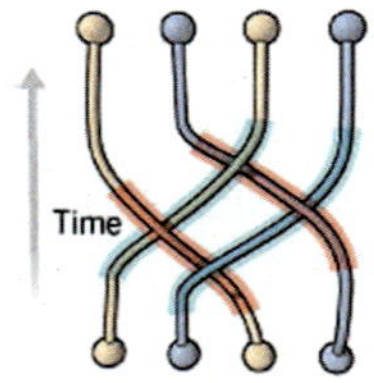

Quasiparticles can be seen in the behavior of electrons channeled through semiconductor structures. Their braided paths can encode quantum information.

Longevity (seconds) N/A
Logic success rate N/A
Number entangled N/A

Company support
Microsoft, Bell Labs

⊕ **Pros**
Greatly reduce errors.

⊖ **Cons**
Existence not yet confirmed.

Diamond vacancies

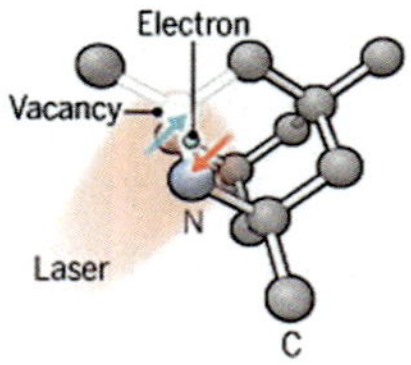

A nitrogen atom and a vacancy add an electron to a diamond lattice. Its quantum spin state, along with those of nearby carbon nuclei, can be controlled with light.

Longevity (seconds) 10
Logic success rate 99.2%
Number entangled 6

Company support
Quantum Diamond Technologies

⊕ **Pros**
Can operate at room temperature.

⊖ **Cons**
Difficult to entangle.

Note: Longevity is the record coherence time for a single qubit superposition state, logic success rate is the highest reported gate fidelity for logic operations on two qubits, and number entangled is the maximum number of qubits entangled and capable of performing two-qubit operations.

Figure 1.1: Although the emphasis is on the photon qubits, the figure highlights the breadth of physical systems that could form platforms for quantum information systems. Reprinted figure with permission from [11], ©AAAS (2020).

followed by discussions of key concepts involved in this context, including quantum entanglement, measurement, decoherence and quantum cloning. We then describe how quantum communication can be encoded using a single photon and the different techniques that are used. This is followed by the presentation of quantum communication protocols including quantum key distribution protocols, teleportation, and superdense coding, before turning to a discussion of the ranges of quantum communication that are already in action.

1.2 Optical polarization

Light is an oscillating electromagnetic field composed of an electric vector field **E** and a magnetic vector field **H**. The state of optical polarization is determined by the relative orientation and phase of these two fields—which, for freely propagating light, are both perpendicular to the propagation direction. There are three common types of polarization: linear, circular, and elliptical.

Linear polarizations

If the electric and magnetic field vectors of light differ in phase by 90°, light/photon can be in either of the following state: vertically polarized; horizontally polarized and slant-polarized which occurs when the electric field vector makes an angle $\pm\theta$ with the horizontal axis. Typically, the most used angle is the $\pm 45°$.

Circular polarization

In circularly polarized light the electric field vector changes direction as it propagates in a circular fashion (alternating between horizontal and vertical polarization). The tip of the vector appears to be moving in a circle as it leaves the light source; hence it is called circular polarization. The electric field vector makes one complete revolution in an axial distance equal to its wavelength. Circular polarization has two types: right-handed and left-handed. Note that this classification is relative to the observer. It can either be right-handed or left-handed by an observer that sees the photon coming toward him/her. Here, we define it from the point of the source (conventions in optics do vary). Right-handed: the photon is to be called right-handed if the electric field vector tip moves in clockwise direction, while, left-handed if it moves in an anticlockwise direction. See Fig. 1.2.

Note that circularly polarized light rotates around the axis of propagation. Its electric field component rotates in time with angular frequency ω, while rotates in space with angular wave-vector k. Hence, circularly polarized light has angular momentum [16–18], but linearly polarized light carries no angular momentum.

6 Chapter 1

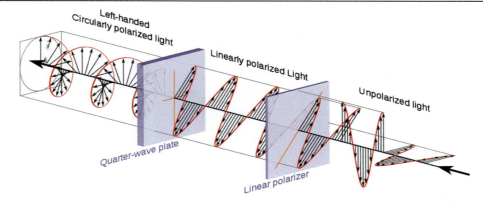

Figure 1.2: Circularly polarized light is left-handed as viewed from the receiver, but right-handed as viewed from the source. Adapted from https://en.wikipedia.org/wiki/User:Dave3457/Sandbox/Circular_Polarizer, under the Creative Commons condition http://creativecommons.org/licenses/by/2.5/.

Elliptical polarization

Elliptical polarization is a more general case than circular polarization, where the amplitude of the x and y components are not equal as shown in the following mathematical representation of the polarization vector for the electric field [18]:

$$\vec{E}(z,t) = E_x \hat{x} \cos(\vec{k}\cdot\vec{r} - \omega t) + E_y \hat{y} \cos(\vec{k}\cdot\vec{r} - \omega t + \phi). \tag{1.1}$$

Here ω is the angular frequency, ϕ is the relative phase between x and y components (E_x and E_y) while k is the wave-vector. This equation describes all forms of polarization as follows: Elliptical polarization: when E_x and E_y are both nonzero and unequal, and the relative phase is nonzero. Here, the polarization changes in magnitude as it rotates and gives an elliptical shape. Linear polarizations: when $\phi = 0, \pm\pi$, and vertical polarization happens when both E_x and ϕ are equal to zero. When both E_y and ϕ are equal to zero we have the horizontal polarization. Circular polarization occurs when $E_x = E_y$ and $\phi = \pm\pi/2$ (for left- and right-handed cases).

1.3 Dirac's notation

Each quantum state can be represented by a vector in a complex Hilbert space using a special notation called the bra-ket notation, which was coined by Paul Dirac (later altered by Max Born into the form we currently know). A column vector in Hilbert space is denoted by the ket state vector $\langle\psi|$, and it has a mirror counterpart in dual space that is called a bra $|\psi\rangle$. A bra is just the Hermitian conjugate of the corresponding ket, and it is a row vector. Using

Table 1.1: Any polarization state can be formulated using a superposition of horizontal and vertical states.

Basis	Polarization	Notation	Representation
Standard basis	Horizontal	$|H\rangle$	n/a
	Vertical	$|V\rangle$	n/a
Hadamard basis	Diagonal	$|+\rangle$	$\frac{1}{\sqrt{2}}(|H\rangle + |V\rangle)$
	Anti-Diagonal	$|-\rangle$	$\frac{1}{\sqrt{2}}(|H\rangle - |V\rangle)$
Circular basis	Right-circular	$|R\rangle$	$\frac{1}{\sqrt{2}}(|H\rangle + i|V\rangle)$
	Left-circular	$|L\rangle$	$\frac{1}{\sqrt{2}}(|H\rangle - i|V\rangle)$

the Dirac notation we now introduce the superposition principle. Since quantum mechanics is linear, any two states (or more) can form a valid quantum state. Focusing on two-state systems say $|\alpha_1\rangle$ and $|\alpha_2\rangle$ we have

$$|\psi\rangle = \frac{1}{\sqrt{2}}(|\alpha_1\rangle + |\alpha_2\rangle) \quad (1.2)$$

where $\frac{1}{\sqrt{2}}$ is the normalization factor when $|\alpha_1\rangle$ and $|\alpha_2\rangle$ are mutually orthonormal $\langle\alpha_1|\alpha_2\rangle = 0$. If the states $|\alpha_1\rangle$ and $|\alpha_2\rangle$ are normalized but not mutually orthogonal, then $|\psi\rangle$ is not normalized and its self-overlap shows interference between these two component states,

$$\langle\psi|\psi\rangle = 1 + \Re\langle\alpha_1|\alpha_2\rangle \quad (1.3)$$

The real part term $\Re\langle\alpha_1|\alpha_2\rangle$ leads to many interesting quantum phenomena [19–21]. Note that the superposition principle is, of course, not limited to just two states, but can involve a Hilbert space of an infinite dimension.

Any arbitrary state of polarization can be formulated as a superposition of horizontal and vertical states but with different phases, as can be seen in Table 1.1.

The photon polarization state can be expressed in terms of different basis sets: (a) The normal computational basis set, sometimes called the standard basis set, where we have a horizontally polarized state $|H\rangle$ corresponding to state $|0\rangle$, while vertically polarized state $|V\rangle$ to state $|1\rangle$. (b) The diagonal basis set, sometime called Hadamard basis set. (c) The circular basis set.

The sets of states (b) and (c) are just superposition of $|H\rangle$ and $|V\rangle$ states but with different phases. For more details see Table 1.1.

1.4 Quantum bits (qubits)

Classically, information comes in the form of units of bits with two possible values, 0 and 1, to encode and process information. For example, a transistor can be in the 0 state (no current) or at the 1 state (with current). In quantum physics, it is not necessary to have the quantum system in one state or the other, but rather in a superposition of both, and that leads us to the quantum counterpart of the classical bit and that is the quantum bit or "qubit" for short, which is the fundamental unit of quantum information: the term was coined by Schumacher [22]. The qubit is mathematically represented by $|\psi\rangle = \alpha_0 |0\rangle + \alpha_1 |1\rangle$, where α_0 and α_1 are complex amplitudes. It is difficult to distinguish between the two polarization states until we make measurements. Performing the measurement results in either having the $|0\rangle$ state with probability $|\alpha_0|^2$ or the $|1\rangle$ state with probability $|\alpha_1|^2$, where $|\alpha_0|^2 + |\alpha_1|^2 = 1$. It is apparent that the act of measurement leads to the collapse of the quantum state to be in one or the other of the two states, so it is not deterministic. We have so far dealt with the superposition of two states, and this can be generalized to qubits in higher dimensions. For example, two qubits can be in a superposition of four states, while n qubits can be in a superposition of 2^n.

Qubits can be physically represented by a variety of physical systems. However, as mentioned at the outset we shall restrict ourselves in this chapter on the degrees of freedom that photons can offer, and there are a few examples. One has so far been discussed, namely the photon polarization (which consists of two orthogonal states). Another degree of freedom is afforded by the two paths that a photon can take in an interferometer. However, photons have other accessible degrees of freedom of higher dimensions as in the case of structured light which is the subject of coverage in the other chapters of this book.

1.5 The Bloch sphere

Unlike algebraic tools, the Bloch sphere is a very good geometrical visualization tool that highlights the difference between the classical bit and the qubit; see Fig. 1.3. Mathematically, a qubit state is represented as a point on the surface or within the interior of the Bloch sphere as follows:

$$|\psi\rangle = e^{i\delta}\left[\cos(\frac{\theta}{2})|0\rangle + e^{i\phi}\sin(\frac{\theta}{2})|1\rangle\right]. \tag{1.4}$$

Here δ is a global phase and can be ignored as it has no observable effects. The angles θ and ϕ are the polar and azimuthal angles, respectively.

All members of the polarization basis set are represented here; the z-axis for the computational basis, the x-axis for diagonal basis, and the y-axis for circular basis. Focusing on the computational basis, the North Pole and the South Pole are two distinct points representing the classical bit, while any pure qubit can be represented by a vector that starts at the center of the sphere and terminates on its surface. Therefore, any point on the surface of the sphere represents a linear combination of $|0\rangle$ and $|1\rangle$ states with complex coefficients.

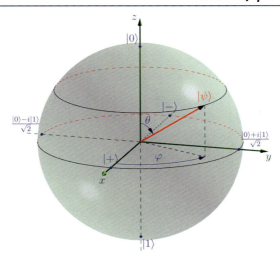

Figure 1.3: Different qubits in their Dirac's notation and the corresponding geometrical Bloch representations.

So far we have identified the qubit representation as a pure state, but how about a mixed state? A mixed qubit state $\hat{\rho}$ cannot be written as a linear combination of state vectors but rather as a statistical combination of two orthogonal pure qubit states $|\psi_i\rangle$,

$$\hat{\rho} = \sum_i P_i |\psi_i\rangle \langle\psi_i| \tag{1.5}$$

with $0 < P_i < 1$, a normalized mixed state when $\Sigma_i = 1$. The Bloch sphere is helpful in the visualization of mixed qubit states; they correspond to points terminated inside the Bloch sphere not on its surface. Note that the Bloch sphere representation is not just useful in the context of qubit states, but also in higher-dimensional systems [23,24].

1.6 Quantum entanglement and nonlocality

Entanglement was first given this name by Erwin Schrödinger and called it "the characteristic trait of quantum mechanics" [25], and it is a property of correlation between two or more quantum systems as a result of the superposition principle. However, in quantum physics, entangled states cannot be factorized as a product of the substates of the individual systems. Strictly, the substate information can be described independently, but the information in the whole state is not the sum of the information in the substates. Substates do carry some information unless the state is maximally entangled, in which case tracing over one subsystem gives a maximally mixed state. There is another important feature of entangled states and that is related to the measurement process. When one of the substates is measured then simultaneously the other one is affected due to the correlation leading to the destruction of the entanglement. This does not imply that information is sent from one place to another, but rather strongly correlated, which can be used to share a random outcome.

As examples, we consider the following cases involving entanglement in two and three dimensions. In the case of 2D qubit states, there are four possible Bell states, named after John S. Bell [26], and written as $\frac{1}{\sqrt{2}}(|00\rangle \pm |11\rangle)$, and $\frac{1}{\sqrt{2}}(|01\rangle \pm |10\rangle)$. These states are maximally-entangled states. The 3D example involves three qubit states $\frac{1}{\sqrt{2}}(|000\rangle \pm |111\rangle)$, which is called the GHZ state after Greenberger, Horne, and Zeilinger [27]. This state is clearly a good example for a maximally-entangled state. Utilizing entanglement in quantum communication is very vital in many applications (e.g. superdense coding and quantum teleportation) as will be seen throughout the book. There have also been reports of quantum entanglement in a biological system [28–30]. Entanglement causes another quantum phenomenon called nonlocality, which is the notion that distant particles (or photons) cannot influence one another in a period of time shorter than that when a light signal is sent between them. Since entanglement causes particles (or photons) to become permanently correlated (dependent on each other's states and properties), these particles will effectively lose their individuality, and hence; cannot be looked at "locally" without considering the other entangled particles, which could be thousands of miles away. To solve the mystery around nonlocality, it is worth going back to Bell, who was stimulated by the Einstein, Podolsky, and Rosen (EPR) paper [31], in which they pointed out the puzzling nonlocal character of the quantum system due to entanglement. Bell was able to derive what has become known as the Bell inequality [26], which is satisfied by local realistic theory but violated by quantum mechanics. Clauser, Horne, Shimony and Holt (CHSH) proposed a game to show such an inequality [32]. Generally, the Bell inequality allows classical players to win with probability $\leq 3/4$, however, the situation is different for the maximally entangled state like EPR where a player is allowed to win with probabilities range from 0 to 0.85. This is called the violation of the Bell inequality. This is not just a theoretical game but there have been several continuing experimental research showing violations of this inequality [33,34], and that is evidence of the quantum nature.

Note that there exist separable, non-entangled states that produce a nonlocal behavior [35] and that there are some cases where entanglement does not produce nonlocality [36]. Both these cases do not serve the quantum communication cause, and such states are not favorable in our context. For clarity, it is worth mentioning that the GHZ experiment [27] demonstrated quantum nonlocality but with one catch and that it contains no inequalities.

The most popular experimental method to generate entangled photons (in polarization) is the spontaneous parametric down conversion, where a single pulse (or pulses) of photons passes through a crystal that produces entangled photons with a different probability amplitude as shown in Fig. 1.4.

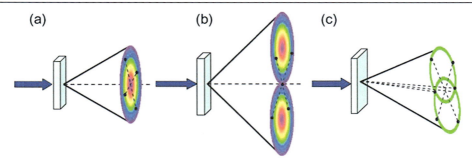

Figure 1.4: Different types of parametric down-conversion schemes to generate entangled photon pairs, where (a) type-I phase matching; (b) collinear degenerate type-II phase-matching; (c) non-collinear degenerate type-II phase-matching. Reprinted figure with permission from [37], ©IOP Publishing Ltd. All rights reserved.

1.7 Measurement, decoherence, and irreversibility

The concepts of superposition, quantum correlations and measurement constitute the areas where quantum schemes differ from classical ones. Unlike classical systems, quantum systems are fragile, and to mitigate this fragility it is important to keep the system isolated from the environment. Hence, observing entanglement and superposition requires a well-isolated quantum system. Coherence is another requirement for entanglement, which can be completely destroyed once it is measured. The act of measurement is equivalent to an interaction with the environment.

Experiments are done with good isolation conditions to reach the best results, however, in reality there is no perfect isolation. This leads to the uncontrolled interaction with the surrounding environment, which leads to noise and error and that is what is meant by decoherence. It is usually cursed for being the key obstacle in damaging information.

There are tools that one can use to avoid decoherence; one of them is the decoherence-free subspaces technique, where information is stored in the many degrees of freedom of an individual quantum system, and so is decoupled and hidden from the outside world [38]. Another tool is the quantum error correction, which is a set of techniques to protect quantum states from the decoherence effects [39], and it can be done by storing information jointly in many qubits, and further make a careful comparison among these qubits. This allows examining whether any of the qubits was distorted as a result of decoherence and then can be corrected.

Although decoherence is an irreversible process, there are some natural phenomena that can be reversed in time. An empty bottle is initially empty (initial state). Once we fill the bottle with perfume (new state), the system (perfume, bottle and the environment) will eventually

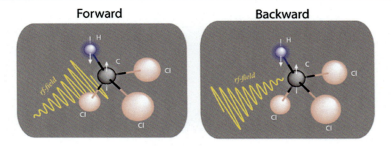

Figure 1.5: The reversal of the arrow of time embedded in a reserved input signal. Adapted with permission from [41,42], ©2020 by the American Physical Society.

return to the initial state since the volatile perfume will evaporate and the bottle will be empty again! However, not all systems behave as such. A natural system may get described as irreversible when the reversed process does not spontaneously occur with the famous example of hot water's tendency to become cold in time and not vise versa. This phenomenon has been called "arrow of time" by the British astronomer Arthur Eddington [40]. Some call it the asymmetry or "one-way direction" of time, and it has been known in a thermodynamical context, and is inherently a macro phenomenon. This irreversibility, that one can notice in thermodynamics, surprisingly can be found elsewhere in physics, namely, in quantum mechanics. It is known fact in quantum mechanics that the act of measurement of a quantum system leads to the collapse of the wave function making the system irreversible to its original state. Having said that, it can be confusing to see some irreversible physical phenomena explained by the very physics laws, which describe revisable phenomena. It is a well-understood fact that these physical laws are unchangeable, and it was Ludwig Boltzmann who proposed a solution to this discrepancy by altering the initial conditions of the physical process. Solving this discrepancy in quantum systems was tried by several research efforts, most notably the experimental effort by Batalhão [41], where the reversibility was introduced to the quantum system by reversing the excitation signal, as shown in Fig. 1.5. Here, in a typical NMR setup, a spin-1/2 carbon atom in a chloroform molecule is driven by an external magnetic field, where reversing the excitation signal in time had indeed reversed the state of the system.

This is unlike conventional thermodynamics, where systems are assumed to be initially uncorrelated and consistent with a single direction of time. However, correlating systems (through initial conditions) will cause the arrow of time to be reversed (heat flows from cold to hot spontaneously), and there have been proposed mechanisms that sought to overcome such problems [43–46]. See Fig. 1.6.

Basics of quantum communication   13

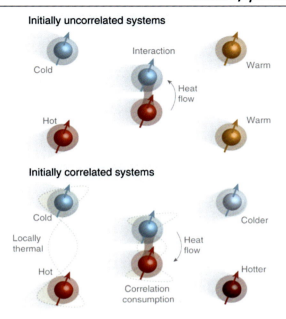

Figure 1.6: Correlation as a mean for reversal of arrow of time. Reprinted figure from [43] under the Creative Commons CC BY License.

1.8 Quantum cloning

Classically, we can clone states without limit, however, it is impossible quantum mechanically due to the so-called no-cloning theorem. The no-cloning theorem was introduced by Wootters and Zurek [9], and separately by Dieks [10], its name is related to its reality. This theorem is very simple; yet it is considered to be a key result in quantum communication. Using contradiction, we present the following proof of this theorem.

To prove the impossibility of cloning, we start by supposing that we have a cloning device that is capable of duplicating an arbitrary qubit state $|\psi\rangle_1$ of system 1 onto the qubit state $|\psi\rangle_2$ of system 2. This means preparing a second qubit in the blank state $|\phi\rangle_2$ and copy the state of the original qubit $|\psi\rangle_1$ into it, ending up with the state $|\psi\rangle_2$. This cloning machine requires having a unitary operator U_{clon} that performs the following transformations:

$$\hat{U}_{clon} |\psi\rangle_1 |\phi\rangle_2 = |\psi\rangle_1 |\psi\rangle_2. \tag{1.6}$$

This unitary operator has to be universal and independent of $|\psi\rangle_1$, and holds for any possible qubit state. Let us start with system 1 in an original qubit state $|0\rangle_1$. The system will evolve using the cloning transformation as follows

$$\hat{U}_{clon} |0\rangle_1 |\phi\rangle_2 = |0\rangle_1 |0\rangle_2 \tag{1.7}$$

and if it is in state $|1\rangle_1$ the cloning transformation is

$$\hat{U}_{clon} |1\rangle_1 |\phi\rangle_2 = |1\rangle_1 |1\rangle_2. \tag{1.8}$$

That is clear and trivial, but now let us consider a qubit in a superposition state $|\psi\rangle_1 = \alpha_0 |0\rangle_1 + \alpha_1 |1\rangle_1$ to be cloned. Applying the cloning transformation yields

$$\begin{aligned}\hat{U}_{clon} (\alpha_0 |0\rangle_1 + \alpha_1 |1\rangle_1) \otimes |\phi\rangle_2 &= \alpha_0 \hat{U}_{clon} |0\rangle_1 |\phi\rangle_2 + \alpha_1 \hat{U}_{clon} |1\rangle_1 |\phi\rangle_2 \\ &= \alpha_0 |0\rangle_1 |0\rangle_2 + \alpha_1 |1\rangle_1 |1\rangle_2 .\end{aligned} \tag{1.9}$$

To our disappointment, this is an unexpected result since we hoped to have a state of the form

$$\begin{aligned}(\alpha_0 |0\rangle_1 + \alpha_1 |1\rangle_1) &\otimes (\alpha_0 |0\rangle_2 + \alpha_1 |1\rangle_2) \\ &= \alpha_0^2 |0\rangle_1 |0\rangle_2 + \alpha_0 \alpha_1 |0\rangle_1 |1\rangle_2 + \alpha_0 \alpha_1 |1\rangle_1 |0\rangle_2 + \alpha_1^2 |1\rangle_1 |1\rangle_2 .\end{aligned} \tag{1.10}$$

This result, therefore, suggests that quantum cloning is indeed impossible, as stated here,

$$\hat{U}_{clon} (\alpha_0 |0\rangle_1 + \alpha_1 |1\rangle_1) \otimes |\phi\rangle_2 \neq (\alpha_0 |0\rangle_1 + \alpha_1 |1\rangle_1) \otimes (\alpha_0 |0\rangle_2 + \alpha_1 |1\rangle_2). \tag{1.11}$$

The above arguments show that a universal cloning procedure is just impossible due to quantum mechanics. This indeed is what distinguishes between classical and quantum information. Another issue to consider is the link between measurement and the no-cloning theorem. The catch here is that measurement cannot be improved by cloning. If cloning were possible we could make many copies of a state and measure them all, obtaining full information about the state. The no-cloning theorem has practical implications in broad areas, among which are quantum cryptography and teleportation, as we will see later on.

1.9 Quantum communication with single photons

It is important to recognize the connections, and the differences, between quantum communication and quantum computation. Quantum computation addresses the challenge of reducing the computation time, where the results of such computation is not necessarily transmitted by a quantum communication network. Nevertheless, quantum communication uses the rules of quantum mechanics to transfer the information between two parties, typically guarding its secrecy—an intrinsic feature that classical communication lacks. One main difference between quantum computation and communication is that quantum computation requires non-local quantum systems interactions [47–51], while quantum communication requires quantum systems that are less prone to decoherence (during signal transmission) [39,52,53]. However, quantum communication is not just data forwarding: it can be called distributed quantum computation.

Among the main ingredients for having quantum communication protocols is to have a single-photon source that can emit each time only one photon, on demand, with high generation rates, and most importantly in well-defined states. Having a well defined photonic state facilitates achieving the encoding and decoding strategies, which can be done by exploiting different degrees of freedom that single photons can offer. Among those strategies are polarization encoding, orbital angular momentum encoding, time-bin encoding, path encoding, and frequency-bin encoding. These strategies can be further mixed to have hybrid configurations.

1.9.1 Polarization encoding

In quantum communication, information is encoded in binary logic 0 and logic 1; logic 0 is represented by a horizontally polarized photon $|H\rangle$, while a vertically polarized photon $|V\rangle$ represents logic 1. As stated earlier, polarization qubits can be encoded either in the computational/normal basis set as $|H\rangle$ and $|V\rangle$, the diagonal/Hadamard basis set as diagonally polarized $|D\rangle$ and anti-diagonally polarized $|A\rangle$, or the circular basis set as right-circularly polarized $|R\rangle$, and left-circularly polarized $|L\rangle$ [54]; see Table 1.1. These three sets of pairs of states form together a set of non-orthogonal and mutually unbiased basis sets (MUB). The mutually unbiased basis set is used when a state is prepared in a given basis set but is measured in another MUB, the outcome bit value that the measurement gives has to be uniformly random. For example, having a state prepared in the Hadamard basis set but measured in the computational basis set. There is then a 1/2 probability of obtaining either a $|H\rangle$ or a $|V\rangle$ state [55]. Fig. 1.7 shows how an arbitrary photonic state can be generated starting with either a $|H\rangle$ or a $|V\rangle$ polarized photon.

1.9.2 Orbital angular momentum (OAM) encoding

Optical angular momentum is divided into two parts: spin and orbital angular momentum. Spin angular momentum is associated with light's circular polarization, while orbital angular momentum arises from the spatial structure of the light's wavefront. Orbital angular momentum comes directly from the spatial distribution of the electromagnetic field, and it has two terms [12,57]: internal, origin-independent, associated to helical wavefront structure, and external, origin-dependent. The helical wavefront structure originates from nothing but the azimuthal phase dependence $e^{i\ell\phi}$, where ϕ is the azimuthal coordinate and ℓ is an integer. The OAM modes are well-defined by the $\hbar\ell$ per photon, producing the OAM eigenstates. Information gets encoded through arranging photons in a specific pattern or structure; hence, it is called structured light [12,13,57,58]. See Fig. 1.8 for an example of such light. This book is about this new growing research area of interest within the quantum communication and more details about this is presented in the coming chapters.

16 Chapter 1

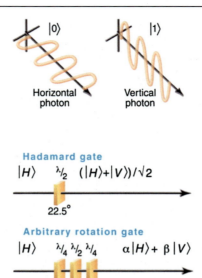

Figure 1.7: A schematic of horizontally and vertically polarized photon (top), and showing how to generate arbitrary photonic state by just going through birefringent wave plates with different orientation and wavelength (bottom). Figures adapted with permissions from [56], ©2020 by AAAS.

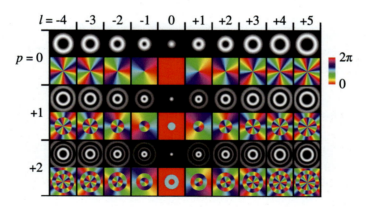

Figure 1.8: Phase and intensity profiles of the first modes of the structured light (Laguerre-Gaussian family) showing azimuthal indices ℓ, and radial indices p. Reproduced figure from [58] under the Creative Commons CC BY License.

1.9.3 Time-bin encoding

The time-bin qubits are very robust against decoherence yet are somewhat difficult to interact with each other; hence, better used in quantum communication rather than quantum computa-

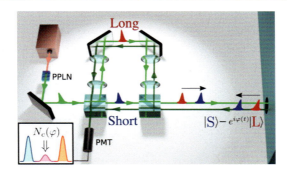

Figure 1.9: The setup of time-bin encoded qubits. The qubit among returning from the receiver (Bob) back to the sender (Alice), is measured by photomultiplier tube (PMT), which gives a third peak which is due to the interaction with the receiver. Figures adapted with permissions from [62], ©2020 by the American Physical Society.

tion. Time-bin qubits are prepared using an unbalanced interferometer [59–61] of two paths, where one path is longer than the other. Therefore, photons may take the short path and arrive early with an "early"/"short" state and this is the first time-bin state, or take the long path and acquire the "late"/"long" state, and this the second time-bin state; see Fig. 1.9. Therefore, a time-bin qubit is formed by the following coherent superposition:

$$|T\rangle = \frac{1}{\sqrt{2}}\left(|S\rangle + e^{i\phi}|L\rangle\right) \tag{1.12}$$

where $\phi = \{-\pi/2, 0, \pi/2, \pi\}$ is the relative phase and based on the value formed as the time-bin analogue to the computational basis, but the time difference between the two paths of these time-basis states must be greater than the photon coherence time.

1.9.4 Path encoding

The path degree of freedom is simply an encoding with the help of optical modes, and the qubit here is taken to be a single photon that can be in one of two different spatial coherent modes; a right mode $|r\rangle_A$ and a left mode $|\ell\rangle_A$, where A is the photon. A qubit with these two spatial modes can be written as

$$|P\rangle = \frac{1}{\sqrt{2}}\left(|r\rangle_A + e^{i\phi}|\ell\rangle_A\right) \tag{1.13}$$

where $\phi = \{-\pi/2, 0, \pi/2, \pi\}$ is the relative phase, and similar to the scenario in the time-bin encoding, we can form a path basis set analogue to the computational basis set based on the

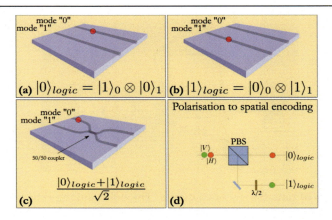

Figure 1.10: Sketch showing how to form qubit from path degree of freedom within waveguide. (a) and (b) There are two spatial paths labeled 0 and 1. Here $|0\rangle_{logic}$ and $|1\rangle_{logic}$ indicates the number of photons that populate these spatial paths. Reproduced with permission from [63], ©2020 by John Wiley and Sons.

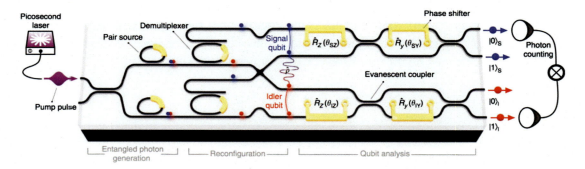

Figure 1.11: Chip comprises three stages, path-entangled state generation, reconfiguration, and qubit analysis. Reprinted figure from [67] under the Creative Commons CC BY License.

value of p. Note that these two spatial modes represent two equivalent channels, hence we can switch between them using polarization beam splitters (see Fig. 1.10).

The concept of path encoding provides a natural choice of qubit states in on-chip waveguide architectures. This is because the interferometric setup is inherently stable within integrated optics, unlike the on-chip polarization encoding, which requires special designs [64–66]. Moreover, path encoding makes higher-dimensional qubits plausible since the photon can spread out easily between multiple waveguides, which makes it a candidate for future developments of quantum optical technologies; for an example see Fig. 1.11.

Basics of quantum communication 19

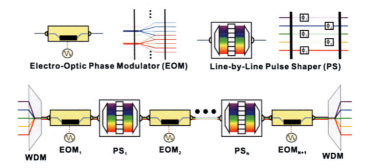

Figure 1.12: Frequency bin, and the many horizontal paths correspond to distinct frequencies, and all belong to a single spatial mode; for more details see [70]. Figures adapted with permissions from [70], ©2020 by IEEE.

1.9.5 Frequency-bin encoding

Qubits can be encoded in different frequency modes while keeping the spatial and polarization degrees of freedom constants over all frequencies. The way to do the encoding is to have every spectral line an information carrier. The available frequency modes normally separated by microwave frequencies and easily controllable by using optical and optoelectronic devices [68,69]. The frequency bins are centered at frequencies $\omega_n = \omega_0 + n\Delta\omega$ where n is integer; see Fig. 1.12. A single photon centered at the bin with frequency ω_{n_0} corresponds to a state $|1_{n_0}\rangle$, whereas a photon at ω_{n_1} corresponds to $|1_{n_1}\rangle$. The corresponding qubit states can be represented using these two modes n_0 and n_1 to form the superposition states as

$$|F\rangle = \alpha_0 |1_{n_0}\rangle + \alpha_1 |1_{n_1}\rangle \tag{1.14}$$

where $|\alpha_0|^2 + |\alpha_1|^2 = 1$.

The frequency-bin encoding has been of great attraction with many proposed theoretical ideas and different experimental setups [70–74].

1.10 Protocols of quantum communications

Quantum communication research covers many areas but we will limit ourselves here to quantum cryptography protocols followed by quantum teleportation and superdense coding protocols.

1.10.1 Quantum key distribution (QKD) protocols

Quantum cryptography is one of the prominent areas of quantum communication since it has a direct impact on the wider society. It utilizes three key principles of quantum physics, namely the no-cloning theorem, state collapse during measurement, and the irreversibility of measurement. These principles help provide unconditional security in optical communications. In this section, we introduce quantum key distribution QKD, which is a key concept of this field. QKD can provide completely secure quantum communication capitalizing on measurement disturbance. However, we know that disturbance for a quantum system was traditionally considered undesirable, so how can disturbance become useful? Most QKD protocols involve two parties, Alice and Bob, who jointly generate a common random key. But there is also an unwanted party and that is Eve being a possible eavesdropper. Here comes the disturbance issue since Eve tries to steal the secret communication between Alice and Bob; this is an act of measurement, which in turn introduces disturbances to the signal, alerting the legitimate users Alice and Bob of Eve's presence.

This field was kicked off by the pioneering work of Wiesner [75] who proposed the idea of quantum money. This was followed by the first protocol BB84 named after Bennett and Brassard in 1984 [7] followed by the E91 protocol by Ekert 1991 [8]. Many other protocols were proposed after that [7,8,76–85], and these can be grouped into three general categories: (a) Device-dependent QKD; (b) source-device-independent QKD [80] and (c) measurement-device-independent QKD [81]. Subsequently, all the above approaches can be divided, based on the nature of the random keys and of the quantum states that represent them, into discrete variable (DV-) [7,8] and continuous variable (CV-) QKD [82,83]. A good overview of these schemes can be found in the recent review paper by Pirandola et al. [86]. In this section, we only present the two well-known protocols BB84 and E91 for illustration.

1.10.1.1 The BB84 protocol

As mentioned earlier, there are many degrees of freedom (DOF) that a physical system can offer to be used to implement QKD, however, our preferable choice here is polarization. In all QKD protocols including the BB84 protocol, the two legitimate users Alice and Bob are connected via two channels; one is quantum and the other one is classical. The quantum channel is not used to exchange messages, but rather to transmit a sequence of random bits between both of them, and that is the key. The eavesdropper Eve has open access to the quantum channel. As for the classical channel, we explain its use shortly.

The BB84 protocol works as follows, see Fig. 1.13: on Alice's side there is a single-photon source that can be polarized in two different polarization bases, which are orthogonal to each other, say the computational and the diagonal bases. Again as a reminder: the computational/normal basis consists of a horizontal polarization $|H\rangle$ and a vertical polarization $|V\rangle$,

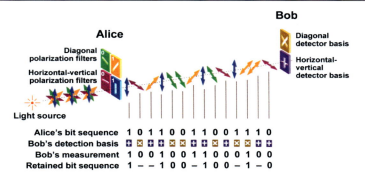

Figure 1.13: Sketch showing how the BB84 protocol works. Originally published in Physics World March 1998 [87].

while the diagonal/Hadamard basis consists of the diagonal polarization $|D\rangle$ and the anti-diagonal polarization $|A\rangle$.

Now, Alice and Bob have to agree to assign logic 0 to $|H\rangle$ and $|D\rangle$ versus logic 1 to $|V\rangle$ and $|A\rangle$. Alice prepares her photons by randomly switching between the computational basis and the diagonal basis and sending individually encoded photons through a quantum channel. At Bob's side, he decides to measure the incoming photon randomly by switching his detection between the two basis sets of states. After a sequence of such actions by both of them, now comes the importance of the classical channel since Alice and Bob need to check the basis of their measurement but not the results. If Alice's basis agrees with that of Bob then they keep their finding and that forms their secret raw key. This takes place when the quantum channel is free from noise and the tampering by Eve.

We mentioned that Eve has full access to the quantum channel during the transmission of photons between Alice and Bob, and she can exploit this by trying have a copy of the photon state, but that is not possible due to the no-cloning theorem. Therefore, she can mount an attack by measuring the incoming photons, and note down their results. Based on that she tries randomly to guess a basis for photons that she needs to regenerate in order to send along to Bob. By doing this she introduces on average 25% errors in the secret key, and that alerts Alice and Bob of Eve's presence.

1.10.1.2 The E91 protocol

Unlike BB84 where a single photon is sent from Alice to Bob, here E91 protocol utilizes entanglement to create a shared secret key between Alice and Bob. Having a maximally entangled state between both of them leads to highly correlated results when both Alice and Bob preform their measurements.

The E91 protocol works as follows: the quantum channel has an entanglement source that can generate and emit many maximally entangled photon pairs in a form like that of the Bell states. The entangled photon pair fly apart with one photon belonging to Alice while the other one to Bob. Similar to the BB84 protocol, Alice and Bob independently and randomly start to measure the incoming photons, by choosing a random basis, out of three possible bases characterized by certain azimuthal angles. We know that those entangled photons are highly correlated with each other, meaning that once Alice measures her photon in a certain polarization, this instantaneously causes the other photon at Bob's side to be perpendicular to the chosen Alice polarization. This only happens if they are measured in the same basis.

In the next step Alice and Bob announce and compare each basis for each measurement that they took through the classical channel. And in order to check for the security of their finding they can check that by using the generalized Bell inequality [26], known as CHSH [32]. In order to calculate that they need to do some house keeping beforehand such as discarding all measurements that either or both of them failed to register the arrival of a photon. Now Alice and Bob can compute the CHSH inequality for the data that they acquire for their measurement results, and they have one of the following two options: (a) Obtaining $2\sqrt{2}$, and that tells them that the CHSH inequality is violated. Having this violation ensures that the shared photon pairs are maximally entangled, and there is no chance that Eve has any information about the key. Now they can be sure that there is no eavesdropper and they have a joint secret raw key. (b) Other than that gives no violation of the CHSH inequality, which is simply an indication of a separable state, meaning that Eve has tampered with the key and it is no longer a secret. One would rightly ask here what will happen if Eve gets control over the source of entanglement? The answer is that she gets no information since the entanglement source is not significant when it comes to encoding processes.

Experimental research on QKD has been very active in many labs around the globe since its discovery, and for basic experimental elements and systems, one can consult the review by Gisin [52], while [53] covers the basic security analysis tools of various QKD protocols. Although QKD is theoretically secure, it is not the case as regards real-life practical implementations. The reason is that QKD suffers from several imperfections leading to a window of security loopholes. Needless to say that Eve has unlimited resources at her disposal and among them is the exploitation of the practical imperfections window without being detected. Among the obvious imperfections targets that Eve can exploit to mount her quantum attack are the practical QKD source and the detection system. For relevant references with more details on this see [88–91].

1.10.2 Quantum teleportation protocol

Quantum teleportation is a unique feature of quantum information, which is a direct consequence of quantum entanglement. It was discovered by Bennett and coworkers [92] and was

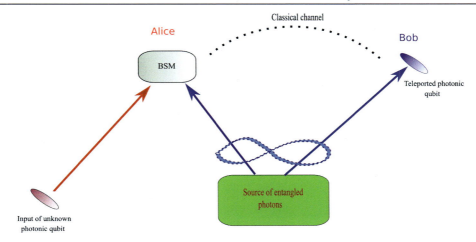

Figure 1.14: Teleportation schematic, where Alice performs a Bell-state measurement between the qubit she wants to teleport and one of the photons of the entangled pair, while the pair help Bob to get the qubit performing the transformation.

soon after realized experimentally [93,94]. It enables the transfer an unknown quantum state of a particle but not the particle itself between two distant locations, say Alice's and Bob's stations, through the main of transmitting classical information.

Let us assume that Alice has a single photon, which is in some quantum state that she wishes to transmit to Bob. A trivial way would be that Alice sends the photon itself directly to Bob, however there are two restrictions; (a) She does not know the state of that photon. (b) She has no idea about the location of the receiver. Adding to these two restrictions, we know that quantum mechanics forbids cloning of a quantum state. Having these limitation makes the teleporting process look impossible! However, we can still accomplish the task using quantum entanglement. The quantum teleportation protocol (see Fig. 1.14), requires three qubits in total including the desired unknown state that Alice wants to teleport. The steps involved in the protocol are as follows:

Let us assume that Alice wants to teleport the unknown following qubit to Bob (first qubit),

$$|\chi\rangle_x = \alpha_H |H\rangle_x + \alpha_V |V\rangle_x \qquad (1.15)$$

where α_H, and α_V are unknown to Alice, and the subscript x indicates 'unknown'.

One of the key ingredients to accomplish this task is to have a source of entanglement to enable the generation of a maximally entangled state that is shared between Alice and Bob, such as

$$|\psi^\pm\rangle = \frac{1}{\sqrt{2}} (|HV\rangle_{AB} \pm |VH\rangle_{AB}),$$
$$|\phi^\pm\rangle = \frac{1}{\sqrt{2}} (|HH\rangle_{AB} \pm |VV\rangle_{AB}),$$

(1.16)

where we have adopted the notation $|HH\rangle_{AB} = |H\rangle_A \otimes |H\rangle_B$. Any one of the above maximally entangled Bell states can be used to complete the quantum teleportation protocol.

Here, we show how to achieve this by just using one of the above Bell states. We choose the state $|\psi^-\rangle_{AB} = \frac{1}{\sqrt{2}}(|HV\rangle_{AB} - |VH\rangle_{AB})$, which is a maximally entangled state between Alice and Bob. Note that $|\psi^-\rangle_{AB}$ contains no information about the unknown qubit $|\chi\rangle_x$.

Alice holds the unknown qubit $|\chi\rangle_x$ that she wants to transfer to Bob, together with her part of the entangled state, A. While Bob has just his part of the entangled pair B. Now, we can write the total state shared by Alice and Bob as

$$|\Psi\rangle_{xAB} = |\chi\rangle_x \otimes |\psi^-\rangle_{AB},$$

(1.17)

which can be expanded as

$$|\Psi\rangle_{xAB} = \frac{1}{\sqrt{2}} \left(\alpha_H |HHV\rangle_{xAB} - \alpha_H |HVH\rangle_{xAB} + \alpha_V |VHV\rangle_{xAB} - \alpha_V |VVH\rangle_{xAB} \right).$$

(1.18)

At this stage, Alice needs to introduce another set of Bell states that involve both the unknown photon states and the photon at her end,

$$|\phi_1\rangle = \frac{1}{\sqrt{2}} \left(|VH\rangle_{xA} - |HV\rangle_{xA} \right),$$ (1.19a)

$$|\phi_2\rangle = \frac{1}{\sqrt{2}} \left(|VH\rangle_{xA} + |HV\rangle_{xA} \right),$$ (1.19b)

$$|\phi_3\rangle = \frac{1}{\sqrt{2}} \left(|HH\rangle_{xA} - |VV\rangle_{xA} \right),$$ (1.19c)

$$|\phi_4\rangle = \frac{1}{\sqrt{2}} \left(|HH\rangle_{xA} + |VV\rangle_{xA} \right).$$ (1.19d)

From Eqs. (1.19a) and (1.19b) we get

$$|VH\rangle_{xA} = \frac{1}{\sqrt{2}} (|\phi_1\rangle + |\phi_2\rangle),$$
$$|HV\rangle_{xA} = \frac{1}{\sqrt{2}} (|\phi_1\rangle - |\phi_2\rangle).$$

(1.20)

Table 1.2: Output states of quantum teleportation and Bob's local operation.

Bell state observed by Alice	Projected state at Bob	Bob's action (local operation)
$\|\phi_1\rangle$	$\alpha_H \|H\rangle_B + \alpha_V \|V\rangle_B$	No action needed
$\|\phi_2\rangle$	$-\alpha_H \|H\rangle_B + \alpha_V \|V\rangle_B$	$\|H\rangle_B \Rightarrow -\|H\rangle_B$ and $\|V\rangle_B \Rightarrow \|V\rangle_B$
$\|\phi_3\rangle$	$\alpha_H \|V\rangle_B + \alpha_V \|H\rangle_B$	$\|V\rangle_B \Rightarrow \|H\rangle_B$ and $\|H\rangle_B \Rightarrow \|V\rangle_B$
$\|\phi_4\rangle$	$\alpha_H \|V\rangle_B - \alpha_V \|H\rangle_B$	$\|V\rangle_B \Rightarrow \|H\rangle_B$ and $\|H\rangle_B \Rightarrow -\|V\rangle_B$

Meanwhile Eqs. (1.19c) and (1.19d) yield

$$|HH\rangle_{xA} = \frac{1}{\sqrt{2}} (|\phi_3\rangle + |\phi_4\rangle),$$
$$|VV\rangle_{xA} = \frac{1}{\sqrt{2}} (|\phi_3\rangle - |\phi_4\rangle). \qquad (1.21)$$

The complete state of the three particles prior to Alice performing any measurement is

$$|\Psi\rangle_{xAB} = \frac{1}{\sqrt{2}} \Big\{ |\phi_1\rangle (\alpha_H |H\rangle_B + \alpha_V |V\rangle_B) + |\phi_2\rangle (-\alpha_H |H\rangle_B + \alpha_V |V\rangle_B)$$
$$+ |\phi_3\rangle (\alpha_H |V\rangle_B + \alpha_V |H\rangle_B) + |\phi_4\rangle (\alpha_H |V\rangle_B - \alpha_V |V\rangle_B) \Big\}. \qquad (1.22)$$

Eq. (1.22) contains all the information one needs to understand quantum teleportation. Alice now performs a measurement on the joint system consisting of her photon A and the unknown state x in order to distinguish each one of the four orthogonal Bell states $|\phi_i\rangle$, where $i = 1, ..., 4$. Measuring $|\phi_i\rangle$ results in Bob's photon being projected into one of the four pure states. For example measuring $|\phi_1\rangle$ leads Bob's photon to be projected into $\alpha_H |H\rangle_B + \alpha_V |V\rangle_B$. Regardless of which state $|\phi_i\rangle$ is being measured, probabilities of each measurement outcome is equal to 1/4, and it is independent of the parameters α_H and α_V of the unknown original state. Once the measurement has been carried out, Alice needs to communicate to Bob via the classical channel which state she has measured (since she has four possibilities, two classical bits are sufficient). Needless to say that the original teleported photon is destroyed in the process due to the act of measurement.

Bob needs to perform a set of different transformations on his entangled photon pair, depending on the outcomes that were communicated to him by Alice in order to reconstruct the original state that is being teleported to him. The transformation that Bob needs to perform is summarized in Table 1.2.

By simply receiving a classical communication from Alice, Bob can further manage to perform the right transformation and hence obtain a copy of the source state $|\chi\rangle$, which is destroyed in the process. Does teleportation violate physical laws, particularly the no faster than light law? The answer is that the realization of the teleportation protocol requires classical communication between Alice and Bob.

Quantum teleportation research, both theoretical and experimental, continues to be very active. On the experimental side, it has been achieved using many different physical systems and technologies [95–97] ranging from on-chip [98] all the way to long distance teleportation [99,100], and there has even been success in demonstrating the transfer of more complex 3D quantum states [101].

1.10.3 Superdense coding protocol

In the teleportation protocol, Alice can "teleport" a qubit to Bob using two classical bits, but that cannot be accomplished without the help of the maximally entangled qubit pair prepared beforehand. Now let us consider the following scenario in which Alice and Bob are positioned at different locations, and they wish to communicate two classical bits using a classical channel. Classically, communicating two bits of information requires two classical bits. It is intriguing to know that Alice can transmit two classical bits of information directly to Bob but using only one qubit, and the key to achieve this is to have shared entanglement between them. Without entanglement, Alice can still send a single qubit to Bob, which is inefficient, since she can only communicate one bit of information. This is the well-known superdense coding, which was invented by Bennett and Wiesner [102]. Zeilinger's group implemented and verified the protocol experimentally [103].

Before Alice starts to think about how to send her message, it is a must that both Alice and Bob share a maximally entangled qubit. The maximally entangled state is independent of the message intended to be communicated, however, it is the main resource behind establishing the communication between Alice and Bob. Bob has the choice to prepare a pair of maximally entangled photon qubits using just one of the Bell states, say $|\psi^-\rangle_{AB}$, where A is for the qubit that belongs to Alice and B belongs to Bob. Clearly the state $|\psi^-\rangle_{AB}$ can be transformed to any of the other three states by simply applying a local unitary transformation using the Pauli operators $\hat{1} = |H\rangle\langle H| + |V\rangle\langle V|$, $\hat{\sigma}_x = |H\rangle\langle V| + |V\rangle\langle H|$, $\hat{\sigma}_y = i(|V\rangle\langle H| - |H\rangle\langle V|)$, and $\hat{\sigma}_z = |H\rangle\langle H| - |V\rangle\langle V|$. Now Alice wishes to send a message to Bob that consists of 2 bits, which can be expressed using binary values 00, 01, 10, and 11. The natural choice for such binary values is to be represented using one of the following four states $|HH\rangle$, $|HV\rangle$, $|VH\rangle$, and $|VV\rangle$. The standard implementation of this protocol is as follows:

i. Bob needs to send one photon from each pair to Alice.
ii. Alice manipulates the received photons based on what message she want to convey to Bob. All she needs is to perform the operation $\hat{1}$, $\hat{\sigma}_x$, $\hat{\sigma}_y$, and $\hat{\sigma}_z$ on her entangled qubit locally by transforming the initial entangled state $|\psi^-\rangle$ into any of the four Bell states. This keeps entanglement intact since there is no act of measurement.

iii. After performing the unitary transformation, Alice now sends her entangled qubit to Bob using a quantum channel.
iv. Bob now possesses both two qubits, and in order for him to find out which classical bits Alice had sent, he needs to perform a Bell-state measurement BSM on both qubits. This way he can disclose the message that Alice sent him.

This is an elegant way for Alice to transfer two bits of classical information by sending only a single qubit to Bob.

1.11 Ranges of quantum communication

The photon is regarded as the best choice for long distance quantum communication since it rarely interacts with the surrounding environment, and there are two transmission media that can be exploited; optical fiber and free space. The main advantage of using an optical fiber is that the fiber technology is already the existing telecommunication infrastructure, however, there is transmission loss that makes it impossible to communicate a quantum state beyond a few 100 kms [104].

The solution to this restriction is to use free-space communication, particularly, satellite-to-earth and inter-satellite communication, which are less developed. There are many early long distance free-space quantum communication experiments [104–111] which paved the way for a series of long distance quantum communication, ranging from earth to a satellite in space (thousands of kilometers) [106–108], in the intermediate distance range between two drones (kilometer(s)) [112], moving trucks [113], planes [114,115] and air balloons [116], down to very short distances (less than a meter) as in ATM transactions. Some application of the latter case is called "hand-held", "on-chip", or "compact", depending on the architecture of the system [117–120].

1.11.1 Long distance quantum communication

Long distance quantum communication is either ground to space or inter-space communication. The inter-space communication offers very low losses in vacuum for the same distance. For 1000 km, as an example, the inter-space communication loss is around 50-80 dB, while the fiber loss is around 200 dB. Another advantage of long distance quantum communication is the opportunity it offers to explore the limits of quantum mechanics and its correlations over long distance. Initially, quantum communication satellite research was put on hold in Europe and USA but then it has assumed a strong presence in various countries like the USA, Canada, Singapore, and some European countries, particularly after the strong push in both Japan and China:

28 Chapter 1

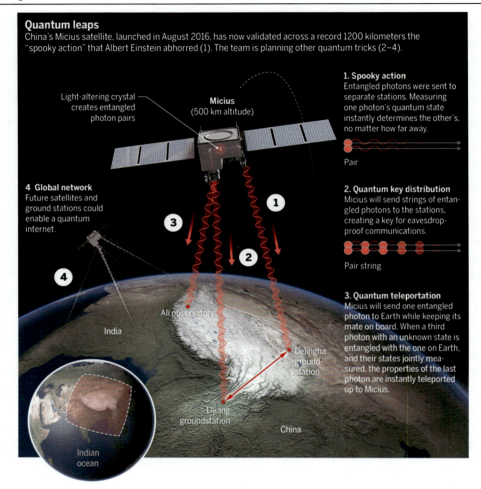

Figure 1.15: Schematics describing the operation of the Chinese (600 km range one way, 1200 km two ways) quantum satellite, which contains a crystal that produces entangled photons. Reprinted with permission from [122], ©2020 by AAAS.

i. In May 2014, the Japanese National Institute of Information and Communications (NICT) developed a small lasercom terminal, called SOTA, and launched it onboard the microsatellite SOCRATES into a 600-km sun-synchronous orbit and had performed some successful experiments [121].

ii. China had followed suit, and had launched Quantum Experiments at Space Scale (QUESS) initiative, see Fig. 1.15, which resulted in the launch of the Micius satellite that carries some experiment; most notably, the Space-Earth quantum key distribution and laser communications experiments; hence, facilitating space-to-ground quantum communication up to 1200 km: a giant step towards the quantum internet [122,123].

Ground to space communications require qubits with high robustness to decoherence, like those encoded in the time-bin [124], where the shorter the distance between the two bins (usually few nanometers) the better the communications as it will counteract the phase shift induced by the satellite motion; conceptually similar to the Doppler shift. However, for interspace communication, polarization encoding is preferred, especially in the context of low earth orbit satellites (LEO).

Quantum communication in free space is an essential step forward since it offers the possibility to perform some fundamental tests in physics. Using free space, it is possible to set up communication not only over very large distances but utilizing the quantum nature of light allowing us not to have any signal to be transmitted at the speed of light. This is what is called the locality loophole, and there have been a number of experiments on this, which have been successfully performed including [125–129], and [130] showing the violation of Bell's inequality over 1200 km.

There are still many challenges for long distance quantum communications, including (but not limited to) the efficient delivery of the quantum state, the range of transmission, the reliability with which the quantum state can be transmitted, and the storage time for entangled states. For state-of-the-art and recent developments on the satellite-based quantum communications, one can consult these references [131–134].

1.11.1.1 Quantum Internet

Another possible manifestation and future development that has been identified arising from long distance quantum communication is the quantum internet, which could be realized through a network of quantum repeaters [135,136]. We know that a quantum state cannot be copied, adding to that the fact that each quantum state faces losses over long distances whether in free space or in fiber. This is not good news for the implementation of the quantum internet since entanglement distribution and storage are of great importance for the network of users [137,138]. One possible solution is using quantum repeaters, which re-establish the qubit state before losing it due to interaction with the transmission medium for long distances. There is a challenging task for experiments that seek to demonstrate quantum repeaters and that the entanglement rates between nearby repeater nodes have to be much faster than the decoherence rate of that entanglement [14]. The development of the quantum internet, however, is still in its infancy, and still requires a great deal of theoretical investigations along with proof-of-principle demonstrations.

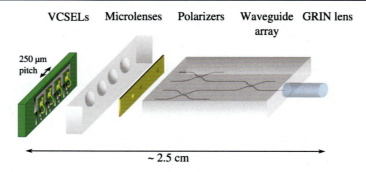

Figure 1.16: An example of a short distance hand-held QKD transmitting device. The compact integrated photonic device consist of lasers, lenses, polarizers and a waveguide array. Reprinted figure with permission from [119], ©2020 by IEEE.

1.11.2 Short distance quantum communications

When thinking of communication devices, usually one thinks of huge cryogenic chambers and devices of that size. However, quantum devices size reduction is an essential step towards the move to real life communication applications such as ATM and points of sale, with transaction time of 1 s despite ambient conditions and hand movements [117]. This implies the need for short distance quantum communication (less than a meter), which can be called either on-chip, hand-held, or compact [118] depending on the size of the technology.

These small on-chip quantum communication devices ought to be very efficient when it comes to handling important and challenging tasks, for example generating single-photon states, manipulating and storing those photon states and then successfully detecting them. Generally, these devices are to be attached to classical devices for any quantum–human interface. The success in this pursuit is mostly measured on how well the new chip uses components and manufacturing processes from the well-established telecommunications industry. There have been very promising experimental demonstrations of quantum teleportation [98], and entanglement distribution [139]. However, the most commercially successful quantum technology is the QKD [120,140–143] with a secret key rate around 30 kb/s over a distance of around 0.5 m. Several research publications in the area have produced such devices including but not limited to [117–119]. The mechanism of such communication range as in the work in Ref. [119], in which quantum a key was transmitted from a hand-held device by using one-side beam steering and a wave-plate rotation in the receiver, as shown in Fig. 1.16, while Fig. 1.17 shows another QKD architecture, which combines several integrated devices. The on-chip quantum communication devices can suffer from losses, which need to be reduced at all system implementation levels in order to have a meaningful practical performance.

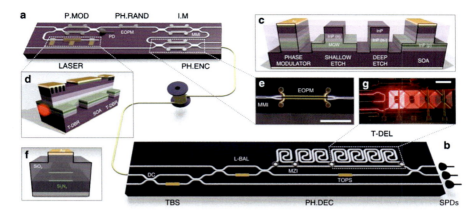

Figure 1.17: Compact integrated photonic design to generate QKD [120]. The system has two main chips; first chip (a, c, d, and e) is built on an InP substrate, and made of several components; laser, optical amplifiers and phase modulators. The second chip (b, f, and g) is SiON platform, which has receiver circuit along with Mach Zehnder interferometers and delay lines. Adapted with permission from [120] under the Creative Commons CC BY License.

1.12 Conclusions

With a more than ever connected world, privacy and security are increasingly becoming of utmost importance for human beings around the globe. Quantum communication, with the dominance of harnessing quantum mechanical effects, has come to the fore to enable secure communication. Looking back a couple of decades up to now, it is clear that quantum communication has come a long way from purely theoretical and fundamental research into becoming a very attractive area of research around the world with the potential to be the backbone of future technology for communication. This has been the main motivation for this chapter, where we have presented, without going into too many details, the set of novel concepts, ideas and the protocols of quantum communication. We have focused primarily on the polarization of single photons to represent the photonic qubit, which is just two-dimensional space. More fascinating and spectacular findings are awaiting to be seen when enlarging the Hilbert space by simply using high-dimensional quantum states or qudits, which are the subject of many of the chapters in this book. Needless to say that there are protocols that are uniquely designed for the high-dimensional spaces and have no equivalent in the two-dimensional world. Therefore, it is of interest to see how to generate high-dimensional entangled quantum states. How far these states can be transferred and then detected and verified in order to be of use for quantum communication protocols. There is no doubt that high-dimensional quantum states, as we will see, will play a crucial role for the next generation of quantum communication technology.

Acknowledgments

We thank M. Babiker and D. L. Andrews for their careful reading of this chapter, and one of us, Z-H. L., acknowledges the financial support from National Natural Science Foundation of China (NSFC) (11704241).

References

[1] C.E. Shannon, Bell Syst. Tech. J. 27 (1948) 379.
[2] G.E. Moore, Cramming more components onto integrated circuits, Reprinted from Electronics, volume 38, number 8, April 19, 1965, pp.114 ff., IEEE Solid-State Circuits Newsl. 11 (3) (2006) 33, https://doi.org/10.1109/N-SSC.2006.4785860.
[3] R.P. Feynman, Int. J. Theor. Phys. 24 (1982) 467.
[4] P. Benioff, J. Stat. Phys. 22 (1980) 563.
[5] Yuri Manin, Computable and Uncomputable, Sovetskoye Radio, Moscow, 1980, p. 128 (in Russian).
[6] D. Deutsch, Proc. R. Soc. A 400 (1985) 97.
[7] C.H. Bennett, G. Brassard, in: Proceedings of IEEE International Conference on Computers, Systems and Signal Processing, 1984, p. 175; reprinted in Theor. Comput. Sci. 560 (2014) 7.
[8] A.K. Ekert, Phys. Rev. Lett. 67 (1991) 661.
[9] W.K. Wooters, W.H. Zurek, Nature 299 (1982) 802.
[10] D. Dieks, Phys. Lett. A 92 (1982) 271.
[11] https://www.sciencemag.org/news/2016/12/scientists-are-close-building-quantum-computer-can-beat-conventional-one.
[12] Andrew Forbes, Isaac Nape, AVS Quantum Sci. 1 (2019) 011701.
[13] Robert Fickler, Radek Lapkiewicz, William N. Plick, Mario Krenn, Christoph Schaeff, Sven Ramelow, Anton Zeilinger, Science 338 (2012) 640.
[14] Daniele Cozzolino, Beatrice Da Lio, Davide Bacco, Leif Katsuo Oxenløwe, Adv. Quantum Technol. 2 (2019) 1900038.
[15] M. Al-Amri, M. El-Gomati, M. Zubairy (Eds.), Optics in Our Time, Springer, 2016, see Chapters 17 and 18.
[16] Masud Mansuripur, Opt. Express 13 (2005) 5315.
[17] Masud Mansuripur, Phys. Rev. A 84 (2011) 033838.
[18] Dennis Goldstein, Polarized Light, 3rd edition, CRC Press, 2017.
[19] J.G. Rarity, P.R. Tapster, Proc. R. Soc. A 355 (1997) 2267–2277.
[20] Marlan O. Scully, Berthold-Georg Englert, Herbert Walther, Nature 351 (1991) 111–116.
[21] Yoon-Ho Kim, Rong Yu, Sergei P. Kulik, Yanhua Shih, Marlan O. Scully, Phys. Rev. Lett. 84 (2000) 1.
[22] B. Schumacher, Phys. Rev. A 51 (1995) 2738.
[23] E. Toninelli, B. Ndagano, A. Vallés, B. Sephton, I. Nape, A. Ambrosio, F. Capasso, M.J. Padgett, A. Forbes, Adv. Opt. Photonics 11 (2019) 67.
[24] A. Forbes, M.d. Oliveira, M.R. Dennis, Nat. Photonics 15 (2021) 253.
[25] E. Schrödinger, Proc. Camb. Philos. Soc. 32 (1935) 446.
[26] J.S. Bell, Physics 1 (1964) 195;
Reprinted in J.S. Bell, Speakable and Unspeakable in Quantum Mechanics, Cambridge University Press, Cambridge, 1987.
[27] D.M. Greenberger, M. Horne, A. Zeilinger, in: M. Kafatos (Ed.), Bell's Theorem, Quantum Theory, and Conceptions of the Universe, Kluwer Academic, Dordrecht, 1989, pp. 69–72;
D.M. Greenberger, M.A. Horne, A. Shimony, A. Zeilinger, Am. J. Phys. 58 (1990) 1131.
[28] Mohan Sarovar, Akihito Ishizaki, Graham R. Fleming, K. Birgitta Whaley, Nat. Phys. 6 (2010) 462.
[29] G.D. Scholes, G.R. Fleming, A. Olaya-Castro, R. van Grondelle, Nat. Chem. 3 (2011) 763.
[30] G.D. Scholes, Nat. Phys. 7 (2011) 448.

[31] A. Einstein, B. Podolsky, N. Rosen, Phys. Rev. 47 (1935) 777.
[32] J.F. Clauser, M.A. Horne, A. Shimony, R.A. Holt, Phys. Rev. Lett. 23 (1969) 880.
[33] M. Giustina, M.A.M. Versteegh, S. Wengerowsky, J. Handsteiner, A. Hochrainer, K. Phelan, F. Steinlechner, J. Kofler, J.-A. Larsson, C. Abellan, W. Amaya, V. Pruneri, M.W. Mitchell, J. Beyer, T. Gerrits, A.E. Lita, L.K. Shalm, S.W. Nam, T. Scheidl, R. Ursin, B. Wittmann, A. Zeilinger, Phys. Rev. Lett. 115 (2015) 250401.
[34] L.K. Shalm, E. Meyer-Scott, B.G. Christensen, P. Bierhorst, M.A. Wayne, M.J. Stevens, T. Gerrits, S. Glancy, D.R. Hamel, M.S. Allman, K.J. Coakley, S.D. Dyer, C. Hodge, A.E. Lita, V.B. Verma, C. Lambrocco, E. Tortorici, A.L. Migdall, Y. Zhang, D.R. Kumor, W.H. Farr, F. Marsili, M.D. Shaw, J.A. Stern, C. Abellan, W. Amaya, V. Pruneri, T. Jennewein, M.W. Mitchell, P.G. Kwiat, J.C. Bienfang, R.P. Mirin, E. Knill, S.W. Nam, Phys. Rev. Lett. 115 (2015) 250402.
[35] Charles H. Bennett, David P. DiVincenzo, Christopher A. Fuchs, Tal Mor, Eric Rains, Peter W. Shor, John A. Smolin, William K. Wootters, Phys. Rev. A 59 (1999) 1070.
[36] Reinhard F. Werner, Phys. Rev. A 40 (1989) 4277.
[37] B. Lounis, M. Orrit, Rep. Prog. Phys. 68 (2005) 1129.
[38] A. Shabani, D.A. Lidar, Phys. Rev. A 72 (2005) 042303.
[39] Dieter Suter, Gonzalo A. Alvarez, Rev. Mod. Phys. 88 (2016) 041001.
[40] A.S. Eddington, The Nature of the Physical World, Cambridge University Press, 1948.
[41] T.B. Batalhão, A.M. Souza, R.S. Sarthour, I.S. Oliveira, M. Paternostro, E. Lutz, R.M. Serra, Phys. Rev. Lett. 115 (2015) 190601.
[42] Alexia Auffèves, Physics 8 (2015) 106.
[43] Kaonan Micadei, John P.S. Peterson, Alexandre M. Souza, Roberto S. Sarthour, Ivan S. Oliveira, Gabriel T. Landi, Tiago B. Batalhão, Roberto M. Serra, Eric Lutz, Nat. Commun. 10 (2019) 2456.
[44] D. Jennings, T. Rudolph, Phys. Rev. E 81 (2010) 061130.
[45] L. Maccone, Phys. Rev. Lett. 103 (2009) 080401.
[46] J. Dressel, A. Chantasri, A.N. Jordan, A.N. Korotkov, Phys. Rev. Lett. 119 (2017) 220507.
[47] L.M. Duan, M.D. Lukin, I. Cirac, P. Zoller, Nature 414 (2001) 413.
[48] J. Majer, J.M. Chow, J.M. Gambetta, Jens Koch, B.R. Johnson, J.A. Schreier, L. Frunzio, D.I. Schuster, A.A. Houck, A. Wallraff, A. Blais, M.H. Devoret, S.M. Girvin, R.J. Schoelkopf, Nature 449 (2007) 443.
[49] C. Monroe, J. Kim, Science 339 (2013) 1164.
[50] Christopher J. Axline, Luke D. Burkhart, Wolfgang Pfaff, Mengzhen Zhang, Kevin Chou, Philippe Campagne-Ibarcq, Philip Reinhold, Luigi Frunzio, S.M. Girvin, Liang Jiang, M.H. Devoret, R.J. Schoelkopf, Nat. Phys. 14 (2018) 705.
[51] F. Borjans, X.G. Croot, X. Mi, M.J. Gullans, J.R. Petta, Nature 577 (2020) 195.
[52] N. Gisin, G. Ribordy, W. Tittel, H. Zbinden, Rev. Mod. Phys. 74 (2002) 145.
[53] V. Scarani, H. Bechmann-Pasquinucci, N.J. Cerf, M. Dušek, N. Lütkenhaus, M. Peev, Rev. Mod. Phys. 81 (2009) 1301.
[54] Daniel F.V. James, Paul G. Kwiat, William J. Munro, Andrew G. White, Phys. Rev. A 64 (2001) 052312.
[55] Zhibo Hou, Guoyong Xiang, Daoyi Dong, Chuan-Feng Li, Guang-Can Guo, Opt. Express 23 (2015) 10018.
[56] Jeremy L. O'Brien, Science 318 (2007) 1567.
[57] D. Bhatti, J. von Zanthier, G.S. Agarwal, Phys. Rev. A 91 (2015) 062303.
[58] Adrien Nicolas, Lucile Veissier, Elisabeth Giacobino, Dominik Maxein, Julien Laurat, New J. Phys. 17 (2015) 033037.
[59] L.D. Franson, Phys. Rev. Lett. 62 (1989) 2205.
[60] J.D. Franson, Phys. Rev. A 44 (1991) 4552.
[61] I. Marcikic, H. de Riedmatten, W. Tittel, V. Scarani, H. Zbinden, N. Gisin, Phys. Rev. A 66 (2002) 062308.
[62] Giuseppe Vallone, Daniele Dequal, Marco Tomasin, Francesco Vedovato, Matteo Schiavon, Vincenza Luceri, Giuseppe Bianco, Paolo Villoresi, Phys. Rev. Lett. 116 (2016) 253601.
[63] S. Tanzilli, A. Martin, F. Kaiser, M.P. DeMicheli, O. Alibart, D.B. Ostrowsky, Laser Photonics Rev. 6 (2012) 115.

[64] F. Monteiro, E. Verbanis, V. Caprara Vivoli, A. Martin, N. Gisin, H. Zbinden, R.T. Thew, Quantum Sci. Technol. 2 (2017) 024008.
[65] P. Walther, K.J. Resch, T. Rudolph, E. Schenck, H. Weinfurter, V. Vedral, M. Aspelmeyer, A. Zeilinger, Nature 434 (2005) 169.
[66] Alexander S. Solntsev, Andrey A. Sukhorukov, Rev. Phys. 2 (2017) 19.
[67] J.W. Silverstone, R. Santagati, D. Bonneau, M.J. Strain, M. Sorel, J.L. O'Brien, M.G. Thompson, Nat. Commun. 6 (2015) 7948.
[68] L. Olislager, J. Cussey, A.T. Nguyen, P. Emplit, S. Massar, J.-M. Merolla, K. Phan Huy, Phys. Rev. A 82 (2010) 013804.
[69] E.H. Huntington, T.C. Ralph, Phys. Rev. A 69 (2004) 042318.
[70] Hsuan-hao Lu, Andrew Weiner, Pavel Lougovski, Joseph M. Lukens, IEEE Photonics Technol. Lett. 31 (2019) 1858.
[71] Joseph M. Lukens, Pavel Lougovski, Optica 4 (2017) 8.
[72] Xianxin Guo, Yefeng Mei, Shengwang Du, Optica 4 (2017) 388.
[73] Benjamin Brecht, Andreas Eckstein, Raimund Ricken, Viktor Quiring, Hubertus Suche, Linda Sansoni, Christine Silberhorn, Phys. Rev. A 90 (2014) 030302(R).
[74] M. Bloch, S.W. McLaughlin, J-M. Merolla, F. Patois, Opt. Lett. 32 (2007) 301.
[75] S. Wiesner, Conjugate coding, ACM SIGACT News 15 (1983) 78.
[76] H. Salih, Z.-H. Li, M. Al-Amri, M.S. Zubairy, Phys. Rev. Lett. 110 (2013) 170502.
[77] T.-G. Noh, Phys. Rev. Lett. 103 (2009) 230501.
[78] C.H. Bennett, G. Brassard, N.D. Mermin, Phys. Rev. Lett. 68 (1992) 557.
[79] C.H. Bennett, Phys. Rev. Lett. 68 (1992) 3121.
[80] J. Barrett, R. Colbeck, A. Kent, Phys. Rev. Lett. 106 (2013) 010503.
[81] H.-K. Lo, M. Curty, B. Qi, Phys. Rev. Lett. 108 (2012) 130503.
[82] T.C Ralph, Phys. Rev. A 61 (1999) 010303.
[83] L.S Madsen, V.C. Usenko, M. Lassen, R. Filip, U.L. Andersen, Nat. Commun. 3 (2012) 1083.
[84] H. Bechmann-Pasquinucci, N. Gisin, Phys. Rev. A 59 (1999) 4238.
[85] K. Boström, T. Felbinger, Phys. Rev. Lett. 89 (2002) 187902.
[86] S. Pirandola, U.L. Andersen, L. Banchi, M. Berta, D. Bunandar, R. Colbeck, D. Englund, T. Gehring, C. Lupo, C. Ottaviani, J. Pereira, M. Razavi, J.S. Shaari, M. Tomamichel, V.C. Usenko, G. Vallone, P. Villoresi, P. Wallden, https://arxiv.org/abs/1906.01645v1, 2019.
[87] W. Tittel, G. Ribordy, N. Gisin, Phys. World (March 1998) 41.
[88] Hoi-Kwong Lo, Marcos Curty, Kiyoshi Tamaki, Nat. Photonics 8 (2014) 595.
[89] Zheng-Hong Li, Luojia Wang, Jingping Xu, Yaping Yang, M. Al-Amri, M. Suhail Zubairy, Phys. Rev. A 101 (2020) 022336.
[90] Eleni Diamanti, Hoi-Kwong Lo, Bing Qi, Zhiliang Yuan, npj Quantum Inf. 2 (2016) 16025.
[91] Feihu Xu, Xiongfeng Ma, Qiang Zhang, Hoi-Kwong Lo, Jian-Wei Pan, arXiv:1903.09051v1, 2019.
[92] C.H. Bennett, G. Brassard, C. Crepeau, R. Jozsa, A. Peres, W.K. Wootters, Phys. Rev. Lett. 70 (1993) 1895.
[93] D. Boschi, S. Branca, F. De Martini, L. Hardy, S. Popescu, Phys. Rev. Lett. 80 (1998) 1121.
[94] D. Bouwmeester, J.W. Pan, K. Mattle, M. Eibl, H. Weinfurter, A. Zeilinger, Nature 390 (1997) 575.
[95] S. Pirandola, J. Eisert, C. Weedbrook, A. Furusawa, S.L. Braunstein, Nat. Photonics 9 (2015) 641.
[96] M. Al-Amri, Jörg Evers, M. Suhail Zubairy, Phys. Rev. A 82 (2010) 022329.
[97] Zheng-Hong Li, M. Al-Amri, Xi-Hua Yang, M. Suhail Zubairy, Phys. Rev. A 100 (2019) 022110.
[98] Benjamin J. Metcalf, Justin B. Spring, Peter C. Humphreys, Nicholas Thomas-Peter, Marco Barbieri, W. Steven Kolthammer, Xian-Min Jin, Nathan K. Langford, Dmytro Kundys, James C. Gates, Brian J. Smith, Peter G.R. Smith, Ian A. Walmsley, Nat. Photonics 8 (2014) 770.
[99] Xiao-Song Ma, Thomas Herbst, Thomas Scheidl, Daqing Wang, Sebastian Kropatschek, William Naylor, Bernhard Wittmann, Alexandra Mech, Johannes Kofler, Elena Anisimova, Vadim Makarov, Thomas Jennewein, Rupert Ursin, Anton Zeilinger, Nature 489 (2012) 269.

[100] Juan Yin, Ji-Gang Ren, He Lu, Yuan Cao, Hai-Lin Yong, Yu-Ping Wu, Chang Liu, Sheng-Kai Liao, Fei Zhou, Yan Jiang, Xin-Dong Cai, Ping Xu, Ge-Sheng Pan, Jian-Jun Jia, Yong-Mei Huang, Hao Yin, Jian-Yu Wang, Yu-Ao Chen, Cheng-Zhi Peng, Jian-Wei Pan, Nature 488 (2012) 185.
[101] Yi-Han Luo, Han-Sen Zhong, Manuel Erhard, Xi-Lin Wang, Li-Chao Peng, Mario Krenn, Xiao Jiang, Li Li, Nai-Le Liu, Chao-Yang Lu, Anton Zeilinger, Jian-Wei Pan, Phys. Rev. Lett. 123 (2019) 070505.
[102] C.H. Bennett, S.J. Wiesner, Phys. Rev. Lett. 69 (1992) 2881.
[103] K. Mattle, H. Weinfurter, P.G. Kwiat, A. Zeilinger, Phys. Rev. Lett. 76 (1996) 4656.
[104] R. Ursin, F. Tiefenbacher, T. Schmitt-Manderbach, H. Weier, T. Scheidl, M. Lindenthal, B. Blauensteiner, T. Jennewein, J. Perdigues, P. Trojek, et al., Nat. Phys. 3 (2007) 481.
[105] C. Kurtsiefer, P. Zarda, M. Halder, H. Weinfurter, P.M. Gorman, P.R. Tapster, J.G. Rarity, Nature 419 (2002) 450.
[106] R.J. Hughes, J.E. Nordholt, D. Derkacs, C.G. Peterson, New J. Phys. 4 (2002) 43.
[107] J.G. Rarity, P.R. Tapster, P.M. Gorman, P. Knight, New J. Phys. 4 (2002) 82.1–82.21.
[108] M. Aspelmeyer, T. Jennewein, M. Pfennigbauer, W. Leeb, A. Zeilinger, IEEE J. Sel. Top. Quantum Electron. 9 (2003) 1541.
[109] X.M. Jin, J.G. Ren, B. Yang, Z.H. Yi, F. Zhou, X.F. Xu, S.K. Wang, D. Yang, Y.F. Hu, S. Jiang, et al., Nat. Photonics 4 (6) (2010) 376.
[110] X.S. Ma, T. Herbst, T. Scheidl, D. Wang, S. Kropatschek, W. Naylor, B. Wittmann, A. Mech, J. Kofler, E. Anisimova, et al., Nature 489 (7415) (2012) 269.
[111] J. Yin, J.G. Ren, H. Lu, Y. Cao, H.L. Yong, Y.P. Wu, C. Liu, S.K. Liao, F. Zhou, Y. Jiang, et al., Nature 488 (7410) (2012) 185.
[112] Hua-Ying Liu, Xiao-Hui Tian, Changsheng Gu, Pengfei Fan, Xin Ni, Ran Yang, Ji-Ning Zhang, Mingzhe Hu, Jian Guo, Xun Cao, Xiaopeng Hu, Gang Zhao, Yan-Qing Lu, Yan-Xiao Gong, Zhenda Xie, Shi-Ning Zhu, Phys. Rev. Lett. 126 (2021) 020503.
[113] J.-P. Bourgoin, B.L. Higgins, N. Gigov, C. Holloway, C.J. Pugh, S. Kaiser, M. Cranmer, T. Jennewein, Opt. Express 23 (26) (2015) 33437.
[114] S. Nauerth, F. Moll, M. Rau, C. Fuchs, J. Horwath, S. Frick, H. Weinfurter, Nat. Photonics 7 (5) (2013) 382.
[115] Christopher J. Pugh, Sarah Kaiser, Jean-Philippe Bourgoin, Jeongwan Jin, Nigar Sultana, Sascha Agne, Elena Anisimova, Vadim Makarov, Eric Choi, Brendon L. Higgins, Thomas Jennewein, Quantum Sci. Technol. 2 (2017) 024009.
[116] J.-Y. Wang, B. Yang, S.-K. Liao, L. Zhang, Q. Shen, X.-F. Hu, J.-C. Wu, S.-J. Yang, H. Jiang, Y.-L. Tang, B. Zhong, H. Liang, W.-Y. Liu, Y.-H. Hu, Y.-M. Huang, B. Qi, J.-G. Ren, G.-S. Pan, J. Yin, J.-J. Jia, Y.-A. Chen, K. Chen, C.-Z. Peng, J.-W. Pan, Nat. Photonics 7 (5) (2013) 387.
[117] Hyunchae Chun, Iris Choi, Grahame Faulkner, Larry Clarke, Bryan Barber, Glenn George, Colin Capon, Antti Niskanen, Joachim Wabnig, Dominic O. Brien, David Bitauld, Opt. Express 25 (6) (2017) 6784.
[118] J.L. Duligall, M.S. Godfrey, K.A. Harrison, W.J. Munro, J.G. Rarity, New J. Phys. 8 (2006) 249.
[119] Gwenaelle Melen, Tobias Vogl, Markus Rau, Giacomo Corrielli, Andrea Crespi, Roberto Osellame, Harald Weinfurter, IEEE J. Sel. Top. Quantum Electron. 21 (2015) 6600607.
[120] P. Sibson, C. Erven, M. Godfrey, S. Miki, T. Yamashita, M. Fujiwara, M. Sasaki, H. Terai, M.G. Tanner, C.M. Natarajan, R.H. Hadfield, J.L.O. Brien, M.G. Thompson, Nat. Commun. 8 (2017) 13984.
[121] Hideki Takenaka, Alberto Carrasco-Casado, Mikio Fujiwara, Mitsuo Kitamura, Masahide Sasaki, Morio Toyoshima, Nat. Photonics 11 (2017) 502.
[122] https://www.sciencemag.org/news/2017/06/china-s-quantum-satellite-achieves-spooky-action-record-distance.
[123] https://www.nature.com/news/chinese-satellite-is-one-giant-step-for-the-quantum-internet-1.20329.
[124] John M. Donohue, Megan Agnew, Jonathan Lavoie, Kevin J. Resch, Phys. Rev. Lett. 111 (2013) 153602.
[125] B. Hensen, H. Bernien, A.E. Dréau, A. Reiserer, N. Kalb, M.S. Blok, J. Ruitenberg, R.F.L. Vermeulen, R.N. Schouten, C. Abellán, W. Amaya, V. Pruneri, M.W. Mitchell, M. Markham, D.J. Twitchen, D. Elkouss, S. Wehner, T.H. Taminiau, R. Hanson, Nature 526 (2015) 682.

[126] Lynden K. Shalm, Evan Meyer-Scott, Bradley G. Christensen, Peter Bierhorst, Michael A. Wayne, Martin J. Stevens, Thomas Gerrits, Scott Glancy, Deny R. Hamel, Michael S. Allman, Kevin J. Coakley, Shellee D. Dyer, Carson Hodge, et al., Phys. Rev. Lett. 115 (2015) 250402.

[127] Marissa Giustina, Marijn A.M. Versteegh, Sören Wengerowsky, Johannes Handsteiner, Armin Hochrainer, Kevin Phelan, Fabian Steinlechner, Johannes Kofler, Jan-Åke Larsson, Carlos Abellán, Waldimar Amaya, Valerio Pruneri, Morgan W. Mitchell, et al., Phys. Rev. Lett. 115 (2015) 250401.

[128] D. Rideout, T. Jennewein, G. Amelino-Camelia, T.F. Demarie, B.L. Higgins, A. Kempf, A. Kent, R. Laflamme, X. Ma, R.B. Mann, et al., Class. Quantum Gravity 29 (2012) 224011.

[129] F. Vedovato, C. Agnesi, M. Schiavon, D. Dequal, L. Calderaro, M. Tomasin, D.G. Marangon, A. Stanco, V. Luceri, G. Bianco, et al., Sci. Adv. 3 (2017) e1701180.

[130] J. Yin, Y. Cao, Y.-H. Li, S.-K. Liao, L. Zhang, J.-G. Ren, W.-Q. Cai, W.-Y. Liu, B. Li, H. Dai, et al., Science 356 (2017) 1140.

[131] S.-K. Liao, W.-Q. Cai, W.-Y. Liu, et al., Nature 549 (2017) 43.

[132] L. Calderaro, C. Agnesi, D. Dequal, F. Vedovato, M. Schiavon, A. Santamato, V. Luceri, G. Bianco, G. Vallone, P. Villoresi, Quantum Sci. Technol. 4 (2019) 015012.

[133] N. Hosseinidehaj, Z. Babar, R. Malaney, S.X. Ng, L. Hanzo, IEEE Commun. Surv. Tutor. 21 (2019) 881.

[134] D. Dequal, L.T. Vidarte, V.R. Rodriguez, G. Vallone, P. Villoresi, A. Leverrier, E. Diamant, npj Quantum Inf. 7 (2021) 3.

[135] A. Trabesinger, Nat. Phys. 8 (2012) 263.

[136] K.M. Svore, M.M. Troyer, Computer 49 (2016) 21.

[137] D.-S. Ding, W. Zhang, Z.-Y. Zhou, S. Shi, J.-s. Pan, G.-Y. Xiang, X.-S. Wang, Y.-K. Jiang, B.-S. Shi, G.-C. Guo, Phys. Rev. A 90 (2014) 042301.

[138] J. Bavaresco, N. Herrera Valencia, C. Klöckl, M. Pivoluska Erker, N. Friis, M. Malik, M. Huber, Nat. Phys. 14 (2018) 1032.

[139] J. Trapateau, J. Ghalbouni, A. Orieux, E. Diamanti, I. Zaquine, J. Appl. Phys. 118 (2015) 143106.

[140] P. Zhang, K. Aungskunsiri, E. Martín-López, J. Wabnig, M. Lobino, R.W. Nock, J. Munns, D. Bonneau, P. Jiang, H.W. Li, A. Laing, J.G. Rarity, A.O. Niskanen, M.G. Thompson, J.L. O'Brien, Phys. Rev. Lett. 112 (2014) 130501.

[141] B. Korzh, N. Walenta, R. Houlmann, H. Zbinden, Opt. Express 21 (2013) 19579.

[142] K.A. Patel, J.F. Dynes, M. Lucamarini, I. Choi, A.W. Sharpe, Z.L. Yuan, R.V. Penty, A.J. Shields, Appl. Phys. Lett. 104 (2014) 051123.

[143] T. Honjo, K. Inoue, H. Takahashi, Opt. Lett. 29 (2004) 2797.

CHAPTER 2

Structured light

M. Babiker[a], V.E. Lembessis[b], Koray Köksal[c], and J. Yuan[a]

[a]Department of Physics, University of York, York, United Kingdom [b]Quantum Technology Group, Department of Physics and Astronomy, College of Science, King Saud University, Riyadh, Saudi Arabia [c]Physics Department, Bitlis Eren University, Bitlis, Turkey

Contents

- 2.1 Introduction 38
- 2.2 Optical angular momentum 40
- 2.3 Helmholtz equation and paraxial regime 40
  - 2.3.1 Linearly polarized 'unstructured' light 40
  - 2.3.2 Elliptically polarized 41
- 2.4 Structured light 42
  - 2.4.1 Phase-structured light 43
  - 2.4.2 Laguerre-Gaussian (LG) light beams 44
- 2.5 Bessel and Bessel-Gaussian vortex beams 46
  - 2.5.1 Bessel vortex beams 46
  - 2.5.2 Bessel-Gaussian vortex beams 47
- 2.6 Non-paraxial LG beams 47
  - 2.6.1 Extracting the paraxial regime 49
- 2.7 Paraxial beams with small waists 51
- 2.8 Chirality and helicity 52
  - 2.8.1 Cycle-averaged fields 52
  - 2.8.2 Effects of the Gouy and curvature phases 55
- 2.9 Multiple vortex beams 57
  - 2.9.1 Linearly polarized LG beams 57
  - 2.9.2 Axial shift 58
    - 2.9.2.1 *Effects of frequency shift* 60
    - 2.9.2.2 *Ferris wheels and conveyor belts* 60
    - 2.9.2.3 *Radial gaps in double rings* 61
- 2.10 No axial shift—polarization gradients 61
  - 2.10.1 Co-propagating LG beams 62
    - 2.10.1.1 *Co-propagating with $\sigma^+ - \sigma^-$* 62
    - 2.10.1.2 *Lin $\perp$ Lin polarizations* 63
  - 2.10.2 Counter-propagating beams 65

         2.10.2.1 $\sigma^+ - \sigma^-$ polarizations 65
         2.10.2.2 Counter-propagating with Lin $\perp$ Lin 66
     2.10.3 Bi-chromatic vortex beams 67
         2.10.3.1 Bi-chromatic co-propagating with $\sigma^+ - \sigma^-$ 68
         2.10.3.2 Bi-chromatic with Lin $\perp$ Lin 70
2.11 **Quantization of optical angular momentum** 71
     2.11.1 Quantized SAM 72
         2.11.1.1 Circularly polarized light 72
         2.11.1.2 Linearly polarized light 72
     2.11.2 Orbital angular momentum 73
2.12 **Conclusions** 73
References 75

2.1 Introduction

We should begin by explaining what specifically we mean by 'structured light' in the context of this book as this term is known to have much wider uses in various other contexts. In particular, it features in vision systems whereby light is projected through specific patterns such as grids or sieves onto objects so enabling the determination of surface information of the objects, as, for example in 3D scanners. This chapter and the rest of the book, however, are focused on a special type of structured light, namely laser light endowed with the property of orbital angular momentum in addition to its spin angular momentum. The specific structured light we are concerned with is phase-structured light in the form of optical vortex beams, which as an area of research, stemmed from the work by Allen et al. [1] who were the first to suggest that light beams can be made to carry orbital angular momentum. The laser beams of such light is often referred to as optical vortices, or twisted light. This area is now well established as an active branch of optical physics and there now exists a good number of sources on optical vortices, which the reader's attention is drawn to, including reviews, edited books and special issues, dealing with fundamentals as well as applications [2–11].

The primary aim in this chapter is to outline some features of structured light of this kind that would be relevant to optical communication. There are two key properties of structured light, namely its spin angular momentum (SAM) associated with its wave polarization and its orbital angular momentum (OAM) associated with its twisted wavefronts. The OAM property has been recognized as providing the means in the form of additional higher degrees of freedom, which allow an enhancement of the capacity for optical information transfer. Using the unbounded set of OAM quantum states allows the realization of a large number of quantum communication channels. Application of structured light in quantum communication is the subject of a number of chapters in this volume.

Structured light of the optical vortex kind comes in a variety of forms and can be generated using many different techniques. Currently the best known techniques for optical vortex generation are spatial light modulators [12], spiral phase plates [13], q-plates [14] and cylindrical lenses [15].

We aim to outline the background theory of optical vortices and their spin and orbital angular momentum properties, highlighting, in particular, situations where both SAM and OAM are involved and when vortex beams interfere. We also briefly describe the essential features of a few of the most widely discussed vortex beams, but elaborate a little on the non-paraxial as well as the paraxial regimes of the Laguerre-Gaussian type and show how the paraxial regime emerges from the non-paraxial formalism.

Each optical vortex, or twisted light beam, is phase-structured in that it is characterized by an azimuthal phase function $\exp(i\ell\phi)$, which endows it with an orbital angular momentum of magnitude $\ell\hbar$ per photon, where ℓ is the winding number and integer $\ell \neq 0$ can be positive or negative. In addition to the Laguerre-Gaussian type of vortex light the Bessel and Mathieu modes have also featured [16–18] and we aim to discuss such types, which are endowed with the OAM property.

The general flow of this chapter is as follows. The standard formalism of the optical angular momentum is briefly stated, emphasizing the formal division into spin angular momentum (SAM) and orbital angular momentum (OAM) [19]. Next we consider, also in general terms, the OAM of a vortex or twisted light beam, the chirality and helicity of such light beams [20–26] and aim to highlight the main features of the most widely discussed vortex beams, namely Bessel beams, and Laguerre-Gaussian beams. For the latter we broaden the discussion to include the non-paraxial regime, before dealing in some lengths with the paraxial regime as a limit of the non-paraxial theory. We explore with details how OAM and optical spin combine in the case of multiple beams to generate novel polarization gradients. We focus only on the two-beam cases, which include co-propagating and counter-propagating configurations and couple those with different types of polarization arrangements, namely circular polarizations, the case of orthogonal linear polarizations and the case of two beams with polarization vectors inclined at angle γ. We also discuss the case of bi-chromatic beams, differing slightly in frequency and also when the focal planes of the beams are shifted so creating novel interference patterns in the region between the focal planes.

Finally, we briefly describe the quantization of the optical spin angular momentum carried by both structured and unstructured light and we also outline the essential quantum formalism for OAM of structured light, as exemplified by the LG beam set.

2.2 Optical angular momentum

The electromagnetic fields associated with optical beams are transverse, designated by the symbol $\mathbf{E}^\perp$ satisfying $\nabla \cdot \mathbf{E}^\perp = 0$. The optical angular momentum density vector is defined as the moment of the linear momentum density $\mathbf{P} = \epsilon_0 \mathbf{E} \times \mathbf{B}$ about a chosen origin $\mathbf{r}$

$$\mathbf{J}(\mathbf{r}) = \epsilon_0 \int d^3\mathbf{r}' (\mathbf{r}' - \mathbf{r}) \times \mathbf{E}^\perp(\mathbf{r}') \times \mathbf{B}(\mathbf{r}'). \tag{2.1}$$

We now express $\mathbf{B}$ in terms of the vector potential $\mathbf{A}^\perp$. It can then be shown that the total optical angular momentum splits into two parts [19]

$$\mathbf{J}(\mathbf{r}) = \mathbf{S} + \mathbf{L}(\mathbf{r}) \tag{2.2}$$

where $\mathbf{S}$ is identified as the spin angular momentum (SAM)

$$\mathbf{S} = \epsilon_0 \int d^3\mathbf{r}' [\mathbf{E}^\perp(\mathbf{r}') \times \mathbf{A}^\perp(\mathbf{r}')] \tag{2.3}$$

and $\mathbf{L}(\mathbf{r})$ is identified as the orbital angular momentum (OAM)

$$\mathbf{L}(\mathbf{r}) = \epsilon_0 \sum_i \int d^3\mathbf{r}' E_i^\perp \{(\mathbf{r}' - \mathbf{r}) \times \nabla'\} A_i^\perp(\mathbf{r}'). \tag{2.4}$$

Note that $\mathbf{L}(\mathbf{r})$, the OAM, depends on the choice of origin $\mathbf{r}$, while $\mathbf{S}$, the SAM, does not. It is for this reason that SAM is identified as an intrinsic angular momentum.

2.3 Helmholtz equation and paraxial regime

2.3.1 Linearly polarized 'unstructured' light

Consider a monochromatic linearly polarized light beam propagating in the z-direction with frequency ω and wavenumber $k_z = \omega^2/c^2$. The electric field vector is written $\mathbf{E} = \hat{\epsilon}\mathcal{E}$ where $\hat{\epsilon}$ is the linear polarization vector and $\mathcal{E}$ is the scalar function, which is a solution of the Helmholtz equation

$$\nabla^2 \mathcal{E} + \frac{\omega^2}{c^2}\mathcal{E} = 0. \tag{2.5}$$

The scalar function $\mathcal{E}$ is a product of an amplitude function $\mathcal{U}$ and a phase function $\exp i\theta$. Thus

$$\mathcal{E} = \mathcal{U} e^{i\theta}. \tag{2.6}$$

Structured light

For a light beam that is not phase-structured, the phase function is $\theta = k_z z - \omega t$. We have, on substituting in the Helmholtz equation,

$$\nabla_\perp^2 \mathcal{U} + \frac{\partial^2 \mathcal{U}}{\partial z^2} + 2ik_z \frac{\partial \mathcal{U}}{\partial z} = 0 \tag{2.7}$$

where $\nabla_\perp^2 = \frac{\partial^2}{\partial x^2} + \frac{\partial^2}{\partial y^2}$. A version of wave equation for the amplitude function $\mathcal{U}$ emerges on dropping the term $\partial^2 U/\partial z^2$, which is regarded as small relative to the term $k\partial U/\partial z$. Thus we obtain from Eq. (2.7) on dropping the second term

$$i\frac{\partial \mathcal{U}}{\partial z} = -\frac{1}{2k_z}\nabla_\perp^2 \mathcal{U}. \tag{2.8}$$

This is the paraxial approximation, which is widely used and identified as providing a reasonable description of optical beam propagation along the z-axis. Physically it emphasizes the fact that the beam profile for most laser beams changes slowly with axial position z relative to its profile at the focal plane. Note that Eq. (2.8) can be identified as the Schrödinger equation in two dimensions. It describes how the variations in the transverse (x–y) plane change with the axial position. Hence in this Schrödinger equation the axial coordinate z assumes the role of time t in the case of quantum mechanics.

The magnetic field associated with Eq. (2.6) can be evaluated and used to evaluate the time-averaged Poynting vector. We find

$$\epsilon_0 \langle \mathbf{E} \times \mathbf{B} \rangle = \frac{i\omega\epsilon_0}{2}[(\mathcal{U}\nabla\mathcal{U}^* - \mathcal{U}^*\nabla\mathcal{U}) - 2ik_z|\mathcal{U}|^2\hat{z}]. \tag{2.9}$$

Eq. (2.9) is the cycle-averaged linear momentum density vector of the light.

2.3.2 Elliptically polarized

The above formalism for linearly polarized unstructured light can be generalized to cover the case of circularly polarized or (in general) elliptically polarized light so that the electric field has components in both x- and y-directions and the general polarization vector is

$$\hat{\epsilon} = \alpha\hat{x} + \beta\hat{y}, \tag{2.10}$$

where α and β are, in general, complex constants. The electric field vector in this case satisfies the equation in Cartesian coordinates

$$\mathbf{E} = \left\{(\alpha\hat{x} + \beta\hat{y})\mathcal{U} + \frac{ic}{\omega}\left(\alpha\frac{\partial \mathcal{U}}{\partial x} + \beta\frac{\partial \mathcal{U}}{\partial y}\right)\hat{z}\right\}e^{i\theta}. \tag{2.11}$$

This general case covers the special scenarios of circularly polarized light for which $\alpha = 1/\sqrt{2}$ and $\beta = \pm i/\sqrt{2}$ in which case we have

$$|\alpha|^2 + |\beta|^2 = 1 \quad (\alpha\beta^* + \alpha^*\beta) = 0. \tag{2.12}$$

The left and right circular polarization are such that

$$i(\alpha\beta^* - \alpha^*\beta) = \sigma \tag{2.13}$$

with $\sigma = \pm 1$.

The above formalism has been applied for a variety of ordinary laser beams such as the Gaussian beams whose phase is unstructured. The situation changes when the light beam is phase-structured as is the case with the Laguerre-Gaussian beams.

2.4 Structured light

Consider a general paraxial field that has the following amplitude and phase function in cylindrical coordinates $\mathbf{r} = (\rho, \phi, z)$:

$$\mathcal{E} = \mathcal{U}(\rho, z) e^{i\ell\phi} \tag{2.14}$$

The appearance of the phase function $\exp(i\ell\phi)$ is a signature of a vortex beam, or structured light carrying orbital angular momentum. Eq. (2.9) is applicable to this field, so on expressing the ∇ operator in cylindrical polar coordinates we find that the azimuthal component of $\epsilon_0 \langle \mathbf{E} \times \mathbf{B} \rangle$ as given in Eq. (2.9) is as follows:

$$\epsilon_0 \langle \mathbf{E} \times \mathbf{B} \rangle_\phi = \frac{\epsilon_0 \omega \ell}{\rho} |\mathcal{U}|^2. \tag{2.15}$$

The angular momentum density j_z is then obtained as the vector product of $\mathbf{r}$ with $\epsilon_0 \langle \mathbf{E} \times \mathbf{B} \rangle_\phi$

$$j_z = \epsilon_0 \mathbf{r} \times \langle \mathbf{E} \times \mathbf{B} \rangle_\phi = \epsilon_0 \omega \ell |\mathcal{U}|^2. \tag{2.16}$$

The energy density of the light carried in the z-direction is $w = c\epsilon_0 \langle \mathbf{E} \times \mathbf{B} \rangle_z$, which is (using $ck = \omega$)

$$w = \epsilon_0 \omega^2 |\mathcal{U}|^2. \tag{2.17}$$

We therefore have

$$\frac{j_z}{w} = \frac{\ell}{\omega}. \tag{2.18}$$

Since the numerator and the denominator of the right-hand side are constants, we may integrate each over a normal cross-section of the light beam. This leads to the ratio of the total angular momentum per unit length J_z to the energy per unit length W

$$\frac{J_z}{W} = \frac{\int\int \rho d\rho d\phi \, j_z}{\int\int \rho d\rho d\phi \, w} = \frac{\ell}{\omega}. \tag{2.19}$$

2.4.1 Phase-structured light

A phase-structured light beam is one in which the scalar function $\mathcal{E}$ has the formal expression in Eq. (2.6), but with a general phase function $\exp(i\Theta)$, which includes $\exp(i\theta)$ as in Eq. (2.14), in addition to other phase functions. We have

$$\mathcal{E} = \mathcal{U} e^{i\Theta} \tag{2.20}$$

The amplitude function $\mathcal{U}$ is such that it is normalized. It also satisfies the condition of the paraxial regime and leads to a finite energy of the beam. Θ is the structured phase function, which includes characteristic terms the most significant of which is $\ell\phi$, in addition to $\theta = (k_z z - \omega t)$. Since Θ is now a function of x and y as well as z the differential operators in Eq. (2.7) now pick up terms involving differentiation of Θ with respect to x and y and we have

$$\mathbf{E} = \left\{ (\alpha\hat{\mathbf{x}} + \beta\hat{\mathbf{y}})\mathcal{U} + \frac{1}{k}\left[i\left(\alpha\frac{\partial\mathcal{U}}{\partial x} + \beta\frac{\partial\mathcal{U}}{\partial y}\right) - \mathcal{U}\left(\alpha\frac{\partial\Theta}{\partial x} + \beta\frac{\partial\Theta}{\partial y}\right) \right]\hat{\mathbf{z}} \right\} e^{i\Theta}. \tag{2.21}$$

A similar evaluation of linear momentum density in this will lead to the analogue of Eq. (2.9) as follows:

$$\epsilon_0 \langle \mathbf{E} \times \mathbf{B} \rangle = \frac{i\omega\epsilon_0}{2}\left[(\mathcal{U}\nabla\mathcal{U}^* - \mathcal{U}^*\nabla\mathcal{U}) - 2ik_z|\mathcal{U}|^2\hat{\mathbf{z}} + (\alpha\beta^* - \alpha^*\beta)\left\{ \hat{\mathbf{x}}\frac{\partial}{\partial y} - \hat{\mathbf{y}}\frac{\partial}{\partial x} \right\}|\mathcal{U}|^2 \right]. \tag{2.22}$$

In view of the definition of the spin σ in Eq. (2.13) we identify the last term in Eq. (2.22) as a spin-dependent contribution. Evaluations of the derivatives in cylindrical polar coordinates lead after, some algebra [2], to the analogue of Eq. (2.19)

$$\frac{J_z}{W} = \frac{\ell + \sigma}{\omega}. \tag{2.23}$$

The above general formalism applies to elliptically polarized *phase-structured light*. We first consider the Laguerre-Gaussian (LG) set of light beams since they are the prototypical phase-structured light beams and are the most discussed and applied in the context of twisted light or optical vortices.

2.4.2 Laguerre-Gaussian (LG) light beams

The analytical forms of the amplitude function and the phase function of the LG set of phase-structured light beams are best displayed in cylindrical polar coordinates $\mathbf{r} = (\rho, \phi, z)$. Each beam propagates along the z-axis with frequency ω and axial wave vector k. Each LG mode is characterized by two integers: the winding number ℓ, which is a non-zero integer and $p \geq 0$ is the radial number. The amplitude function is

$$U_{k_z l p}(\rho, \phi, z) = \frac{\mathcal{U}_{k00} C_{|\ell|, p}}{\sqrt{1 + z^2/z_R^2}} \left(\frac{\rho \sqrt{2}}{w_0 \sqrt{1 + z^2/z_R^2}} \right)^{|\ell|}$$
$$\times \exp\left[-\frac{2\rho^2}{w_0^2(1 + z^2/z_R^2)} \right] L_p^{|\ell|} \left(\frac{2\rho^2}{w_0^2(1 + z^2/z_R^2)} \right) \quad (2.24)$$

Here $L_p^{|\ell|}$ are the associated Laguerre polynomials; $\mathcal{U}_{k00}$ is the amplitude for a corresponding plane wave of wave vector k; $C_{|\ell|,p} = \sqrt{\frac{p!}{(|\ell|+p)!}}$ and $w(z)$ is the beam waist at axial position z such that $w^2(z) = 2(z^2 + z_R^2)/k_z z_R$, where $z_R = w_0^2 k_z/2$ is the Rayleigh range with $w_0 = w(0)$ the beam width at the focal plane at $z = 0$. For the phase function we have

$$\Theta_{k_z l p} = k_z z + l\phi + \Theta_{Gouy} + \Theta_{curv} - \omega t, \quad (2.25)$$

where

$$\Theta_{Gouy} = -(2p + |l| + 1)\tan^{-1}(z/z_R) \quad \Theta_{curv} = \frac{k_z \rho^2 z}{2(z^2 + z_R^2)}. \quad (2.26)$$

The first term in the phase function is the usual term representing plane wave propagation with axial wave vector k_z and the second term is the azimuthal phase, which gives rise to ℓ intertwined helical wavefronts and is the basis for the orbital angular momentum content of the beam. The third term is the Gouy phase and the final term is a phase contribution due to the variation of the beam curvature with both ρ and z. The characteristic phase term $\exp(i\ell\phi)$ indicates that we have a phase vortex everywhere on the beam axis $\rho = 0$. The azimuthal phase function is in fact an eigenfunction of the orbital angular momentum operator $\hat{L}_z = -i\hbar\partial/\partial\phi$ such that

$$\hat{L}_z(e^{i\ell\phi}) = \ell\hbar(e^{i\ell\phi}). \quad (2.27)$$

The eigenvalues are $\ell\hbar$. This is the angular momentum carried by a single photon of a Laguerre-Gaussian beam $LG_{\ell,p}$.

Structured light 45

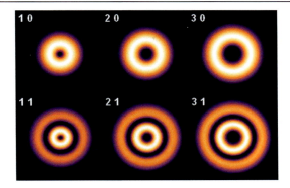

Figure 2.1: The intensity distributions at the focal plane of some Laguerre-Gaussian beams $LG_{\ell p}$. The single-ring modes ℓ, p on the first row are doughnut modes for which $\ell = 1, 2, 3$ and $p = 0$. The second row are two-ring modes for which $\ell = 1, 2, 3$ and $p = 1$.

Fig. 2.1 displays the intensity distribution of the LG modes $LG_{\ell,p}$ showing only the lowest values of ℓ and p. On the first row are single-ring modes $(\ell, p) = (1, 0), (2, 0), (3, 0)$ called the doughnut modes and $(\ell, p) = (1, 1), (2, 1), (3, 1)$ with two rings. In general for an $LG_{\ell p}$ beam there are $p + 1$ rings. Note that the doughnut modes are much discussed in optical vortex applications and are often highlighted as the simplest modes displaying the OAM properties of the Laguerre-Gaussian modes. The ring of a doughnut mode has a radius at high intensity given by $\rho_{max} = w_0 \sqrt{|\ell|/2}$.

In addition to the value and sign of the winding number ℓ and value of the radial number $p \geq 0$ the Laguerre-Gaussian set is characterized by a set of other beam parameters. These can be identified from the analytical form given above and they are as follows. In addition to the frequency of the light ω and the axial wave vector k_z, a key parameter is the beam waist w_0, which is related to the Rayleigh range $z_R = \pi w_0^2 / \lambda$, where λ is the wavelength.

Small values of z_R or w_0 indicate strong focusing when the LG beam is generated from an ordinary laser beam using, for example, a spatial light modulator (SLM). Tightly focused laser light can now be generated with spots in the sub-wavelength scale. This should not be confused with a situation where an LG light of a relatively large beam waist once produced can itself be focused further to tight spots of sub-wavelength dimensions (see [27] and references therein). As we explain later, tightly focused beams, including linearly polarized beams, have been shown to possess a substantial longitudinal electric field component and associated magnetic field components [27]. These additional components are negligibly small for beams with relatively larger waists, and become significant only when the beam waists are sufficiently small. The new field components have important consequences for the properties of the beam itself and its interaction with matter.

Laguerre-Gaussian beams are also characterized by their convergence phase functions, namely the Gouy phase and the curvature phase defined in Eq. (2.26). These additional phases functions have significant variations in the vicinity of the focal plane. Beside optical beams the Gouy phase has featured in other focused beams, including acoustic and electron vortex beams. Feng and Winful [28] provided a transparent interpretation of the Gouy phase in terms of the spatial confinement of the beam, while Hariharan and Robinson [29] interpreted it as a geometrical effect in terms of the uncertainty principle. It is clear in the case of LG beams that its magnitude increases with increasing winding number $|\ell|$ and/or the radial number p and both the Gouy and the curvature phase terms increase in magnitude with tighter focusing corresponding to smaller w_0 or z_R.

2.5 Bessel and Bessel-Gaussian vortex beams

2.5.1 Bessel vortex beams

The intensity distribution of a Bessel vortex beam forms a set of concentric rings and are distinguished by being non-diffracting on propagation. For example, it is known that the beam experiences no change in its cross-section as it propagates, in contrast with the case of other beams such as the Laguerre-Gaussian beam, where the beam radius increases with axial distance from the focal plane.

The electric field vector of a Bessel vortex beam of frequency ω as an exact solution of the Helmholtz equation $\nabla^2 \mathbf{E} + (\omega^2/c^2)\mathbf{E} = 0$ in free space can be written in cylindrical polar coordinates $\mathbf{r} = (\rho, \phi, z)$ as follows:

$$\mathbf{E}(\mathbf{r}) = E_0 \hat{\epsilon} J_\ell(\kappa\rho) e^{i\ell\phi} e^{ik_z z}. \tag{2.28}$$

Here, $\hat{\epsilon}$ is the unit vector of the wave polarization, E_0 is a normalization factor, J_ℓ is the Bessel function of order ℓ; k_z and κ are the axial (longitudinal) and radial (transverse) components of the total wave vector $\mathbf{k}$ such that $k = \sqrt{(k_z^2 + \kappa^2)}$. Fig. 2.2 shows the intensity distribution of a Bessel beam of order $\ell = 1$.

Bessel beams are characterized by a number of features. They have a stable profile and in their propagation axially, they extend indefinitely. They are self-healing in the sense that on encountering an obstruction while propagating, the beam reconstructs and self-heals. However, only an ideal beam can extend indefinitely along its axis, while having an infinite number of rings so the Bessel beams as defined in Eq. (2.28) cannot be generated in practice. Real Bessel beams are localized beams in the sense that the light emerges from an aperture using, for example, an annular slit and confined to a finite axial length.

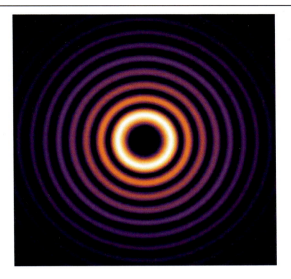

Figure 2.2: The intensity distribution of a Bessel beam on a normal cross-section. Here $\ell = 1$.

2.5.2 Bessel-Gaussian vortex beams

These are orbital angular momentum carrying vortex beam with both Bessel and Gaussian features. They automatically ensure finiteness of energy and radial extent due to a Gaussian factor. The electric field vector of a Bessel-Gaussian mode is as follows:

$$\mathbf{E}(\mathbf{r}) = E_0 \hat{\epsilon} J_\ell \left(\frac{2\pi\rho}{\rho_0} \right) e^{-\frac{\rho^2}{w_0^2}} e^{i\ell\phi} e^{ik_z z}, \tag{2.29}$$

where ρ_0 is the Gaussian radius. These Bessel-Gaussian beams have served as OAM states in optical communication. In addition to non-diffraction characteristics, they also have the ability to self-heal after encountering an obstruction on propagation, which is one of the desirable features in optical communication.

There are other beams characterized by the non-diffraction property of which we mention Mathieu beams, which constitute a family of elliptical non-diffracting beams with variations following the Mathieu functions [17,18].

2.6 Non-paraxial LG beams

The non-paraxial regime is the full theory of optical vortices formally as the solutions of Maxwell's equations without resort to the paraxial approximation. The main properties of a non-paraxial light beam of frequency ω propagating in the z-direction in free space are W

the energy, **P** the linear momentum and **J** the angular momentum (both orbital and spin). The cycle-averaged forms of these beam properties are obtainable as integrals over the (x–y) plane as follows [2]:

$$W = \frac{1}{2}\epsilon_0 \int\int dxdy \{\mathbf{E}^* \cdot \mathbf{E}\}, \tag{2.30}$$

$$\mathbf{P} = \left(\frac{\epsilon_0}{i\omega}\right) \int\int dxdy \{\mathbf{E}^* \times (\nabla \times \mathbf{E})\}, \tag{2.31}$$

$$\mathbf{J} = \frac{1}{2}\left(\frac{\epsilon_0}{i\omega}\right) \int\int dxdy \{\mathbf{r} \times \mathbf{E}^*[\times (\nabla \times \mathbf{E})]\}. \tag{2.32}$$

In order to account for both optical spin and OAM, we write the general form of the electric field associated with a non-paraxial vortex beam in cylindrical coordinates $\mathbf{r} = (\rho, \phi, z)$ as follows:

$$\mathbf{E}(\mathbf{r}) = (\alpha\hat{\mathbf{x}} + \beta\hat{\mathbf{y}})E^T(\rho, z)e^{i\ell\phi} + \hat{\mathbf{z}}E^z(\rho, \phi, z), \tag{2.33}$$

where E^T indicates transverse components and E^z longitudinal components. In the above carets denote unit vectors and, as earlier, the inclusion of the terms involving the factors α and β allows for optical spin (wave polarization, which is in general elliptical) to be taken into account. This prescription also ensures spin-orbit coupling as encountered below. The last term in Eq. (2.33) is the electric field z-component, also called the longitudinal or the axial electric field component of the beam. This component is fixed by the transversality condition $\nabla \cdot \mathbf{E} = 0$ [30].

The starting point in the description of non-paraxial LG light beams is the Barnett–Allen formulation [31–33]. The main characteristics of the LG light beam are its frequency ω; its wave vector **k** with κ and k_z its transverse and longitudinal components such that $k_z = \sqrt{k^2 - \kappa^2}$. The key point is to exploit the Bessel function set $\{J_\ell(\kappa\rho)\}$ as a complete orthonormal set of functions, which are exact solutions of the Helmholtz equation. In the Barnett–Allen formulation the LG field components are written as integral superpositions of the Bessel set. This leads to the definition of the non-paraxial electric field components as Hankel transforms as follows:

$$E^x = \alpha \int_0^k d\kappa \, d_{\ell p}(\kappa) e^{i\ell\phi} e^{i\sqrt{k^2-\kappa^2}z} J_\ell(\kappa\rho); \quad E^y = \frac{\beta}{\alpha}E^x, \tag{2.34}$$

$$E^z = i \int_0^k d\kappa \, \frac{d_{\ell p}(\kappa)}{\sqrt{(k^2-\kappa^2)z}} e^{i\ell\phi} e^{i\sqrt{k^2-\kappa^2}z}$$
$$\times \left\{ F_{\alpha\beta}(\phi) [J_{\ell-1}(\kappa\rho) - J_{\ell+1}(\kappa\rho)] + \frac{i2\ell}{\kappa\rho} G_{\alpha\beta}(\phi) J_\ell(\kappa\rho) \right\}, \tag{2.35}$$

where

$$F_{\alpha\beta}(\phi) = (\alpha\cos\phi + \beta\sin\phi); \quad G_{\alpha\beta}(\phi) = (\beta\cos\phi - \alpha\sin\phi). \tag{2.36}$$

As earlier, z_R is the Rayleigh range, which is related to beam radius at axial coordinate z by $w^2(z) = (z^2 + z_R^2)/kz_R$. Also, $w_0 = w(0)$ is the beam radius at the focal plane $z = 0$. The amplitude functions $d_{\ell p}(\kappa)$ are obtained for LG$_{\ell p}$ modes in the form

$$d_{\ell p}(\kappa) = \left(\frac{\kappa}{k}\right)^{1+|\ell|+2p} \exp\left(-\frac{\kappa^2}{2k}z_R\right). \tag{2.37}$$

Note that the form of the longitudinal field component E^z shown in Eq. (2.35) is the same as and differs only in appearance from the expression presented by Götte and Barnett, Chapter 1 in [6]. The version presented here highlights the appearance of the ℓ-dependence term, which changes sign when the sign of ℓ changes. In the paraxial regime the effects of this field component are negligible for LG beams of beam waists w_0 greater than a wavelength. However, its presence in Eq. (2.35) is imposed by the transversality condition and, as we shall see, its effects do indeed become important for LG light of small beam waists.

2.6.1 Extracting the paraxial regime

The electric field components shown in Eq. (2.34) and (2.35) are the non-paraxial solutions of the Helmholtz equation $\nabla^2 \mathbf{E} + k^2 \mathbf{E} = 0$ and also conform with the transversality condition $\nabla \cdot \mathbf{E} = 0$. The factors α and β are in general complex numbers and determine the wave polarization such that $|\alpha|^2 + |\beta|^2 = 1$.

For any given k and for arbitrary ℓ and p, there are no known analytical forms of the integrals in Eq. (2.34) and (2.35) of the non-paraxial regime. The paraxial regime, however, is amenable to evaluations leading to analytical forms. Formally this regime corresponds to $k \gg \kappa$. A Taylor expansion in powers of κ/k allows us to write

$$\sqrt{k^2 - \kappa^2} \approx k - \frac{\kappa^2}{2k}; \quad \frac{\kappa}{2\sqrt{k^2 - \kappa^2}} \approx \frac{\kappa}{2k}. \tag{2.38}$$

This step, together with replacing the upper limit in the κ integrals by ∞ facilitate the extraction of the LG field components in the paraxial regime. Consider a LG beam traveling along the positive z-axis. We use the cylindrical coordinates $\mathbf{r} = (\rho, \phi, z)$ and assume that the light is linearly polarized along $\hat{\epsilon}$ which is transverse to the direction of propagation with two in-plane components E^x and E^y. The above procedure of extracting the paraxial formalism from the non-paraxial theory leads to the following transverse electric field components:

$$E^x = \alpha \mathcal{E}_{\ell,p} L_p^{|\ell|}\left(\frac{2\rho^2}{w^2(z)}\right),$$
$$E^y = \beta \mathcal{E}_{\ell,p} L_p^{|\ell|}\left(\frac{2\rho^2}{w^2(z)}\right),$$
(2.39)

where

$$\mathcal{E}_{\ell,p} = E_0 \frac{C_{\ell,p}}{\sqrt{1+\frac{z^2}{z_R^2}}} e^{-\frac{\rho^2}{w^2(z)}} \left(\frac{\sqrt{2}\rho}{w^2(z)}\right)^{|\ell|} e^{-i\omega t} e^{i\Theta}.$$
(2.40)

Here, $C_{\ell,p} = \sqrt{\frac{p!}{(|\ell|+p)!}}$ is a constant factor. The other parameters, $w(z)$, w_0, z_R emerge from the procedure precisely as defined earlier. The phase function Θ in Eq. (2.40) is as follows:

$$\Theta = k_z z + \ell\phi - (2p + |\ell| + 1)\tan^{-1}(z/z_R) + \frac{k_z \rho^2 z}{2(z^2 + z_R^2)}.$$
(2.41)

This phase function, too, is identical to the one encountered earlier for LG modes with the first term representing plane wave propagation with axial wave vector k_z, while the second term is the azimuthal term bearing the winding number ℓ. The third term is the Gouy phase and the final term enters as a phase contribution due to the variation of the beam curvature with axial position z. Note that these expressions have not just been quoted. They have emerged from the procedure involving the extraction of the paraxial regime from the non-paraxial formulation.

In addition to the transverse components, the procedure leads to the longitudinal component E^z, which was originally determined in the Barnett–Allen formulation and evaluated using the transversality condition $\nabla \cdot \mathbf{E} = 0$. We find

$$E^z(\ell \neq 0) = \mathcal{E}_{\ell,p}\frac{1}{k\rho}\left[-i\left\{|\ell|L_p^{|\ell|}\left(\frac{2\rho^2}{w(z)^2}\right) - 2(p+1)L_{p+1}^{|\ell|-1}\left(\frac{2\rho^2}{w(z)^2}\right)\right\}\right.$$
$$\left. \times (\alpha\cos\phi + \beta\sin\phi) - \epsilon_\chi\right],$$
(2.42)

where ϵ_χ in Eq. (2.42) is given by

$$\epsilon_\chi = \ell L_p^{|\ell|}\left(\frac{2\rho^2}{w(z)^2}\right)(\beta\cos\phi - \alpha\sin\phi).$$
(2.43)

All the components of the electric field vector have now been determined in analytical forms. These will now facilitate the evaluation of the magnetic field components using Maxwell's equation $\nabla \times \mathbf{E} = -\frac{\partial \mathbf{B}}{\partial t}$. Koksal et al. [30,34] have managed to derive the magnetic field components in analytical, albeit involved, forms and proceeded to discuss the helicity and chirality of the Laguerre-Gaussian light.

2.7 Paraxial beams with small waists

As a specific application of the above formalism we consider paraxial LG$_{\ell,0}$ modes for which $p = 0$ (so-called doughnut modes) and take integer ℓ to be positive or negative. Recent work by Koksal et al. [30] has shown that when a transversely polarized doughnut beam is produced with a sufficiently small beam waist w_0, the longitudinal field component becomes comparable to the transverse components. This leads to a number of effects, including enhanced chirality and helicity features as well as spin-orbit coupling. The above analytical paraxial formalism applies to these modes and it is possible to proceed to explore the properties of doughnut beams of any value of $|\ell|$. However, for illustration Koksal et al. [30] considered the case of two separate counter-rotating doughnut beams of the lowest order namely LG$_{1,0}$ and LG$_{-1,0}$. Each of these two beams can have circular transverse polarization (optical spin $\sigma = \pm 1$) or linear polarization (optical spin $= 0$). We concentrate on the linear polarization case because, as we shall see the emergence of the longitudinal component leads to the chiral behavior of these modes.

Using cylindrical coordinates $\mathbf{r} = (\rho, \phi, z)$ we note that the amplitude function $U_{k10}(\rho, z)$ for $\ell = +1$, coincides with the amplitude function $U_{k(-1)0}(\rho, z)$ for $\ell = -1$. The corresponding phase functions $\Theta_{k\pm 10}(\mathbf{r})$ are not the same, however. The electric field vector for the doughnut mode LG$_{\ell=+1}$ can be found from the above general paraxial forms. We have

$$E^x_{k\ell 0}(\mathbf{r}) = U_{k\ell 0}(\rho, z) e^{i\Theta_{k\ell 0}(\mathbf{r})}, \tag{2.44}$$

where E^x indicates that the doughnut mode is linearly polarized along $\hat{\mathbf{x}}$, so $E^y = 0$. The amplitude functions are well known [11,35],

$$U_{k\pm 10}(\rho, z) = \mathcal{E}_{k00} \frac{1}{(1 + z^2/z_R^2)^{1/2}} \left(\frac{\rho\sqrt{2}}{w(z)}\right) e^{-\frac{\rho^2}{w(z)^2}}, \tag{2.45}$$

where $\mathcal{E}_{k00}$ is a constant amplitude. The phase functions for the beams LG$_{\pm 1,0}$ are given by

$$\Theta_{k\pm 10}(\mathbf{r}) = k_z z \pm \phi + \theta_G + \theta_C, \tag{2.46}$$

where θ_G is the Gouy phase and θ_C is the curvature phase; together they constitute the convergence phase and here they are explicitly given by

$$\theta_G = -2\arctan\left(\frac{z}{z_R}\right); \quad \theta_C = \frac{k_z \rho^2 z}{2(z^2 + z_R^2)}. \tag{2.47}$$

In addition to the transverse component of the electric field given in Eq. (2.44) each of the doughnut beams possesses a longitudinal component E^z, which can be computed from the requirement that $\mathbf{E}$ must be divergence-less, i.e. $\nabla \cdot \mathbf{E} = 0$, where

$$\mathbf{E} = E^x \hat{x} + E^z \hat{z}. \tag{2.48}$$

The longitudinal component for $LG_{1,0}$ emerges as follows:

$$E_{k10}^z(\mathbf{r}) = -i\mathcal{E}_{k00}\left(\frac{4\rho^2 \cos\phi - w(z)^2 e^{-i\phi}}{\xi k w(z)^4}\right) e^{-\frac{\rho^2}{w(z)^2} + i\Theta_{k10}}, \tag{2.49}$$

where $\xi = \sqrt{k/4z_R}$. The same procedure is followed to evaluate $E_{k(-1)0}^z$. We now have all the electric field components of the two doughnut beams. First, we must demonstrate that for small beam waists w_0 the longitudinal component does indeed become comparable to the transverse components, in contrast to the case of larger values of w_0, where the amplitude function of longitudinal component becomes negligibly small. It is sufficient to demonstrate this for one of the beams, as the other beam differs only in the phase function. This is shown in Fig. 2.3, which displays the variations of $|E_z|^2$ along with $|E_x|^2$ for $LG_{1,0}$ for a small width w_0 and compares those variations with the case of a larger w_0. It is clear that the small w_0 case shows that the longitudinal component is comparable in magnitude to that of the transverse component, and is very small for larger beam waists w_0.

2.8 Chirality and helicity

2.8.1 Cycle-averaged fields

One of the effects that become manifest due to the increase in the significance of the longitudinal field for a beam with small w_0 is that this type of light is endowed with the property whereby two beams which are identical except for the sign of the winding number ℓ are distinguishable. This is because one beam is a phase-inverted mirror image of the other and so the beams possess different helicity and chirality [23–26]. Both these properties are negligibly small for large w_0. The chirality and the helicity of a light field can be defined in terms of the

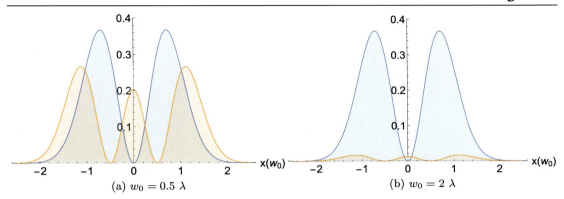

Figure 2.3: The in-plane variations of the modulus squared of the electric field components for the $\ell = 1$, $p = 0$ Laguerre-Gaussian (doughnut) mode in the focal plane $z = 0$. Figures (a) and (b) show the variations of the longitudinal field E^z (orange) along with the transverse component E^x (blue). (a) A beam with a small beam waist $w_0 = 0.5\lambda$, where E^z is comparable with E^x and (b) a relatively large beam waist $w_0 = 2\lambda$, where E^z is seen to be negligibly small in comparison with E^x. These results also apply for the case $\ell = -1$, $p = 0$, as the amplitude functions are the same for $\ell = \pm 1$, $p = 0$, but the phase functions are different.

dual fields **C** and **D**, in addition to the usual electromagnetic fields. The reader is referred to [21,22] for recent accounts on this. In the Coulomb gauge we have for the chirality density

$$\chi = \frac{1}{2}\left(-\epsilon_0 \mathbf{E} \cdot \dot{\mathbf{B}} + \mathbf{B} \cdot \dot{\mathbf{D}}\right), \tag{2.50}$$

and for the helicity density

$$\eta = \frac{1}{2}\left(\sqrt{\frac{\epsilon_0}{\mu_0}}\mathbf{A} \cdot \mathbf{B} - \sqrt{\frac{\mu_0}{\epsilon_0}}\mathbf{C} \cdot \mathbf{D}\right). \tag{2.51}$$

Here $\mathbf{D} = \epsilon_0 \mathbf{E}$ is the electric displacement field and **C** is a dual vector potential distinct from vector potential **A** such that $\nabla \times \mathbf{C} = -\mathbf{D}$ and $\dot{\mathbf{C}} = -\mathbf{B}/\mu_0 = -\mathbf{H}$. However, we are interested only in the cycle-averaged version of the monochromatic fields. Following Barnett,[1] we assume that the fields, represented by the vector field **F**, have the following complex format:

$$\mathbf{F} = \frac{1}{2}\left(\mathcal{F}e^{-i\omega t} + \mathcal{F}^* e^{i\omega t}\right). \tag{2.52}$$

Substituting the complex field format Eq. (2.52), we have for the cycle-averaged helicity $\bar{\eta}$

$$\bar{\eta} = \frac{1}{8}\left[\sqrt{\frac{\epsilon_0}{\mu_0}}\left(\mathcal{A} \cdot \mathcal{B}^* + \mathcal{A}^* \cdot \mathcal{B}\right) - \sqrt{\frac{\mu_0}{\epsilon_0}}\left(\mathcal{C} \cdot \mathcal{D}^* + \mathcal{C}^* \cdot \mathcal{D}\right)\right]. \tag{2.53}$$

---

[1] S.M. Barnett: private communication.

This can be written in terms of the complex amplitudes $\mathcal{E}$ and $\mathcal{H}$ as follows:

$$\bar{\eta} = -\frac{i}{8\omega c}\left[\mathcal{E}\cdot\mathcal{H}^* - \mathcal{E}^*\cdot\mathcal{H} - \mathcal{H}\cdot\mathcal{E}^* + \mathcal{H}^*\cdot\mathcal{E}\right]$$

$$= -\frac{1}{2\omega c}\Im[\mathbf{E}^*\cdot\mathbf{H}]. \tag{2.54}$$

where $\Im[..]$ stands for the imaginary part of $[..]$. Similar evaluations lead to the result that the cycle-averaged chirality density $\bar{\chi}$ is proportional to $\bar{\eta}$

$$\bar{\chi} = -\frac{i\omega}{4c^2}\left(\mathcal{E}\cdot\mathcal{H}^* + \mathcal{E}^*\cdot\mathcal{H}\right)$$

$$= -\frac{\omega}{2c^2}\Im[\mathbf{E}^*\cdot\mathbf{H}]$$

$$= \frac{\omega^2}{c}\bar{\eta}. \tag{2.55}$$

Actual calculations of the helicity and the chirality distributions for a given LG beam of small w_0 will have to make direct use of Eq. (2.54). However, we need first to obtain expressions for the corresponding magnetic field components, which follow by direct use of Maxwell's equation $\mathbf{B} = \nabla\times\mathbf{E}/(i\omega)$ and the evaluations have to be carried out separately for both $LG_{1,0}$ and $LG_{-1,0}$. The expressions for the magnetic field components will not be presented here [27]. We obtain the following expression for the cycle-averaged chirality density for linearly polarized $LG_{1,0}$ and $LG_{-1,0}$ modes (with $\alpha=1$, $\beta=0$) [30]:

$$\bar{\chi}(\ell=\pm 1) = \pm\frac{e^{-\frac{1}{2}\rho^2\left(\frac{k}{z_R}+\frac{2}{w(z)^2}\right)}\left\{w(z)^2\sin^2(\phi)(k\rho^2-z_R)+z_R\cos^2(\phi)(4\rho^2-w(z)^2)+\rho^2 z_R(3\cos(2\phi)+1)\right\}}{2c^2 z_R^2 w(z)^2}. \tag{2.56}$$

Fig. 2.4 shows the variations of $\bar{\chi}$, Eq. (2.56), evaluated in the focal plane $z=0$ for the doughnut beams $LG_{1,0}$ and $LG_{-1,0}$. It can be seen that the bright regions in (a) (positive values of chirality) correspond to dark regions in (b) (negative values) and vice versa. These chirality density distributions are basic properties of the optical vortex beams. The beams are linearly polarized; hence have zero optical spin and so there are no contributions from the spin chirality. It is worth emphasizing again the fact that since we are dealing with cycle-averaged fields, the helicity is proportional to the chirality, as in Eq. (2.55).

The chirality of beams carrying OAM has been experimentally investigated by Wozniak et al. [36]. They also compared their experimental results with numerical evaluations using vectorial diffraction theory (see also [37] and the references therein). In the experiment by Wozniak et al. the two (already formed) doughnut beams, differing only on the signs of their winding number ℓ were subsequently passed through an additional lens and the measured chirality was

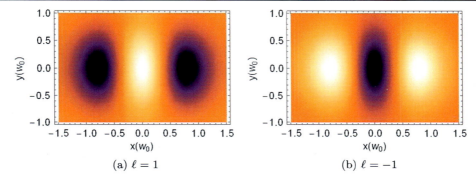

Figure 2.4: Chirality distributions in the focal plane $z = 0$ for two doughnut beams for which the waist is $w_0 = 0.5\lambda$ (a) doughnut beam $\ell = 1$ and (b) doughnut beam $\ell = -1$. The color code is such that white/black are $+/-$ extremes on either side of zero.

observed in the back focal plane of the lens. Their work confirmed the difference in the chiral distributions of the focused doughnut beams, one with $\ell = 1$ and the second had $\ell = -1$. Their chirality distribution in the back focal plane of the lens differs from the results shown above in which no additional lens is involved. Clearly the differences are due to the fact that the doughnut beams are influenced by the additional lens in [36]. The possible effects of additional focusing on already formed optical vortex light were pointed out by Dorn et al. [38] and Levy et al. [39]. The chirality and helicity properties discussed above concern vortex beams and are the properties of such beams without passing them through an additional lens. Earlier theoretical and experimental work on allied properties of vortex beams was done by Zamberana-Puyalto et al. [20,40,41]

2.8.2 Effects of the Gouy and curvature phases

We have discussed the chirality and helicity of vortex beams in relation to the focal plane at $z = 0$, where both the Gouy and the curvature phases are identically zero. These phase functions change sign and their gradients become significant in the vicinity of the two sides of the focal plane. The question is what effects these additional phase functions impose on the chirality and helicity in planes to the left ($z < 0$) and to the right ($z > 0$) of the focal plane. In these planes, the variations with axial and radial positions of the Gouy phase $\theta_{Gouy}(z)$ and the curvature phase $\theta_{curv}(\rho, z)$ become important as well as the variation of the beam radius $w(z)$.

To illustrate the roles of these phase functions on the chirality distributions we consider the planes $z = +2w_0$ and $z = -2w_0$, to the right and to the left of the focal plane, respectively.

56 Chapter 2

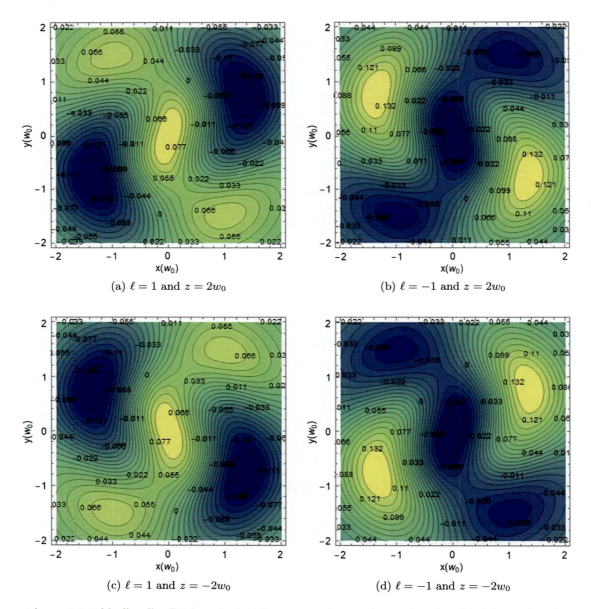

Figure 2.5: Chirality distributions in the planes $z = +2w_0$ and $z = -2w_0$ (to the right and to the left, respectively, of the focal plane $z = 0$) for two doughnut beams each of which has a waist $w_0 = 0.5\lambda$: Left panel (a and c): the case of a doughnut beam with winding number $\ell = 1$ (a) plane $z = +2w_0$ and (c) plane $z = -2w_0$. The right panel (b and d) is the same as the left panel, but for negative winding number $\ell = -1$.

The results are distinctly different distributions as shown in Fig. 2.5. We see that the distributions bear some resemblance to those in the focal plane. In Figs. 2.5 (a and b) for $z = +2w_0$ the same main chirality features pointed out in relation to Fig. 2.4 in the focal plane persist between the doughnut modes $\ell = 1$ and $\ell = -1$. However, the distributions are rotated relative to those in the focal plane as in Fig. 2.4, and Figs. 2.5 (c and d) show the distributions for $z = -2w_0$ for the doughnut modes $\ell = 1$ and $\ell = -1$. Once again the patterns are rotated relative to those in the focal plane; however, the rotation is opposite to that for the same ℓ but on equidistant planes on different sides of the focal plane. Figure (b) is a reflection plus sign inversion of (a), so too (d) with respect to (c). Furthermore, (c) is a mirror image of (a) and (d) a mirror image of (b), plus sign inversion. These features reflect the change in the sign of the variation of the convergence phases across the focal plane.

2.9 Multiple vortex beams

Light fields arising from the superposition of two or more optical vortex beams have novel total amplitude functions and total phase functions. The beams can be co-propagating or counter-propagating in pairs and can in general have any state of wave polarization each. The most studied cases are those in which pairs of beams are superimposed when arranged to propagate along a common axis and have the same magnitudes but different signs of winding numbers, i.e. $\ell_1 = \ell$ and $\ell_2 = -\ell$. The possible scenarios we discuss next include co-propagating and counter-propagating configurations and couple those with different types of polarization arrangements, namely circular polarizations, the case of orthogonal linear polarizations and the case of two beams with polarization vectors inclined at angle γ. We also discuss the case of bi-chromatic beams, differing slightly in frequency and also when the focal planes of the beams are shifted so creating novel interference patterns in the region between the focal planes.

2.9.1 Linearly polarized LG beams

Consider the case of two co-axial vortex beams with the same linear polarization $\hat{\epsilon}$. The total field is simply the sum of the field vectors. For beams 1 and 2 we write

$$\mathbf{E}_1 = U_1 e^{i\Theta_1}\hat{\epsilon}; \quad \mathbf{E}_2 = U_2 e^{i\Theta_2}\hat{\epsilon}, \tag{2.57}$$

where the amplitude functions U_1 and U_2 and phase function Θ_1 and Θ_2 are appropriate for LG beams, as given above. Writing U and Θ as the total amplitude function and the total phase function of the interfering beams, we have for the total electric field vector

$$\mathbf{E} = U e^{i\Theta}\hat{\epsilon} = \left(U_1 e^{i\Theta_1} + U_2 e^{i\Theta_2}\right)\hat{\epsilon}. \tag{2.58}$$

It is straightforward to show that the total amplitude function is given by

$$U = \left\{U_1^2 + U_2^2 + 2U_1U_2\cos(\Theta_1 - \Theta_2)\right\}^{1/2}. \tag{2.59}$$

The total phase function is

$$\Theta = \tan^{-1}\left\{\frac{U_1\sin\Theta_1 + U_2\sin\Theta_2}{U_1\cos\Theta_1 + U_2\cos\Theta_2}\right\}. \tag{2.60}$$

Eqs. (2.59) and (2.60) are applicable to any two LG beams provided the beams are co-axial and both linearly polarized in the same direction. Although assumed co-axial, their focal planes need not be coinciding and can either be co-propagating, or counter-propagating. They can also be co-rotating, i.e. with the same sign of winding number or counter-rotating, i.e. with different signs of winding number. If the beams are identical in amplitude $U_1 = U_2$, then the total amplitude function (2.59) and the phase function (2.60) take simpler forms. The case in which the beam amplitude functions are different is exemplified by the simple case in which the focal planes are shifted axially by a finite distance d in which case both the amplitude and the phase feed dependence into each other.

2.9.2 Axial shift

For illustration, we focus on the case of shifted counter-propagating LG modes and consider specifically the doughnut beams $LG_{\ell_1,0}$ and $LG_{\ell_2,0}$. The focal plane of beam 1 is the plane $z = -d/2$, while that for beam 2 is the plane $z = +d/2$ and the central plane is at $z = 0$. The amplitude functions as well as the phase functions in this case are different and are such that [42]

$$U_{1,2} = U(z \to z \mp d/2); \quad \Theta_{1,2} = \Theta(z \to z \mp d/2). \tag{2.61}$$

Note that in addition to the axial shift the two beams are assumed counter-propagating. Once the expressions for U_1, U_2, Θ_1 and Θ_2 have been determined the total phase function follows by direct substitution in Eq. (2.60). The total amplitude function is given by Eq. (2.59) on substituting for U_1 and U_2, but the phase difference is as follows:

$$\Theta_1 - \Theta_2 = 2k_z z + (\ell_1 + \ell_2)\phi + \Delta\Theta_{Gouy} + \Delta\Theta_{curv}, \tag{2.62}$$

where

$$\Delta\Theta_{Gouy} = -(|\ell_1|+1)\tan^{-1}\left(\frac{z-d/2}{z_R}\right) - (|\ell_2|+1)\tan^{-1}\left(\frac{z+d/2}{z_R}\right). \tag{2.63}$$

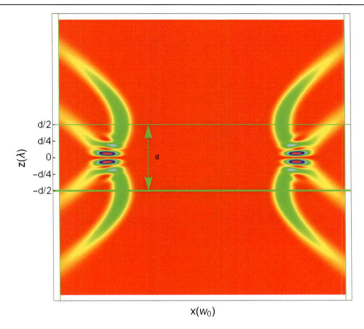

Figure 2.6: Intensity profile due to the interference of two axially shifted counter-propagating doughnut beams LG$_{l0}$. The variations are in the (x–z) plane at $y = 0$ and the parameters are as follows: $\ell_1 = \ell_2 = 40$, $w_0 = 6\lambda$ and $d = 24w_0$.

For $\ell_1 = \ell_2 = \ell$ this simplifies to

$$\Delta\Theta_{Gouy} = -(|\ell| + 1)\tan^{-1}\left\{\frac{2zz_R}{z_R^2 - (z^2 + d^2/4)}\right\}, \quad (2.64)$$

while $\Delta\Theta_{curv}$ is as follows:

$$\Delta\Theta_{curv} = \frac{k_z \rho^2 d}{2(z^2 + z_R^2)}. \quad (2.65)$$

Thus we have for $\Theta_1 - \Theta_2$

$$\Theta_1 - \Theta_2 = 2k_z z + 2l\phi - (|\ell| + 1)\tan^{-1}\left\{\frac{2zz_R}{z_R^2 - (z^2 + d^2/4)}\right\} + \frac{k_z \rho^2 d}{2(z^2 + z_R^2)}. \quad (2.66)$$

The above formalism enables the evaluation of the total power density and its variations with position in the space between the focal planes. The results displayed in Figs. 2.6 and 2.7 illustrate these variations for a specific set of parameters as stated in the captions. These figures show that intensity variations of electromagnetic fields in the region between the focal planes

60 Chapter 2

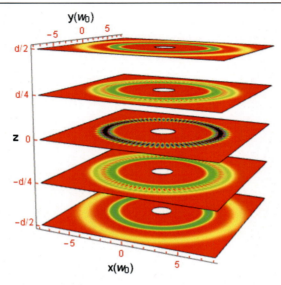

Figure 2.7: Intensity pattern projected on the $(x-y)$ planes due to the interference of two axially shifted counter-propagating doughnut beams LG_{l0}. The variations are shown only in the region between the focal planes $-d/2 < z < d/2$. The parameters are as in Fig. 2.6.

are in the form of a finite ring lattice with the variations determined by the phase difference $\cos(\Theta_1 - \Theta_2)$ in the amplitude function. We now explore how this phase difference can have different forms depending on the set-up and leads to different physical scenarios. We illustrate this by exploring the effects of the beams having slightly different frequencies.

2.9.2.1 Effects of frequency shift

In addition to the axial shift, we now assume that the frequencies of the beams differ by $\Delta\omega$. The total amplitude function is then given by

$$U = \left\{ U_1^2 + U_2^2 + 2U_1 U_2 \cos(\Theta_1 - \Theta_2 - (\Delta k)z - (\Delta\omega)t) \right\}^{1/2}, \quad (2.67)$$

where $(\Delta\omega) = \omega_1 - \omega_2$ and $(\Delta k) = k_{z1} - k_{z2}$. The term proportional to (Δk) causes axial beating/dephasing effects of the envelope function, but these effects are small since the dephasing length $2\pi/\Delta k$ is much longer than the coherence length. The argument of the cosine function in Eq. (2.67) is now time-dependent leading to optical Ferris wheels and lifts (conveyor belts).

2.9.2.2 Ferris wheels and conveyor belts

The argument of the cosine function in Eq. (2.67) is now time-dependent by virtue of the frequency shift $\Delta\omega$. This means that the interference pattern is now time-dependent. There is an angular rotational motion of all the rings with an angular frequency v_ϕ given by

$$v_\phi = \frac{d\phi}{dt} = \frac{\Delta\omega}{2l}. \tag{2.68}$$

The angular motion is a rotating optical Ferris wheel, as first described by Franke-Arnold et al. [43]. Another effect of the frequency difference $\Delta\omega$ is an axial translation in which the whole pattern travels along the z-axis at a speed v_z given by

$$v_z = \frac{dz}{dt} \approx \frac{\Delta\omega}{2k_z}. \tag{2.69}$$

This is the conveyor belt effect.

2.9.2.3 Radial gaps in double rings

Fig. 2.7 shows that the intensity distribution consists of a number of double rings and there is a single ring at $z = 0$. On either side of this central ring the double rings are separated radially by a distance $\delta\rho$, depending on the double ring. For the double ring at $z = (d/2 - \delta)$, where δ is the fringe separation, one of the rings is at an axial distance $(d/2 - \delta)$ from the focal plane of beam 1, and the second ring is at an axial distance of $(d/2 + \delta)$ from the focal plane of beam 2. Thus the radii of the rings are given by

$$w_1 = w_0\sqrt{|\ell|/2}\sqrt{[1 + (d/2 - \delta)^2/z_R^2]}; \quad w_2 = \sqrt{|\ell|/2}w_0\sqrt{[1 + (d/2 + \delta)^2/z_R^2]}. \tag{2.70}$$

The radial separation is the difference. We have

$$\Delta\rho = w_0\sqrt{|\ell|/2}\left(\sqrt{1 + (d/2 + \delta)^2/z_R^2} - \sqrt{1 + (d/2 - \delta)^2/z_R^2}\right). \tag{2.71}$$

This is approximately $\Delta\rho \approx w_0\sqrt{|\ell|/2}(d\delta/z_R^2)$ for small d or large z_R. It is thus possible for the parameters to be chosen so that the rings can be made to be closely separated. This result can be compared with the double rings in an LG beam with the same ℓ, but with $p = 1$, where the rings are separated by a radial distance of w_0. However, in the current case emerging from the shifted beam scenario it may be possible to choose the parameters in order for the factor $\sqrt{|\ell|/2}(d\delta/z_R^2)$ to be much smaller or much larger than unity.

2.10 No axial shift—polarization gradients

We now consider the no axial shift case $d = 0$ and begin with the scenario in which the two vortex beams have circular polarizations giving rise to spatio-temporal polarization gradients [44]. Once more the beams are assumed to have sufficiently large beam waists so that the longitudinal component is negligible.

2.10.1 Co-propagating LG beams

The two co-propagating LG beams, labeled 1 and 2 have the same axial wave vector k_z and frequency ω and with their focal planes coinciding at $z = 0$. The winding numbers are $\ell_1 = -\ell_2 = \ell$, i.e. they differ only in the sign of the winding number. The total electric field vector given by

$$\mathbf{E}(\rho, \phi, z, t) = \mathcal{U}(\rho, \phi, z) e^{-i\omega t} e^{ik_z z}, \tag{2.72}$$

where $\mathcal{U}(\rho, \phi, z)$ is given by

$$\mathcal{U}(\rho, \phi, z) = U_1(\rho, z) e^{i\ell\phi} \hat{e}_1 + U_2(\rho, z) e^{-i\ell\phi} \hat{e}_2, \tag{2.73}$$

where $U_{1,2}$ are the amplitude functions of the beams. The unit vectors $\hat{e}_1$ and $\hat{e}_2$ are the polarization vectors, which can both be circular polarizations (σ^+ or σ^-) or can be two different linear polarizations.

2.10.1.1 Co-propagating with $\sigma^+ - \sigma^-$

Consider the case in which the co-propagating beams are such that $\ell_1 = \ell$ and $\ell_2 = -\ell$. Furthermore, we assume that the beams have counter-rotating circular polarizations as follows:

$$\hat{e}_1 = \sigma^+ = -\frac{1}{\sqrt{2}} (\hat{e}_x + i\hat{e}_y), \tag{2.74}$$

$$\hat{e}_2 = \sigma^- = \frac{1}{\sqrt{2}} (\hat{e}_x - i\hat{e}_y), \tag{2.75}$$

where $\hat{e}_x$ and $\hat{e}_y$ are the polarization vectors along the Cartesian x and y axes. Eq. (2.73) then becomes

$$\mathcal{U}(\rho, \phi, z) = \frac{1}{\sqrt{2}} \left\{ [U_2(\rho, z) - U_1(\rho, z)] \hat{\Sigma}_-(\phi) - i [U_2(\rho, z) + U_1(\rho, z)] \hat{\Sigma}_+(\phi) \right\}, \tag{2.76}$$

where $\hat{\Sigma}_\pm(\phi)$ are ϕ-dependent polarization vectors,

$$\hat{\Sigma}_-(\phi) = \hat{e}_x \cos(\ell\phi) - \hat{e}_y \sin(\ell\phi), \tag{2.77}$$

$$\hat{\Sigma}_+(\phi) = \hat{e}_x \sin(\ell\phi) + \hat{e}_y \cos(\ell\phi). \tag{2.78}$$

The polarization vectors of the total fields due to the co-propagating beams is elliptical with a position-dependent ellipticity given by

$$e_{1,2} = \frac{(U_2(\rho, z) - U_1(\rho, z))}{(U_2(\rho, z) + U_1(\rho, z))}. \tag{2.79}$$

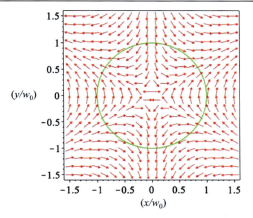

Figure 2.8: Polarization distribution with directions represented by arrows in the (x–y) plane for $\sigma_+ - \sigma_-$ co-propagating doughnut beams $\ell = \pm 2$. The circle indicates the locus of the maximum intensity and distances are in units of w_0.

Since the two beams differ only on the sign of ℓ we may write $U_1 = U_2 = U$ and the ellipticity $e_{1,2} = 0$. Therefore Eq. (2.76) yields

$$\mathcal{U}(\rho, \phi, z) = -i\sqrt{2} U(\rho, z) \hat{\mathbf{\Sigma}}_+(\phi)$$
$$= -i\sqrt{2} U(\rho, z) \left\{ \hat{\mathbf{e}}_x \sin(\ell\phi) + \hat{\mathbf{e}}_y \cos(\ell\phi) \right\}. \quad (2.80)$$

Thus, the polarization vector of the two identical co-propagating beams we are focusing on is locally linear everywhere, but it depends on the in-plane position (ρ, ϕ). Fig. 2.8 shows how the direction of the polarization vector varies on the transverse plane for the case $\ell = \pm 2$. The green circle coincide with the radial positions of maximum intensity.

2.10.1.2 Lin ⊥ Lin polarizations

Once more we consider co-propagating beams with winding numbers $\ell_1 = \ell$ and $\ell_2 = -\ell$, but we first assume that their polarizations are both linear and orthogonal—a situation referred to as the Lin ⊥ Lin case. We have

$$\mathcal{U}(\rho, \phi, z) = U_1(\rho, z) e^{i\ell\phi} \hat{\mathbf{e}}_1 + U_2(\rho, z) e^{-i\ell\phi} \hat{\mathbf{e}}_2. \quad (2.81)$$

We take beam 1 to be polarized along x, i.e. $\hat{\mathbf{e}}_1 = \hat{\mathbf{e}}_x$ and beam 2 polarized along y, i.e. $\hat{\mathbf{e}}_2 = \hat{\mathbf{e}}_y$. Once again we set $U_1 = U_2$ and Eq. (2.81) yields

$$\mathcal{U}(\rho, \phi, z) = \Sigma_\pm(\phi) U(\rho, z) \quad (2.82)$$

Table 2.1: Variation with azimuthal angle ϕ of the type of polarization $\Sigma_\pm$ for two co-propagating doughnut modes, with Lin $\perp$ Lin polarizations, at fixed radial position $\rho_0 = w_0\sqrt{(l/2)}$. It is seen that, as ϕ changes from 0 to π/l, the type of polarization switches between linear (Lin) and circular (Circ).

ϕ	$\Sigma_\pm$
0	Lin
$\pi/4\ell$	Circ σ^-
$\pi/2\ell$	Lin
$3\pi/4\ell$	Circ σ^+
π/ℓ	Lin

where $\Sigma_\pm$ is the overall polarization,

$$\Sigma_\pm(\phi) = (\hat{e}_x + \hat{e}_y)\cos(\ell\phi) + i(\hat{e}_x - \hat{e}_y)\sin(\ell\phi). \tag{2.83}$$

We find that $\Sigma_\pm$ varies with the azimuthal angle as can also be seen in Table 2.1, which concerns doughnut modes $LG_{\ell,0}$ and $LG_{-\ell,0}$. The maximum intensity of the field is located at $\rho_{max} = v_0\sqrt{\ell/2}$. The table shows that $\Sigma_\pm$ varies from linear to circular along the arc of length $\Delta s = \pi w_0/4\sqrt{2\ell}$. Thus, the spatial distribution of the polarization depends on the value of w_0 and the winding number ℓ.

Another interesting interference scheme of co-propagating beams with linear polarizations arises when the two polarizations are tilted at an angle γ with respect to each other [45]. In this case we have for the two fields

$$\mathbf{E}_1(\rho, \phi, z) = \mathcal{U}\exp(ik_z z + i\ell\phi)\hat{e}_x, \tag{2.84}$$

$$\mathbf{E}_2 = \mathcal{U}\exp(ik_z z - i\ell\phi)[\hat{e}_x\cos(\gamma) + \hat{e}_y\sin(\gamma)]. \tag{2.85}$$

The net electric field is then given by

$$\mathbf{E}(\rho, \phi, z) = \mathcal{U}e^{(ik_z z)}[e^{-i\gamma/2}(\hat{e}_x + i\hat{e}_y)\cos(\ell\phi + \gamma/2) + e^{i\gamma/2}(\hat{e}_x - i\hat{e}_y)\cos(\ell\phi - \gamma/2)]. \tag{2.86}$$

The total field is a combination of two circularly polarized Ferris wheel fields [43] one with σ^+ and one with σ^-. By changing the angle γ, we can rotate the intensity maxima on the transverse plane and thus achieve optical conveyor belts on the transverse plane. This provides new opportunities in the transportation of cold atomic samples.

2.10.2 Counter-propagating beams

2.10.2.1 $\sigma^+ - \sigma^-$ polarizations

Next we consider the same scenario in which we have two doughnut counter-propagating (rather than co-propagating) beams LG$_{\pm\ell,0}$. It is easy to see that the individual beams have the same amplitude function $U(\rho, z)$ but only differ in their phase functions. Furthermore we now assume that the beams have opposite circular polarization $\sigma^+ - \sigma^-$. We obtain for the total electric field vector

$$\mathbf{E}(\rho, \phi, z, t) = \left\{ \mathcal{U}(\rho, \phi, z) e^{-i\omega t} + c.c. \right\}, \tag{2.87}$$

where now we have

$$\mathcal{U}(\rho, \phi, z) = U_1(\rho, z) e^{i(k_z z + \ell\phi)} \hat{\mathbf{e}}_1 + U_2(\rho, z) e^{-i(k_z z + \ell\phi)} \hat{\mathbf{e}}_2, \tag{2.88}$$

where the unit vectors $\hat{\mathbf{e}}_1$ and $\hat{\mathbf{e}}_2$ are as in Eqs. (2.74) and (2.75), respectively. Thus, we have

$$\mathcal{U}(\rho, \phi, z) = \frac{1}{\sqrt{2}} \left\{ [U_2(\rho, z) - U_1(\rho, z)] \mathbf{\Sigma}_\beta(\phi, z) - i[U_2(\rho, z) + U_1(\rho, z)] \mathbf{\Sigma}_\alpha(\phi, z) \right\}, \tag{2.89}$$

where $\mathbf{\Sigma}_{\alpha/\beta}(\phi, z)$ are the z- and ϕ-dependent polarization vectors.

$$\mathbf{\Sigma}_\alpha(\phi, z) = \hat{\mathbf{e}}_x \cos(k_z z + \ell\phi) - \hat{\mathbf{e}}_y \sin(k_z z + \ell\phi), \tag{2.90}$$

$$\mathbf{\Sigma}_\beta(\phi, z) = \hat{\mathbf{e}}_x \sin(k_z z + \ell\phi) + \hat{\mathbf{e}}_y \cos(k_z z + \ell\phi). \tag{2.91}$$

Since the amplitude functions are identical ($U_1 = U_2 = U$) we can now write

$$\begin{aligned}\mathcal{U}(\rho, \phi, z) &= -i\sqrt{2} U(\rho, z) \mathbf{\Sigma}_\alpha(\phi, z) \\ &= -i\sqrt{2} U(\rho, z) \left\{ \hat{\mathbf{e}}_x \sin(k_z z + \ell\phi) + \hat{\mathbf{e}}_y \cos(k_z z + \ell\phi) \right\}.\end{aligned} \tag{2.92}$$

Note again that the overall polarization is both z- and ϕ-dependent and can be shown to be linear and constant along spiral curves, which satisfy the equation $k_z z + \ell\phi =$ constant. For example, the field is y-polarized along $\hat{\mathbf{e}}_y$ on points lying on the spiral curve $k_z z + \ell\phi = 0$ and is x-polarized along $\hat{\mathbf{e}}_x$ on points lying on the spiral curve $k_z z + \ell\phi = \pi/2$, as shown in Fig. 2.9. Fig. 2.10 displays the polarization distributions on the planes $z = 0; \lambda/4; \lambda/2; 3\lambda/4$ and λ in the case where $\ell = |2|$.

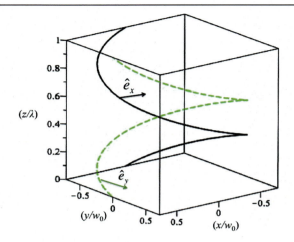

Figure 2.9: Helices of constant linear polarization. The green dashed line corresponds to points where $k_z z + \ell\phi = 0$, where the polarization is $\hat{e}_y$, while the black solid line corresponds to points where $k_z z + \ell\phi = \pi/2$, where the polarization is $\hat{e}_x$. The helices correspond to points of maximum intensity and $\ell = 2$.

2.10.2.2 Counter-propagating with Lin ⊥ Lin

Another interesting scenario is the one in which the two beams have orthogonal linear polarizations (Lin ⊥ Lin). The total electric field vector is

$$\mathbf{E}(\rho, \phi, z, t) = \left\{ \mathcal{U}(\rho, \phi, z) e^{-i\omega t} + c.c. \right\}, \tag{2.93}$$

where now we have

$$\mathcal{U}(\rho, \phi, z) = U_1(\rho, z) e^{i(k_z z + \ell\phi)} \hat{e}_x + U_2(\rho, z) e^{-i(k_z z + \ell\phi)} \hat{e}_y. \tag{2.94}$$

Since we have $U_1 = U_2 = U$ we obtain

$$\mathcal{U}(\rho, \phi, z) = \sqrt{2} U(\rho, z) \left\{ \cos(\ell\phi + k_z z) \frac{\hat{e}_x + \hat{e}_y}{\sqrt{2}} + i \sin(\ell\phi + k_z z) \frac{\hat{e}_x - \hat{e}_y}{\sqrt{2}} \right\}. \tag{2.95}$$

As shown in Table 2.2, the total polarization is in general elliptical, varying between linear and circular as the argument $k_z z + \ell\phi$ varies.

Fig. 2.11 displays two sets of helices in separate plots, one set representing two helices of constant linear polarization and the other set represents two of opposite circular polarizations. The helices correspond to points of maximum intensity.

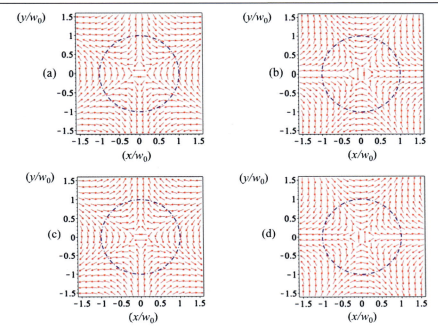

Figure 2.10: Polarization distributions on different z planes, but for $\sigma_+ - \sigma_-$ counter-propagating Laguerre-Gaussian (LG) beams with the same magnitude and sign of ℓ ($|\ell| = 2$). Distances are in units of the beam waist w_0 and arrows represent polarization direction. The blue circles correspond to points of maximum intensity. The figures are labeled (a), (b), (c), and (d) corresponding to the planes at axial positions $z = 0$, $\lambda/4$, $\lambda/2$, and $3\lambda/4$, respectively.

Table 2.2: Variations with (z, ϕ) of total polarization due to two counter-propagating doughnut beams in the Lin $\perp$ Lin configuration evaluated at the radial position $\rho = w_0\sqrt{|\ell|/2}$. As $(k_z z + \ell\phi)$ changes from 0 to π, the polarization switches between linear (Lin) and circular (Circ).

$k_z z + \ell\phi$	Σ_α
0	Lin
$\pi/4$	Circ $\hat{\sigma}_-$
$\pi/2$	Lin
$3\pi/4$	Circ $\hat{\sigma}_+$
π	Lin

2.10.3 Bi-chromatic vortex beams

Bi-chromatic beams, whether co-propagating or counter-propagating, involve superposition of two vortex beams, which have different frequencies ω_1 and ω_2. We focus on the case in

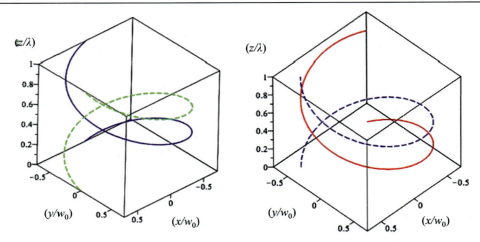

Figure 2.11: Polarization helices for two counter-propagating beams in the Lin ⊥ Lin configuration with $\ell_1 = -\ell_2 = 2$. (Left): helices of constant linear polarization. The green dashed line corresponds to points where $k_z z + \ell\phi = 0$, while the blue solid line corresponds to points where $k_z z + \ell\phi = \pi/2$. (Right): helices of constant circular polarization. The blue dashed line corresponds to points where $k_z z + \ell\phi = \pi/4$, while the red solid line to points where $k_z z + \ell\phi = 3\pi/4$. The helices correspond to points of maximum intensity.

which the difference $\Delta\omega$ in frequencies is small $\Delta\omega \ll \omega_1.\omega_2$. This, along with the different winding numbers, gives rise to a total wave polarization distribution that exhibits both spatial and temporal polarization gradients.

2.10.3.1 Bi-chromatic co-propagating with $\sigma^+ - \sigma^-$

Consider two bi-chromatic co-propagating $LG_{\pm\ell,0}$ doughnut beams, which differ slightly in their frequencies. Assume that beam 1 is right-circularly polarized ($\hat{e}_1 = \sigma^+$) and beam 2 is left-circularly polarized ($\hat{e}_2 = \sigma^-$). The electric field vector is given by the analogues of Eqs. (2.72) and (2.73). We have

$$\mathbf{E}(\rho,\phi,z,t) = \mathcal{U}(\rho,\phi,z,t)e^{ik_z z}, \qquad (2.96)$$

where

$$\mathcal{U}(\rho,\phi,z,t) = U_1(\rho,z)e^{i(\ell\phi-\omega_1 t)}\hat{e}_1 + U_2(\rho,z)e^{-(i\ell\phi+\omega_2 t)}\hat{e}_2, \qquad (2.97)$$

expressing the circular polarization vectors in terms of $\hat{e}_x$ and $\hat{e}_y$ and using $U_1 = U_2 = U$, we have the analogue of Eq. (2.80),

Structured light 69

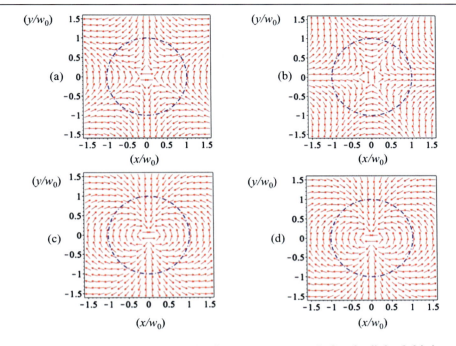

Figure 2.12: The rotation of the polarization vector, at $z = 0$, for the light field due to the interference of two co-propagating LG beams with opposite winding numbers, opposite circular polarizations and slightly different frequencies. (Upper panel) The case where the first beam of frequency ω_1 has $\ell =$ and the second beam of frequency ω_2 has $\ell = -1$ at two different times $t = 0$ (a) and $t = \pi/\Delta$ (b). (Lower panel) The case where the first beam of frequency ω_1 has $\ell = -1$ and the second beam of frequency ω_2 has $\ell = 1$ at two different times $t = 0$ (c) and $t = \pi/\Delta$ (d). The circles correspond to points of maximum intensity.

$$\mathcal{U} = -i\sqrt{2}U e^{-i\frac{(\omega_1+\omega_2)}{2}} \left\{ \hat{e}_x \sin\left[\frac{(2l\phi + (\Delta\omega)t)}{2}\right] + \hat{e}_y \cos\left[\frac{(2l\phi + (\Delta\omega)t)}{2}\right] \right\}, \quad (2.98)$$

where $\Delta\omega = \omega_2 - \omega_1$. This shows that the polarization is spatio-temporal, i.e. a function of both space and time. It is straightforward to show that, on any plane parallel to the focal plane, the polarization distribution rotates around the axis at a rate which depends on $\Delta\omega$ and the value of ℓ. Fig. 2.12 illustrates this pattern rotation first for the case, where beam 1 of frequency ω_1 has $\ell = 2$ and the second beam of frequency ω_2 has $\ell = -2$ at two different times $t = 0$ (left) and $t = \pi/\Delta$ (right). The second case when beam 1 of frequency ω_1 has $\ell = -2$ and beam 2 of frequency ω_2 has $\ell = 2$ at two different times $t = 0$ (left) and $t = \pi/\Delta$ (right). Note that in the latter case the polarization changes from azimuthal to radial. Note also that when $\Delta\omega = 0$, Eq. (2.98) reduces to Eq. (2.73).

2.10.3.2 Bi-chromatic with Lin ⊥ Lin

Finally, we deal with the above case but with the Lin ⊥ Lin polarization arrangement. Assume beam 1 has polarization $\hat{e}_x$ and beam 2 has polarization $\hat{e}_y$. We find

$$\mathbf{E}(\rho, \phi, z, t) = \mathcal{U}(\rho, \phi, z, t) e^{ik_z z}, \tag{2.99}$$

where

$$\mathcal{U}(\rho, \phi, z, t) = U_1(\rho, z) e^{i(\ell\phi - \omega_1 t)} \hat{e}_x + U_2(\rho, z) e^{-(i\ell\phi + \omega_2 t)} \hat{e}_y. \tag{2.100}$$

Once more setting $U_1(\rho, z) = U_2(\rho, z) = U(\rho, z)$ we find

$$\mathcal{U}(\rho, \phi, z, t) = U(\rho, z) e^{-i\frac{(\omega_1 + \omega_2)}{2}} \left\{ e^{i(\ell\phi - (\Delta\omega)t/2)} \hat{e}_x + e^{i(-\ell\phi + (\Delta\omega)t/2)} \hat{e}_y \right\}. \tag{2.101}$$

Fig. 2.13 shows the polarization vector distribution in the focal plane $z = 0$ due to the bi-chromatic $LG_{\pm\ell,0}$ beams in the Lin ⊥ Lin configuration evaluated at two different times. Note how the polarization changes from linear to circular and vice versa at different angular positions.

We briefly summarize this part which has dealt primarily with interfering co-axial LG light beams. We have explored the spatial and temporal polarization of light fields created by the interference of either co-propagating or counter-propagating LG beams when they have opposite winding numbers. Other cases considered include the case when the beams possess opposite circular polarizations, and when they have mutually orthogonal linear polarizations, as in the Lin-Lin configuration.

When the LG beams are co-propagating and possess opposite circular polarizations we have found that, for a fixed value of ℓ, the polarization is independent of the axial position z, so that within a normal beam cross-section, it is everywhere locally linear but the direction changes, depending on its polar position (ρ, ϕ). When the beams have mutually orthogonal polarizations the total polarization again does not depend on the radial position, but now it can change from linear to σ_- and then back to linear and σ_+ as the azimuthal angle changes from 0 to 2π.

When the LG beams are counter-propagating, the polarization distribution depends on the axial position as well as the in-plane polar position (ρ, ϕ). Note the interesting symmetry in which the distribution for $z = \lambda/4$ is the mirror reflection of that at $z = 3\lambda/4$. Similarly, the polarization distribution for $z = \lambda/2$ is the mirror reflection of that for $z = 0$. The distributions at $z = 0$ and $z = \lambda$ are identical.

A case in which the polarization shows both temporal and spatial variations is that of interfering beams with slightly different frequency and opposite signs of the winding number. We may again consider two different cases when the beams have opposite circular polarizations

Structured light 71

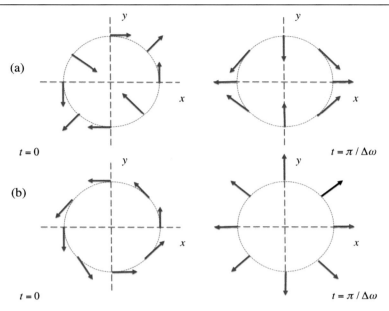

Figure 2.13: The evolution of the polarization vector, at $z = 0$, for a light field due to the interference of two co-propagating LG beams with mutually orthogonal polarizations, opposite signs of ℓ and slightly different frequencies. (Upper panel) The first beam of polarization along the x-axis has frequency ω_1 and $\ell = |\ell|$ and the second beam of polarization along the y-axis has frequency ω_2 and $\ell = -|\ell|$ at two different times $t = 0$ (a) and $t = \pi/\Delta\omega$ (b). (Lower panel) The first beam of polarization along the x-axis has frequency ω_1 and $\ell = -|\ell|$ and the second beam of polarization along the y-axis has a frequency ω_2 and $\ell = |\ell|$ at times $t = 0$ and $t = \pi/\Delta\omega$. The dotted circles correspond to points of maximum intensity.

and mutually orthogonal polarizations. This difference in frequency has been shown to give rise to a rotation of the polarization pattern. The temporal evolution is very similar to the sequence exhibited by the spatial dependence and so similar patterns will emerge, except that position-dependence is now replaced by a time-dependence.

2.11 Quantization of optical angular momentum

As pointed out at the outset the main properties of twisted light that are relevant to quantum information are its spin angular momentum (SAM) and its orbital angular momentum (OAM). It is the OAM property that provides the additional higher degrees of freedom, which leads to an increase in the capacity for optical information transfer. As the OAM states form an unbounded set of states this brings to our disposal a large number of quantum communication channels. Here, we briefly describe the quantization of optical spin carried out by both structured and unstructured light and we also outline the quantum formalism for OAM of structured light exemplified by the LG beam set.

2.11.1 Quantized SAM

The quantized SAM emerges from a standard procedure in which the optical fields are expressed as a sum of normal modes each of wave vector $\mathbf{k}$ and unit polarization vector $\hat{\boldsymbol{\epsilon}}_{\mathbf{k},s}$, where s is the polarization index. Associated with the normal mode are the annihilation and creation operators $\hat{a}_{\mathbf{k},s}$ and $\hat{a}^\dagger_{\mathbf{k},s}$ satisfying the usual commutation rules $[\hat{a}_{\mathbf{k},s}, \hat{a}^\dagger_{\mathbf{k}',s'}] = \delta_{\mathbf{k},\mathbf{k}'}\delta_{s,s'}$. In terms of the transverse fields the spin density operator is

$$\hat{\mathbf{S}} = \epsilon_0 \int d^3\mathbf{r}' [\hat{\mathbf{E}}^\perp(\mathbf{r}') \times \hat{\mathbf{A}}^\perp(\mathbf{r}')]. \tag{2.102}$$

The usual procedure leads to quantized SAM in the following form [19]:

$$\hat{\mathbf{S}} = i\hbar \sum_{\mathbf{k}} \sum_{s,s'} \left\{ \hat{a}^\dagger_{\mathbf{k},s'} \hat{a}_{\mathbf{k},s} + \frac{1}{2}\delta_{s,s'} \right\} (\hat{\boldsymbol{\epsilon}}_{\mathbf{k},s} \times \hat{\boldsymbol{\epsilon}}^*_{\mathbf{k},s'}). \tag{2.103}$$

2.11.1.1 Circularly polarized light

In the case of circularly polarized light the polarization vectors are $\hat{\boldsymbol{\epsilon}}_{\mathbf{k},\pm 1} = \frac{1}{\sqrt{2}}(\hat{\boldsymbol{\epsilon}}_{\mathbf{k},1} \pm i\hat{\boldsymbol{\epsilon}}_{\mathbf{k},2})$, which satisfy $\hat{\boldsymbol{\epsilon}}_{\mathbf{k},s} \times \hat{\boldsymbol{\epsilon}}^*_{\mathbf{k},s'} = -is\hat{\kappa}\delta_{ss'}$ where $s, s' = \pm 1$ and $\hat{\kappa}$ is a unit vector in the direction of $\mathbf{k}$. We obtain finally

$$\hat{\mathbf{S}} = \sum_{\mathbf{k}} \hbar \left(\hat{N}_{\mathbf{k},+1} - \hat{N}_{\mathbf{k},-1} \right) \hat{\kappa}, \tag{2.104}$$

where $\hat{N}$ is the number operator. Thus, the spin angular momentum involves a summation over all $\mathbf{k}$ of the difference between the number of the right-handed and the number of left-handed circularly polarized photons. Circular polarizations represent radiation states that are eigenfunctions of the operator for optical spin angular momentum. As such, each circularly polarized photon conveys a well-defined quantum spin, precisely $\sigma\hbar$, where $\sigma = \pm 1$ according to there being left/right circular polarization [19].

2.11.1.2 Linearly polarized light

In the case of linear polarization the unit vectors $\hat{\boldsymbol{\epsilon}}_{\mathbf{k}1}$ and $\hat{\boldsymbol{\epsilon}}_{\mathbf{k}2}$ are real and orthogonal to each other, then $\hat{\boldsymbol{\epsilon}}_{\mathbf{k},s} \times \hat{\boldsymbol{\epsilon}}^*_{\mathbf{k},s'} = \pm\hat{\kappa}(1 - \delta_{ss'})$. On substituting this in Eq. (2.103) we have

$$\hat{\mathbf{S}} = i \sum_{\mathbf{k}} \hbar(\hat{a}^\dagger_{k2}\hat{a}_{k1} - \hat{a}^\dagger_{k1}\hat{a}_{k2})\hat{\kappa}. \tag{2.105}$$

This shows that $\hat{\mathbf{S}}$ is off-diagonal and suggests that linearly polarized light has zero spin.

2.11.2 Orbital angular momentum

It is convenient to consider only the quantization of the paraxial Laguerre-Gaussian set of modes to illustrate their quantized OAM. Since the LG set, as described above, are orthonormal, their quantized total electromagnetic Hamiltonian is such that

$$\hat{\mathcal{H}}_{em} = \frac{1}{2}\epsilon_0 \int d^3\mathbf{r} \left\{ \hat{E}^2(\mathbf{r}) + c^2\hat{B}^2(\mathbf{r}) \right\}$$
$$= \sum_Q \hbar\omega_Q \left[\hat{a}_Q^\dagger \hat{a}_Q + \frac{1}{2} \right]. \quad (2.106)$$

Here $\hat{a}_Q$ and $\hat{a}_Q^\dagger$ are the annihilation and creation operators of the LG mode Q, satisfying the boson commutation relations

$$[\hat{a}_Q, \hat{a}_{Q'}^\dagger] = \delta_{QQ'}. \quad (2.107)$$

The summation is over all LG modes $|Q\rangle \equiv |(klp)\rangle$. The mode label $Q = (klp)$ stands for the set of degrees of freedom of the mode which are ℓ, the azimuthal winding number, p, the radial number, $\hat{\epsilon}_Q$, the wave polarization and $\mathbf{k}$, the axial wave vector. The quantized electric vector field $\hat{E}$ consistent with Eq. (2.106) can be written as

$$\hat{E}(\rho, \phi, z) = \sum_Q \hat{\epsilon}_Q \left[\mathcal{E}_Q(\rho, \phi, z)\hat{a}_Q + h.c. \right]. \quad (2.108)$$

The electric vector field $\mathcal{E}$ for the LG mode $Q = (klp)$ is as given in Eqs. (2.20) and (2.24). The one-photon state $|Q\rangle \equiv |(klp)\rangle$ is an eigenstate of the Hamiltonian with eigenvalue $\hbar\omega$ and the OAM operator $\hat{L}_z$ with eigenvalue $\hbar\ell$

$$\hat{L}_z |(klp)\rangle = \hbar\ell |(klp)\rangle; \quad \hat{\mathcal{H}} |(klp)\rangle = \hbar\omega |(klp)\rangle \quad (2.109)$$

and as pointed out earlier, the winding number is such that $\ell = \pm|\ell|$ where $|\ell| = 1, 2, 3, \ldots$, i.e. ℓ takes both negative and positive integer values. Vortices for which $\ell = +|\ell|$ can be assumed to have right-handed winding of its wavefronts, while $\ell = -|\ell|$ have left-hand winding of their wavefronts.

The quantum aspects of optical vortices and their significance for quantum communication using structured light are discussed in Chapter 3.

2.12 Conclusions

In this chapter the aim has been to provide a brief introduction to the subject of structured light, emphasizing that the type of structured light to be focused on here is the one associated with twisted light, or optical vortices. Much of what has been discussed here has been based on the LG optical vortex set of modes. We have shown that members of this set of vortex modes are characterized by orbital angular momentum so that each photon carries an OAM of magnitude $\hbar|\ell|$ with $\ell \neq 0$ an integer that takes positive and negative values. The discrete nature of the winding number ℓ in principle suggests using the OAM property as a platform for implementing a high-dimensional generalization of the two-bit quantum states associated with SAM. The higher degrees of freedom afforded by the OAM in turn enhances the capacity for optical information transfer.

The role of optical spin angular momentum when coupled to its orbital angular momentum is discussed at some length when more than one vortex beam is involved, but we have restricted the discussions to the case of two co-axial beams. We have shown that there are a number of realizable scenarios in which the beams can be co-propagating or counter-propagating and can have a combination of linear and/or circular polarizations. Furthermore, they can be bichromatic, differing only slightly in their frequencies, or their focal planes are shifted in space by a finite axial distance d while the axes of the beams remain the same.

We have highlighted the paraxial approximation as the much discussed regime of LG beams of moderate beam waists w_0. We have also considered the Barnett–Allen formulation of the non-paraxial regime, which is essentially the full theory of LG light beams and shown how the paraxial regime emerges from the full theory.

The beam waist is a measure of the tightness of focus and is such that, for w_0 relatively large, the vortex light beam is essentially characterized by field components transverse to the direction of propagation, with a practically negligible longitudinal component. However, as soon as the beam waist becomes sufficiently small a longitudinal field develops with a magnitude that is comparable to the transverse components. A manifestation of the emergence of a substantial longitudinal field component is the demonstration of the chiral nature of vortex light beams due to its orbital (rather than spin) angular momentum. A vortex beam such as a doughnut beam (for which the winding number is ℓ, but the radial number $p = 0$) such as those shown in the first row in Fig. 2.1 has a different helicity and chirality from an identical beam for which the only difference is that its winding number is $-\ell$. This means that vortex beams are characterized by a handedness. We have seen that this intriguing property of vortex light, although exists in principle, could not be demonstrated for a vortex beam with a larger beam waist because it would show diminishing magnitudes. For beams with small waists we have clarified the differences in the chirality and helicity density distributions for doughnut beams that are identical except for the sign of their winding numbers.

Finally we briefly outlined the steps leading to quantizing the OAM, formally identifying vortex modes as eigenstates of the electromagnetic Hamiltonian $\hat{\mathcal{H}}$ and the OAM operator $\hat{L}_z$. The discreteness of the winding number ℓ is again evident in the quantized formulation and confirms the realization of using the OAM property as a platform for implementing a high-dimensional generalization of the two-bit quantum states associated with the SAM.

References

[1] L. Allen, M.W. Beijersbergen, R.J.C. Spreeuw, J.P. Woerdman, Orbital angular momentum of light and the transformation of Laguerre-Gaussian laser modes, Phys. Rev. A 45 (Jun 1992) 8185–8189.

[2] L. Allen, M.J. Padgett, M. Babiker, IV The orbital angular momentum of light, in: Progress in Optics, Vol. 39, Institute of Physics, 1999, pp. 291–372.

[3] Sonja Franke-Arnold, Les Allen, Miles Padgett, Advances in optical angular momentum, Laser Photonics Rev. 2 (4) (2008) 299–313.

[4] Juan P. Torres, Lluis Torner (Eds.), Twisted Photons: Applications of Light with Orbital Angular Momentum, John Wiley & Sons, 2011.

[5] David L. Andrews (Ed.), Structured Light and Its Applications: an Introduction to Phase-Structured Beams and Nanoscale Optical Forces, Elsevier, New York, 2011.

[6] David L. Andrews, Mohamed Babiker (Eds.), The Angular Momentum of Light, Cambridge University Press, Cambridge, 2012.

[7] L. Allen, Stephen M. Barnett, Miles J. Padgett, Optical Angular Momentum, Institute of Physics Pub., Bristol, 2003.

[8] Alison M. Yao, Miles J. Padgett, Orbital angular momentum: origins, behavior and applications, Adv. Opt. Photonics 3 (2) (jun 2011) 161.

[9] Halina Rubinsztein-Dunlop, Andrew Forbes, M.V. Berry, M.R. Dennis, David L. Andrews, Masud Mansuripur, Cornelia Denz, Christina Alpmann, Peter Banzer, Thomas Bauer, Ebrahim Karimi, Lorenzo Marrucci, Miles Padgett, Monika Ritsch-Marte, Natalia M. Litchinitser, Nicholas P. Bigelow, C. Rosales-Guzmán, A. Belmonte, J.P. Torres, Tyler W. Neely, Mark Baker, Reuven Gordon, Alexander B. Stilgoe, Jacquiline Romero, Andrew G. White, Robert Fickler, Alan E. Willner, Guodong Xie, Benjamin McMorran, Andrew M. Weiner, Roadmap on structured light, J. Opt. 19 (1) (jan 2017) 013001.

[10] Stephen M. Barnett, Mohamed Babiker, Miles J. Padgett, Optical orbital angular momentum, Philos. Trans. R. Soc. A, Math. Phys. Eng. Sci. 375 (2087) (feb 2017) 20150444.

[11] Mohamed Babiker, David L. Andrews, Vassilis E. Lembessis, Atoms in complex twisted light, J. Opt. 21 (1) (jan 2019) 013001.

[12] Andrew Forbes, Angela Dudley, Melanie McLaren, Creation and detection of optical modes with spatial light modulators, Adv. Opt. Photonics 8 (2) (Jun 2016) 200–227.

[13] M.W. Beijersbergen, R.P.C. Coerwinkel, M. Kristensen, J.P. Woerdman, Helical-wavefront laser beams produced with a spiral phaseplate, Opt. Commun. 112 (5–6) (1994) 321–327.

[14] L. Marrucci, C. Manzo, D. Paparo, Optical spin-to-orbital angular momentum conversion in inhomogeneous anisotropic media, Phys. Rev. Lett. 96 (Apr 2006) 163905.

[15] Marco W. Beijersbergen, Les Allen, H.E.L.O. Van der Veen, J.P. Woerdman, Astigmatic laser mode converters and transfer of orbital angular momentum, Opt. Commun. 96 (1–3) (1993) 123–132.

[16] D. McGloin, K. Dholakia, Bessel beams: diffraction in a new light, Contemp. Phys. 46 (1) (jan 2005) 15–28.

[17] Chaowei Wang, Liang Yang, Yanlei Hu, Shenglong Rao, Yulong Wang, Deng Pan, Shengyun Ji, Chenchu Zhang, Yahui Su, Wulin Zhu, Jiawen Li, Dong Wu, Jiaru Chu, Femtosecond Mathieu beams for rapid controllable fabrication of complex microcages and application in trapping microobjects, ACS Nano 13 (4) (2019) 4667–4676.

[18] J.C. Gutiérrez-Vega, M.D. Iturbe-Castillo, G.A. Ramírez, E. Tepichín, R.M. Rodríguez-Dagnino, S. Chávez-Cerda, G.H.C. New, Experimental demonstration of optical Mathieu beams, Opt. Commun. 195 (1–4) (2001) 35–40.

[19] Leonard Mandel, Emil Wolf, Optical Coherence and Quantum Optics, Cambridge University Press, 1995.

[20] Ivan Fernandez-Corbaton, Xavier Zambrana-Puyalto, Gabriel Molina-Terriza, Helicity and angular momentum: a symmetry-based framework for the study of light-matter interactions, Phys. Rev. A 86 (Oct 2012) 042103.

[21] Frances Crimin, Neel Mackinnon, Jörg Götte, Stephen Barnett, Optical helicity and chirality: conservation and source, applied sciences, Appl. Sci. 9 (02 2019) 828.

[22] Frances Crimin, Neel Mackinnon, Jörg Götte, Stephen Barnett, On the conservation of helicity in a chiral medium, J. Opt. 21 (08 2019) 094003.
[23] Matt M. Coles, David L. Andrews, Chirality and angular momentum in optical radiation, Phys. Rev. A 85 (Jun 2012) 063810.
[24] Kayn A. Forbes, David L. Andrews, Optical orbital angular momentum: twisted light and chirality, Opt. Lett. 43 (3) (Feb 2018) 435–438.
[25] Kayn A. Forbes, David L. Andrews, Spin-orbit interactions and chiroptical effects engaging orbital angular momentum of twisted light in chiral and achiral media, Phys. Rev. A 99 (Feb 2019) 023837.
[26] Kayn A. Forbes, David L. Andrews, Enhanced optical activity using the orbital angular momentum of structured light, Phys. Rev. Res. 1 (Nov 2019) 033080.
[27] Koray Koksal, Mohamed Babiker, Vassilis E. Lembessis, Jun Yuan, Chirality and helicity of optical vortices of small beam waists, arXiv:2010.06338, 2020.
[28] Simin Feng, Herbert G. Winful, Physical origin of the Gouy phase shift, Opt. Lett. 26 (8) (apr 2001) 485.
[29] P. Hariharan, P. Robinson, The Gouy phase shift as a geometrical quantum effect, J. Mod. Opt. 43 (1996) 219–221.
[30] K. Koksal, M. Babiker, V.E. Lembessis, J. Yuan, Chirality and helicity of linearly-polarised Laguerre-Gaussian beams of small beam waists, Opt. Commun. 490 (2021) 126907.
[31] Stephen M. Barnett, L. Allen, Orbital angular momentum and nonparaxial light beams, Opt. Commun. 110 (5–6) (sep 1994) 670–678.
[32] Jorge Bernhard Gotte, Integral and fractional orbital angular momentum of light, PhD thesis, University of Strathclyde, 2006.
[33] J.B. Götte, S.M. Barnett, Light beams carrying orbital angular momentum, in: D.L. Andrews, M. Babiker (Eds.), The Angular Momentum of Light, Cambridge University Press, Cambridge, 2012, Chapter 1.
[34] Koray Koksal, Vassilis E. Lembessis, Jun Yuan, Mohamed Babiker, JOSA B 37 (2020) 2570.
[35] Smail Bougouffa, Mohamed Babiker, Quadrupole absorption rate and orbital angular momentum transfer for atoms in optical vortices, Phys. Rev. A 102 (6) (2020) 063706.
[36] Paweł Woźniak, Israel De Leon, Katja Höflich, Gerd Leuchs, Peter Banzer, Interaction of light carrying orbital angular momentum with a chiral dipolar scatterer, Optica 6 (8) (2019) 961–965.
[37] Yiqiong Zhao, J. Scott Edgar, Gavin D.M. Jeffries, David McGloin, Daniel T. Chiu, Spin-to-orbital angular momentum conversion in a strongly focused optical beam, Phys. Rev. Lett. 99 (Aug 2007) 073901.
[38] Ralf Dorn, Susanne Quabis, Gerd Leuchs, The focus of light—linear polarization breaks the rotational symmetry of the focal spot, J. Mod. Opt. 50 (12) (2003) 1917–1926.
[39] Uri Levy, Yaron Silberberg, Nir Davidson, Mathematics of vectorial Gaussian beams, Adv. Opt. Photonics 11 (4) (dec 2019) 828.
[40] Xavier Zambrana-Puyalto, Xavier Vidal, Gabriel Molina-Terriza, Angular momentum-induced circular dichroism in non-chiral nanostructures, Nat. Commun. 5 (1) (dec 2014) 4922.
[41] Xavier Zambrana-Puyalto, Xavier Vidal, Ivan Fernandez-Corbaton, Gabriel Molina-Terriza, Far-field measurements of vortex beams interacting with nanoholes, Sci. Rep. 6 (1) (apr 2016) 22185.
[42] K. Koksal, V.E. Lembessis, J. Yuan, M. Babiker, Interference of axially-shifted Laguerre–Gaussian beams and their interaction with atoms, J. Opt. 21 (2019) 104002.
[43] S. Franke-Arnold, J. Leach, M.J. Padgett, D. Lembessis, D. Ellinas, A.J. Wright, J.M. Girkin, P. Ohberg, A.S. Arnold, Optical ferris wheel for ultracold atoms, Opt. Express 15 (14) (June 2007) 8619–8625.
[44] V.E. Lembessis, M. Babiker, Spatiotemporal polarization gradients in phase-bearing light, Phys. Rev. A 81 (Mar 2010) 033811.
[45] V.E. Lembessis, A. Lyras, O.M. Aldossary, Optical ferris wheels as a platform for collisional quantum gates, J. Opt. Soc. Am. B 38 (1) (Jan 2021) 233–240.

CHAPTER 3

Quantum features of structured light

David L. Andrews
University of East Anglia, Norwich Research Park, Norwich, United Kingdom

Contents

3.1 Introduction 77
3.2 Basis for the quantization of structured light 79
3.3 Quantum issues in measurement and localization 83
3.4 Quantized angular momentum: light and matter 86
3.5 Entanglement 88
3.6 Conclusion 89
References 90

3.1 Introduction

Since the earliest inception of the photon as a simple packet of electromagnetic energy, it has become increasingly evident that a simple wave-particle concept is an entirely insufficient representation of a far deeper reality. Following the identification of pivotal roles for light itself in relativity theory, and later in gauge aspects of the electromagnetic interaction in quantum field theory, developments in the laser era have increasingly harnessed the distinct attributes of the photon, especially in the burgeoning field of photonics. In his landmark treatise on the quantum theory of light, Rodney Loudon astutely observed: 'it is no longer so straightforward to explain what is meant by a 'photon' [1]. In the years that have since elapsed, we now find ourselves in the widely heralded 'century of the photon', well beyond the centenary of Einstein's original concept [2]. Nonetheless, despite the incontrovertible status of photons and the necessity of casting a great deal of theory in such terms, the underlying concept has become no less difficult to fully appreciate. Part of the reason undoubtedly lies in the special features that have become evident in the propagation and interactions of structured light.

In modern understanding, photons are discrete quanta of the propagating optical modes that represent monochromatic solutions of Maxwell's equations. As such, each quantum in principle possesses five degrees of freedom: these are the parameters that uniquely define each mode. For light freely radiating from a point source, three of these degrees of freedom may readily be associated with components of a wave-vector (which together define the optical wavelength), while the other two define a state of optical polarization. The latter may correspond to coordinates on the Poincaré sphere, denoting a major axis of polarization, and a degree of ellipticity. These two degrees of freedom that account for conventional polarization afford a binary basis for conveying information, since the surface of a sphere maps to a two-dimensional plane. Polarization states that are orthogonal occupy polar opposite positions on the Poincaré sphere – commonly chosen as either vertical/horizontal or left/right circular polarizations.

In the realm of optical communication and informatics, where light is generally delivered by laser sources emitting in a predetermined direction, the longitudinal component of the wave-vector now accounts for just one of the three spatial degrees of freedom. The other two degrees of freedom, representing wave-vector components in a plane transverse to the propagation direction, often remain unharnessed. At the photon level, this signifies a residual redundancy in the capacity to convey information; it is here that spatially structured modes of light introduce distinctly new possibilities [3]. Such possibilities have long been latent even in the familiar Hermite-Gaussian modes. However, it is with 'twisted' modes that the opportunities have become more substantially realized.

With the dawning of the era of structured light, enticing possibilities to increase the bandwidth of frequency multiplex communication has led to fresh consideration of the latent information-conveying capacity of individual photons – and also attempts to identify the ultimate limitations of such a capacity. In some of these respects, the pace of experimental developments has exposed a surprisingly wide gulf between foundational theory in the spheres of quantum and physical optics. For example, little thought was given, until quite recently, to the implications of considering that a simple Gaussian beam might meaningfully be considered to comprise 'Gaussian photons'. Yet photons are quite properly regarded as quanta of a defined optical mode, so the question might almost have been considered a semantic nicety. Conversely, it is commonplace to find quantum aspects of radiation cast in terms of plane-wave descriptions, despite the intrinsic lack of physicality associated with an infinite transverse extent.

In the broad realm of structured light a wide variety of modal forms is known beyond Hermite-Gaussian beams. It is simplest to characterize and exploit the flexibility afforded by such modes with two independent degrees of freedom in their transverse structure by considering cylindrically symmetric beams, such as Laguerre-Gaussian (LG) and Bessel beams, where one degree of freedom can be captured in a topological charge, ℓ, and the other is determined by a radial distribution of the fields. Any proposition that a beam with transverse structure might simply comprise a distribution of plane-wave or 'unstructured' photons has been unequivocally quashed by counterexample, primarily through careful experimentation on Laguerre-Gaussian (LG) beams [4,5]. It has also been observed that such a notion is at variance with the principles of modal orthogonality [6]. In consequence it has been clearly established that, in a structured beam, individual photons can indeed convey more information than simply the energy and linear momentum established by their wave-vector, or in a simple polarization-related attribute such as spin angular momentum.

Considerations of information content are of crucial relevance in connection with optical communication and data transmission, and it is evident that an initial focus on the azimuthal structure associated with topological charge ℓ and quantized orbital angular momentum $\ell\hbar$ is now being more actively extended to the parametrization of radial form [7,8]. Moreover, it proves possible by superposition techniques to affix spatial degrees of freedom to the polarization structure, such that the polarization may vary modally across the beam, even spanning the whole space of the Poincaré sphere [9]. The realization of an expanded dimensionality for the modal state space affords the possibility of harnessing intrinsically quantum mechanical superposition features, extending well beyond simple qubits [10–13]. Inevitably such findings prompt new questions: what is the limit, if any, on the information capacity of a single photon – although, for high volume applications, it is primarily the rate of data transfer that matters [14]. The cutting edge of such questions lies in its significance for the search for an unequivocally optimum basis for conveying information in this form [15].

3.2 Basis for the quantization of structured light

It is appropriate to begin by concisely reviewing the principles, and the common assumptions, in developing the formal quantum theory of structured light. In particular we highlight the key features in the 'twisted' form of light, i.e. an optical vortex with a helicoidal wavefront structure, exemplified and most frequently implemented in the form of LG modes. Following an early foray into the theory which laid ground for the concept of an optical vortex [16], theory for wavefront-structured light was primarily developed for paraxial (essentially zero divergence) beams of light, whose electromagnetic modal structures satisfy the Helmholtz equation. Its origins in the constructs for orbital angular momentum have been described in a landmark review by Allen et al. [17]. Much more recently, in numerous studies attention has increasingly been focused on the additional scope afforded by features arising as a result of departures from paraxiality.

It is interesting to observe that the quantization of orbital angular momentum (OAM) was in fact first established, quickly and simply, through a deduction from an essentially classical formulation [18]. In paraxial twisted fields with an azimuthal phase structure $\exp(i\ell\varphi)$, where φ is the azimuthal angle about the propagation direction, a ratio of the densities of orbital angular momentum along the propagation direction, j_z, and energy w, proves to satisfy the condition $j_z/w = \ell/\omega$, where ℓ is the topological charge (also called the winding number), and ω is the circular frequency of the radiation. From this it immediately follows that, given an energy per photon $\hbar\omega$, the orbital angular momentum per photon is $\ell\hbar$. Indeed this correlates with a naïve application of the quantum angular momentum operator $\hat{j}_z$ to the azimuthal dependence of the optical phase. A clear distinction from the unit spin angular momentum associated with circular polarization is readily evident. It is also important to recognize that optical modes produced by more intricate azimuth-scaling can also engender non-integer, rational fraction values of the orbital angular momentum [19,20]. Although such modes lack the complete orthogonality of the integer modes, precise values can be experimentally ascertained through their interference with integer modes [21].

Full quantization of the optical fields for twisted or other forms of structured light has to proceed on a more formal basis. Photons are field bosons whose every interaction must engage one or more operators for creation and annihilation. Not only the optical fields are cast in terms of these operators; so too are energy, linear momentum and total angular momentum. As shown by Mandel and Wolf [22], while a paraxial approximation affords a compartmentalization of the total AM into spin and orbital parts (SAM and OAM, respectively) it is only circular polarization states that are eigenfunctions of both the spin angular momentum and Hamiltonian operators: these states alone have sharp eigenvalues for the axial spin. The general quantum formulation of the operators for the spin and orbital components of the optical angular momentum in fact admits two distinct forms – one based on the Poynting vector and the other a canonical form based on quantum electrodynamics [23], inviting experimentation to determine which is the physically most relevant.

The SAM rotates the electromagnetic field vectors locally in space, and for freely propagating paraxial light it is also directly connects to a scalar known as *optical helicity*, κ, defined as a volume integral of the scalar product of the vector potential and magnetic field [24]. In the paraxial approximation, it transpires that for any freely propagating beam, irrespective of topological charge, κ is a conserved property which can be explicitly cast in terms of a difference between the number of left- and right-handed circularly polarized photons [25,26]

$$\kappa = \frac{\hbar}{\varepsilon_0 c} \sum_k \left[N_{(k)}^{(L)} - N_{(k)}^{(R)} \right] \tag{3.1}$$

where the sum is taken over all modes of wavenumber k propagating in the same direction, and $N_{(k)}^{(C)}$ is the photon number operator for each such mode with left/right circular polarization C. Any other polarization state may be resolved into a fractionally weighted superposition of these two circular polarizations. In contrast, the polarization-independent OAM is unequivocally quantized in integer values $\ell \in \mathbb{Z}$, with no upper bound on the magnitude. It is a global property signifying rotations of the spatial distributions within optical modes.

Closely related to the optical helicity is the optical chirality density χ, defined as [27]

$$\chi = \frac{\varepsilon_0}{2} \boldsymbol{E} \cdot \nabla \times \boldsymbol{E} + \frac{1}{2\mu_0} \boldsymbol{B} \cdot \nabla \times \boldsymbol{B}, \tag{3.2}$$

which satisfies the continuity equation

$$\frac{\partial \chi}{\partial t} + \nabla \cdot \boldsymbol{\varphi} = 0. \tag{3.3}$$

Here, $\boldsymbol{\varphi}$ is the corresponding flux, or optical chirality flow. It has been shown that, for a monochromatic, though not necessarily paraxial beam, $\boldsymbol{\varphi}$ is directly proportional to the spin and the optical helicity κ [25].

As we have seen, for a paraxial beam [28] where the electromagnetic fields are perpendicular to the direction of propagation, the intrinsic spin and orbital parts of the angular momentum are well defined [29,30]. For non-paraxial fields, the local values of spin and orbital angular momentum are not so readily separable [24], and the field distributions contain longitudinal components that are strongly dependent on the detailed beam structure [31,32]. For Bessel or Laguerre-Gaussian beam, for example, this generally leads to an OAM-dependent spatial distribution of the helicity density. Spin-orbit interactions in non-paraxial fields generate spin-to-orbital angular momentum conversion for circularly polarized beams [33], the corresponding individual operators no longer having sharp eigenvalues. The non-integer expectation values of the spin and orbital components are expressible as

$$\langle S_z \rangle = \sigma \hbar \cos \theta, \qquad \langle L_z \rangle = [\ell + \sigma (1 - \cos \theta)] \hbar, \tag{3.4}$$

where θ is the apex half-angle of the cone formed by the emergent wave-vectors [34,35]; see Fig. 3.1. Even this formulation fails to account for anomalies that occur at a beam focus, where the handedness of the spin component may itself be reversed [36]. The quantum implications of such features have as yet to receive detailed study. However, the malleability of division into spin and orbital angular momenta becomes much more evident and controllable when interactions with matter occur, as in various kinds of optical element, including suitably tailored metasurfaces [37,38].

Notwithstanding the quantization of angular momentum, it is the transverse electric and magnetic fields of the radiation, $\boldsymbol{E}^\perp$ and $\boldsymbol{B}$, respectively – the latter intrinsically transverse – that engage with matter, and are therefore involved in all detection processes. (Of course, even the registration of the interference patterns that are often created in beam combinations, as a means to characterize structured light, require light absorption by matter for their observation.) It is worth noting a potential risk of confusion here. The property of field transversality for the two fields $\boldsymbol{E}^\perp$ and $\boldsymbol{B}$ is specifically one that is defined with respect to the local disposition of these fields against the local wave-vector, i.e. the normal to the wavefront. As

82   Chapter 3

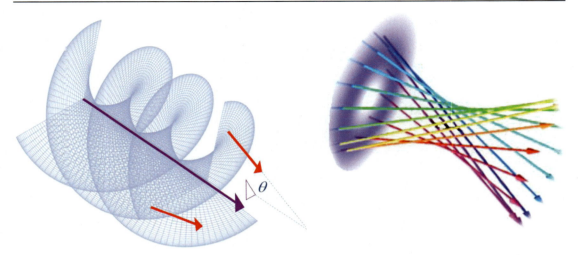

Figure 3.1: *Left*: For a helically structured beam, here illustrated for an LG beam with $\ell = 3$, the beam propagates in the direction shown by the purple arrow, the wavefront surface comprising three intertwined helicoidal surfaces indicated in blue: the wave-vector normal to this surface has an orientation that is position-dependent, as indicated at two different locations by the red arrows. In the paraxial approximation all such vectors are approximated as being aligned with the beam direction. *Right*: Bundle of wave-vectors on a transverse section of an $\ell = 1$ beam, the colors of the arrows signifying a full cycle of the optical phase $\exp(i\ell\phi)$. [Right-hand image adapted from an original by S. Franke-Arnold].

illustrated in Fig. 3.1, in structured light the orientation of this vector varies across the beam, having a major component parallel to the beam axis, and a minor component pointing inwards; the latter is in consequence associated with *longitudinal* components of the orthogonal electromagnetic field vectors, where the term 'longitudinal' is conventionally used to denote the beam axis.

In registering the detection of structured light, with optical elements that are usually formed from dielectric materials, it is the electric field of the photon flux that plays the dominant role. In the quantum formulation, both position-dependent electromagnetic fields become operators, cast as Fourier expansions over a modal set;

$$\begin{aligned} \boldsymbol{E}^{\perp}(\mathbf{r}) &\sim \sum_{\Omega} \mathrm{i} k_{\Omega}^{1/2} \boldsymbol{e}_{\Omega} a_{\Omega} \exp\left(\mathrm{i}\boldsymbol{k}_{\Omega}.\mathbf{r} + \mathrm{i}\ell\varphi\right) + h.c. : \\ \boldsymbol{B}(\mathbf{r}) &\sim \sum_{\Omega} \mathrm{i} k_{\Omega}^{1/2} \left(\hat{\boldsymbol{k}}_{\Omega} \times \boldsymbol{e}_{\Omega}\right) a_{\Omega} \exp\left(\mathrm{i}\boldsymbol{k}_{\Omega}.\mathbf{r} + \mathrm{i}\ell\varphi\right) + h.c. \end{aligned} \quad (3.5)$$

where Ω is a generic mode label, defined in terms of all five degrees of freedom, and $h.c.$ denotes hermitian conjugate. On the left of these two expressions, the equation for the electric field at position **r** is cast in terms of e_Ω, the complex polarization vector, and a_Ω, the corresponding photon annihilation operator – while the creation operator counterpart features in the Hermitian conjugate term written as $h.c$. For *plane waves*, the five degrees of freedom are three associated with Cartesian components of the wave-vector k_Ω (with wavenumber k_Ω) and two more for the polarization η. However, for *vortex* structured beams, the wave-vector space is defined in terms of parameters: k, ℓ, p, represent a partition into longitudinal, polarization, angular and radial functions [39]. It is this disparity between plane-wave and structured radiation which forms the basis for the conveyance of additional information density for a photon with the character of the latter form of light, i.e. a photon possessing OAM can carry more information than its plane-wave counterpart [40,41].

3.3 Quantum issues in measurement and localization

In considering the local measurement of any physical property of individual photons in a structured beam, it is first important to recognize a serious limitation: absolute measurement is fundamentally compromised by the impossibility of defining a position operator for any relativistic quantum particle [22,42–44]. A case has been made that the transverse Riemann-Silberstein vector $\boldsymbol{F} = \boldsymbol{E}^\perp + i\boldsymbol{B}$ might serve as a proxy for a 'photon wavefunction' of sorts [45]. Overlooking the distinction between this and a true quantum mechanical wavefunction is partly defensible when single photons are involved, and the distinction from the state vector presents less of a problem. But for states with two or more identical photons, there is no sense in which each one could be considered to have its own wavefunction – not least because of the character of a boson field [46]. Nonetheless, the lower limit on spatial registration of a photon is not constrained by the wavelength, since material particles of much smaller dimension can measurably interact with them. This is, of course, another aspect of the principle upon which STED (stimulated-emission depletion) subwavelength imaging, defeating Abbe limitations, is based [47].

In the dawning era of structured light experimentation, based on the theory developed by Allen et al. [48], it was already widely inferred that the detection of orbital angular momentum in a beam would of necessity indicate the possession of this property by its constituent photons – see for example an early review [17]. Interferometric measurements proved the truth of this assumption, though some residual concepts of a classical framework led others to still conceive of OAM as a beam property. The significance of this difference in perspective was highlighted by the arrival of experiments that enabled the efficient sorting of light with different OAM values into spatially separated channels [49,50]. However, a quantum uncertainty principle recognized only recently in fact precludes *precise* registration of the orbital angular momentum content of a single photon at a position through any angle-specific aperture;

$$\Delta L \Delta \psi \geq \hbar/2, \tag{3.6}$$

where ψ is the angle – determined in the experiment by suitably encoding a spatial light modulator [51].

Much interest attaches to the process of spontaneous parametric down-conversion (SPDC, also known as degenerate down-conversion, DDC, in which each input photon of an optical frequency ω produces two photons, each of frequency $\omega/2$). For photon pairs produced by this means it has been shown possible to establish a violation of a measurement condition concerning uncertainties in topological charge, $\ell = L/\hbar$ and azimuthal angle ϕ;

$$(\Delta \ell)^2 (\Delta \phi)^2 \geq 1/4. \tag{3.7}$$

The demonstration of quantum correlations serves as a proof of principle that angular disposition and angular momentum are indeed suitable variables for applications in quantum information [52]. Moreover, through an analysis of position correlation within the azimuthal region, it has been discovered how to manipulate the degree of entanglement between the DDC output photons, through control over the orbital angular momentum and radial intensity profiles of the input light [53].

There is still an issue over modal orthogonality. In the current generation of experimentation on sorting OAM modes, it has become customary to accept incomplete or imperfect separation as a reflection of practical limitations. However, it has also emerged that an often-quoted orthogonality relation needs correction, properly reflecting the fact that modes with very marginally different directions of propagation cannot be discriminated with arbitrary precision. Unpacking the generic mode label Ω deployed for simplicity in Eq. (3.3), the commutation relation between the photon creation and annihilation operators for two LG modes of the same indices (ℓ, p), within a sufficiently large quantization volume V, may be expressed as follows;

$$\left[a_\Omega, a_{\Omega'}^\dagger \right] = \left(8\pi^3 V \right)^{-1} \delta \left(k_z - k_z' \right) \delta \left(k_\rho - k_\rho' \right) \delta \left(k_\phi - k_\phi' \right) \delta_{\eta \eta'} \tag{3.8}$$

The important distinction from earlier textbook representations is primarily the casting in terms of Dirac, as opposed to Kronecker, deltas. Here, the local wave-vector is expressed in cylindrical coordinates, as shown in the right-hand panel of Fig. 3.2.

Detailed analysis reveals that the differential measurability of the mode at radially different positions (z, ρ, ϕ) and $(z, \rho + \delta \rho, \phi)$, scales as $k^2 \ell^{-1} \delta \rho$ at positions close to the beam axis, consistent with the angle-angular momentum uncertainty principle of Eq. (3.4). Conversely, for positions remote from the axis, the measurability scales with $k^{-1} \ell^2 \rho^{-3} \delta \rho$, representing a much more demanding limit on the degree of non-collinearity [54].

Quantum features of structured light 85

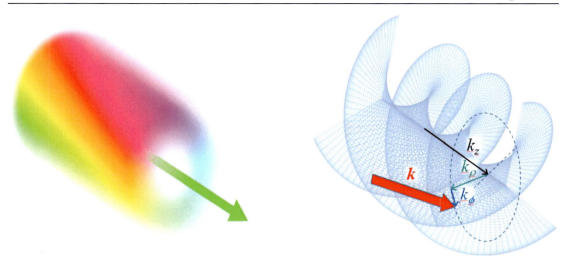

Figure 3.2: *Left*: Phase rotation of a propagating $\ell = 1$, $p = 0$ LG beam, hue denoting phase and color intensity representing the local level of irradiance; the green arrow identifies the propagation axis and there is a phase singularity along this axis. *Right*: Mutually orthogonal components of a local wave-vector in cylindrical coordinates.

The cycling of the optical phase $\exp(i\ell\phi)$ as an LG beam propagates is illustrated in the left-hand panel of Fig. 3.2. The infinitesimal singularity at its core is an essentially classical concept; Berry and Dennis have shown that in the quantum formulation there is an effective core of radius R given by

$$R = 2\pi w_0^2 \left(\hbar c \Delta\omega / P\lambda^3 \right), \tag{3.9}$$

where the beam has beam waist w_0, power P, wavelength λ and linewidth $\Delta\omega$ [55]. This facet of quantum uncertainty may prove significant in developments that depend on the high phase gradient near the core, in order to harness quadrupole interactions; it provides for a higher residual intensity in that region, than the classical formulation would suggest.

Another perspective on the connection between the number of photons and phase resolution is a quantum uncertainty principle of much older provenance [56], which has seldom received the attention it deserves in connection with quanta of structured light: this is the number-phase uncertainty relation. There have been many attempts to formulate a robust statement of this uncertainty, complicated (as also with the angle-angular-momentum relation) by the cyclic continuity in one of the variables – here the optical phase, φ [57]. A clear statement in Heitler's authoritative text is expressed as

$$\Delta n \Delta \varphi \geq 1/2. \tag{3.10}$$

In applications involving the sorting of OAM modes, then when the mode occupancy is restricted to a binary basis, requiring $\Delta n_{k,\eta,l,p} \ll 1$, the optical phase is indeterminate. With a higher flux it would matter less; with large numbers of photons present the phase uncertainty for the same mode detection would be negligible. Contrary to the lessons of 'Young's slits' plane-wave experiments at low intensities [58] single-photon optics with structured light cannot be entirely identified with beam or wave optics. Thus, whereas the principle of achieving heightened information content per photon can be exploited for data transmission at lower intensities, it is conceivable that at some point a trade-off will be met, where increasing the topological charge ceases to deliver an advantage [59]. The first identification of such a feature has been secured in an analysis by Coles [14].

3.4 Quantized angular momentum: light and matter

The basis for compartmentalization of optical angular momentum into spin and orbital components, along with the physical attributes and connotations of each kind, is neat, and it provides the simplicity of allowing wave-vector and polarization structures to be treated as separable. However, it affords considerable scope for misapprehension. Certainly, there is no direct analogue in the familiar spin and orbital angular momentum of electronic states. As mentioned earlier, Mandel and Wolf [21] showed that only a paraxial approximation affords a definitive separation, where integer spin is uniquely associated with circular polarization states. However, although the net angular momentum affords two dynamically separable observables, the simple spin and orbital AM operators do not conform to the necessary quantum mechanical structure. In fact, as shown by Van Enk and Nienhuis [60,61], these relations apply:

$$[S_i, S_j] = 0; \quad [L_i, L_j] = i\hbar(L_k - S_k)\varepsilon_{ijk}; \quad [L_i, S_j] = i\hbar S_k \varepsilon_{ijk} \tag{3.11}$$

where the subscript indices denote Cartesian components, epsilon denotes the Levi-Civita antisymmetric tensor, and the Einstein convention for implied summation over repeated tensor indices applies. Some of the precepts familiar from the standard theory of quantized angular momentum therefore cease to be valid; the commutability of different spin components is especially striking.

There is an inexorable interplay of spin and orbital angular momentum whenever non-paraxial conditions apply [62]. Indeed, even beams whose waist exceeds the reduced wavelength can display such features. One immediate consequence is that the electric and magnetic fields have to be recognized as having both transverse and longitudinal components [63]; transverse spin also arises [64], although their volume integrals vanish. The physical separability of orbital and spin angular momentum is indeed rendered more complex by observations that it is possible to convert between, either directly through passage of the light through a passive optical medium [65] or at a beam focus produced by a high numerical aperture lens [66], for example. The latter instance reflects the fact that this is the realm of *spin-orbit coupling* [34,35,67], a topic that is finding ever-increasing significance in subwavelength photonics. Such coupling is primarily manifest through interactions with matter. As exhibited in Fig. 3.2, the phase gradient of structured light is not purely longitudinal, as it is for plane waves, and the extent of transverse character grows with the topological charge. This becomes significant for a variety of interactions in which electric quadrupole or higher order multipole coupling is involved. For example in Bessel beam interactions with atoms, circular dichroism may thereby occur in atoms, through spin coupling with non-zero OAM [68].

The universality of conservation laws for energy, linear momentum and angular momentum are associated with the temporal, translational and orientational uniformity of space, in accordance with Noether's theorem [69]. Indeed, this is only the most familiar subset of a whole family of conserved quantities [70–72]. In the application of such principles to the entirety of systems comprising matter and light of any form it has to be borne in mind that, in contrast to energy, linear and angular momentum, other quantities do not generally have material counterparts to these optical properties. There is no quantum operator for chirality density of matter, in contrast to the optical quantity χ defined for light in Eq. (3.2). Therefore one cannot meaningfully appropriate conservation laws for representing the transfer of optical chirality to or from matter. Quantum uncertainty conditions on these conservation laws also impose limitations, though these are seldom critical for validation measurements – namely an assumption of timescales that are above an optical cycle, linear dimensions greater than a reduced wavelength, and a sufficient solid angle to provide for detection through a finite aperture – as we have seen in Eq. (3.8). However, in at least one aspect, the quantum issues of angular momentum conservation, in optical vortex interactions with matter, may escape sufficient scrutiny.

There is a large body of work on the interaction of structured light with matter, focusing – in both senses – on atoms. Some studies have assumed a single atom placed along the beam axis, its nucleus neatly coinciding with a singularity in optical phase. As a result of the finite core given by Eq. (3.10), and the physical extent of the electronic orbitals, such an atom will nonetheless experience a small field [73]. Most studies have focused on off-axis interaction, to exploit greater field strengths, though in both cases a field gradient can couple with quadrupole transitions; in both cases a common direction can be assumed for the axis about which the optical and electronic orbital angular momentum are quantized. In the earliest work on the theory [74], it was shown that a quadrupole transition is the lowest electric multipolar process in which an exchange of orbital angular momentum can occur between the light, the internal motion, and the center of mass motion. Near to a resonance, quadrupole effects may become especially prominent [75], and the detailed interplay of photon OAM with atomic transitions continues to present intriguing physics [76,77].

For commercial applications an atom-based scenario does present major technical challenges, quite apart from the energy cost of sustaining cold vacuum conditions. To stably position an atom off-axis requires a strongly structured optical or magnetic field, either of which must strongly perturb the atomic interactions with the structured beam itself (through ac Stark or Zeeman effects). Upscaling also demands a significant amount of data processing to control and sustain atoms in specific nanoscale free-space locations, undermining any virtue of implementation for data communication.

However most matter is not composed of free, individual atoms, and in the pursuit of channels for optical communication and information, the interactions of structured light in, and with, optical components, optical fibers and detectors are of most direct relevance. Such media are commonly anisotropic, and there is no guarantee that any particular material axis, whose rotational symmetry provides for electronic states with sharp eigenvalues for quantized angular momentum, will necessarily be aligned with the axis of the optical beam. Certainly, one cannot achieve definitive quantization of material angular momentum about two macroscopically distinguishable axes. Therefore the conservation of a summed orbital angular momentum, comprising the addition of optical and material elements, can only be measurably ascertained in terms of expectation values; no individual measurement can assume direct correlation between the OAM imparted along one direction, and uptake through an electronic transition with angular momentum values along another.

Another distinct difference arises, again because molecular or dielectric materials do not possess the high, spherical symmetry of atoms; electronic transitions in such media are quite commonly mediated by more than one type of multipole. Many electric dipole transitions are also electric quadrupole allowed, and in general there is not the 1:1 correlation between transition multipolarity and photon angular momentum that atomic transitions provide. Multipolar decay does not produce photons measurably imprinted with a corresponding OAM; this is a topic whose consequences have been comprehensively addressed in references [78,79]. Electronic transitions that are simultaneously allowed by electric dipole and either electric quadrupole or magnetic dipole transitions also provide for quantum interference between them, which is the long-established basis for all chiroptical phenomena [80]. In connection with structured light it is once again the electric quadrupole form of coupling that introduces new features. For example it can introduce a capacity for chiral discrimination – a differential response in physical media of left- or right-handed structure [81,82]. One example is the capacity for angle-resolved scattering of optical vortex light to elicit Raman signals that can distinguish optically active molecular enantiomers [83].

3.5 Entanglement

Generating photon pairs by SPDC affords a well-established and widely investigated means of securing quantum entanglement between two separately resolvable beams [84,85]. Using a twisted beam input allows for the generation of heralded photons with enhanced quality as single-photon source [86]. Given a dimensionality d, maximally entangled states comprise superpositions of product states with identical coefficients $d^{-1/2}$. To achieve the quantum entanglement of more than two photons presents a substantial challenge, though with sufficiently high-intensity input, a proof-of-principle study has shown the possibility, through the interference of two proximal down-conversion events, to deliver a four-photon state entangled in orbital angular momentum [87].

The orthogonality of optical states of a given beam wavelength and direction, but different OAM values, itself provides extensive opportunities for quantum communication [88] and the prospects of exploiting quantum entanglement become especially inviting [10–13]. In general, the use of spatial modes as a basis for encryption can greatly extend the possibilities already provided by polarization states. Based specifically on orbital angular momentum there have been classic illustrations of high-dimension information space [89], and intriguing studies of open-ended opportunities to achieve logical operations, suggesting applications to conflate multiplexed channels [90].

The encoding of high-dimensionality transverse structure opens the doors to the conveyance of more information per photon, nominally running with the base-2 logarithm of the dimensionality d, but subject to both fundamental quantum and technical constraints. The no-cloning theorem, which applies to prepared quantum states [91], presents obstacles to both the security and the cloaked interception of individually relayed photon measurements in a data stream. However, through the use of multiple measurements, spatially entangled states present means for enhanced fidelity of secure transmission [92]. Non-orthogonal basis states necessarily introduce error rates, compromising the amount of secure information that can be conveyed. However, if a fractional error rate Q is deemed admissible, the deployment of a d-dimensional basis affords a secret key rate R (essentially the mean secure datum per photon) that is a negatively-sloped monotonic function of Q, given by [93]

$$R = \log_2 d + \log_2 \left[Q^{2Q} (1-Q)^{2(1-Q)} (d-1)^{-2Q} \right]. \tag{3.12}$$

For modest losses, the first term in this equation dominates and the information content clearly exceeds the conventional binary basis for all $d > 2$. The exploitation of entanglement amongst spatial modes of light thus provides a basis for many emerging applications in quantum cryptography [94], as detailed in later chapters.

3.6 Conclusion

The concept of the photon has for well over a century played a pivotal role in the understanding of huge areas of physics, from nuclear processes to astrophysical spectroscopy. In the broad sphere of optics, the quantized nature of light has often appeared to retreat to the background, with not only familiar optical processes but also a major part of laser physics and nonlinear optics explicable in classical terms. Quantum optics has over long periods often become treated almost as if it were a separate discipline. The arrival of the field of structured light has brought to the fore numerous new issues concerned with the properties and associated information content of photons, establishing new linkages to be forged between classical electrodynamics, quantum optics, nanoscale photonics, informatics and optical communications. It is especially interesting to see how issues of fundamental quantum mechanics now play so readily into rapidly emerging technological applications. In this respect it is hard to see any field progressing so fast as the quantum aspects of structured light.

References

[1] R. Loudon, The Quantum Theory of Light, 3rd edition, Oxford University Press, Oxford, 2000, p. 2.
[2] A. Einstein, Über Einen Die Erzeugung Und Verwandlung Des Lichtes Betreffenden Heuristischen Gesichtspunkt (On a Heuristic Viewpoint Concerning the Production and Transformation of Light), Ann. Phys. (Berlin) 17 (1905) 132–148.
[3] J. Wang, Twisted optical communications using orbital angular momentum, Sci. China, Phys. Mech. Astron. 62 (2019) 034201.
[4] J. Leach, M.J. Padgett, S.M. Barnett, S. Franke-Arnold, J. Courtial, Measuring the orbital angular momentum of a single photon, Phys. Rev. Lett. 88 (2002) 257901.
[5] E.J. Galvez, L.E. Coyle, E. Johnson, B.J. Reschovsky, Interferometric measurement of the helical mode of a single photon, New J. Phys. 13 (2011) 053017.
[6] D.L. Andrews, Conceptualization of the photon for quanta of structured light, Proc. SPIE 11297 (2020) 1129702.
[7] E. Karimi, R.W. Boyd, P. de la Hoz, H. de Guise, J. Řeháček, Z. Hradil, A. Aiello, G. Leuchs, L.L. Sánchez-Soto, Radial quantum number of Laguerre-Gauss modes, Phys. Rev. A 89 (2014) 063813.
[8] Y. Zhou, M. Mirhosseini, D. Fu, J. Zhao, S.M. Hashemi Rafsanjani, A.E. Willner, R.W. Boyd, Sorting photons by radial quantum number, Phys. Rev. Lett. 119 (2017) 263602.
[9] E.J. Galvez, S. Khadka, W.H. Schubert, S. Nomoto, Poincaré-beam patterns produced by nonseparable superpositions of Laguerre–Gauss and polarization modes of light, Appl. Opt. 51 (2012) 2925–2934.
[10] R. Fickler, R. Lapkiewicz, W.N. Plick, M. Krenn, C. Schaeff, S. Ramelow, A. Zeilinger, Quantum entanglement of high angular momenta, Science 338 (2012) 640–643.
[11] S. Franke-Arnold, S.M. Barnett, M.J. Padgett, L. Allen, Two-photon entanglement of orbital angular momentum states, Phys. Rev. A 65 (2002) 033823.
[12] J. Romero, D. Giovannini, S. Franke-Arnold, S.M. Barnett, M.J. Padgett, Increasing the dimension in high-dimensional two-photon orbital angular momentum entanglement, Phys. Rev. A 86 (2012) 012334.
[13] E.J. Galvez, New Methods of Entanglement with Spatial Modes of Light, Colgate University, Hamilton, New York, 2014.
[14] M.M. Coles, An upper bound on the rate of information transfer in optical vortex beams, Laser Phys. Lett. 15 (2018) 095202.
[15] M. Chen, K. Dholakia, M. Mazilu, Is there an optimal basis to maximise optical information transfer?, Sci. Rep. 6 (2016) 22821.
[16] P. Coullet, L. Gil, F. Rocca, Optical vortices, Opt. Commun. 73 (1989) 403–408.
[17] L. Allen, M.J. Padgett, M. Babiker, The orbital angular momentum of light, Prog. Opt. 39 (1999) 291–372.

[18] L. Allen, M.W. Beijersbergen, R.J.C. Spreeuw, J.P. Woerdman, Orbital angular momentum of light and the transformation of Laguerre-Gaussian laser modes, Phys. Rev. A 45 (1992) 8185–8189.
[19] J. Leach, E. Yao, M.J. Padgett, Observation of the vortex structure of a non-integer vortex beam, New J. Phys. 6 (2004) 71.
[20] Y. Wen, I. Chremmos, Y. Chen, Y. Zhang, S. Yu, Arbitrary multiplication and division of the orbital angular momentum of light, Phys. Rev. Lett. 124 (2020) 213901.
[21] D. Deng, M. Lin, Y. Li, H. Zhao, Precision measurement of fractional orbital angular momentum, Phys. Rev. Appl. 12 (2019) 014048.
[22] L. Mandel, E. Wolf, Optical Coherence and Quantum Optics, Cambridge University Press, Cambridge, New York, 1995.
[23] E. Leader, A proposed measurement of optical orbital and spin angular momentum and its implications for photon angular momentum, Phys. Lett. B 779 (2018) 385–387.
[24] S.M. Barnett, L. Allen, R.P. Cameron, C.R. Gilson, M.J. Padgett, F.C. Speirits, A.M. Yao, On the natures of the spin and orbital parts of optical angular momentum, J. Opt. 18 (2016) 064004.
[25] M.M. Coles, D.L. Andrews, Chirality and angular momentum in optical radiation, Phys. Rev. A 85 (2012) 063810.
[26] F. Crimin, N. Mackinnon, J.B. Götte, S.M. Barnett, Optical helicity and chirality: conservation and sources, Appl. Sci. 9 (2019) 828.
[27] K.Y. Bliokh, F. Nori, Characterizing optical chirality, Phys. Rev. A 83 (2011) 021803.
[28] M. Lax, W.H. Louisell, W.B. McKnight, From Maxwell to paraxial wave optics, Phys. Rev. A 11 (1975) 1365.
[29] A. Bekshaev, M. Soskin, M. Vasnetsov, Paraxial Light Beams with Angular Momentum, Nova Science Publishers, New York, 2008.
[30] A. Zangwill, Modern Electrodynamics, Cambridge University Press, Cambridge, New York, 2013.
[31] L.W. Davis, Theory of electromagnetic beams, Phys. Rev. A 19 (1979) 1177.
[32] L. Novotny, B. Hecht, Principles of Nano-Optics, Cambridge University Press, Cambridge, New York, 2012.
[33] S. Nechayev, J.S. Eismann, G. Leuchs, P. Banzer, Orbital-to-spin angular momentum conversion employing local helicity, Phys. Rev. B 99 (2019) 075155.
[34] K.Y. Bliokh, M.A. Alonso, E.A. Ostrovskaya, A. Aiello, Angular momenta and spin-orbit interaction of nonparaxial light in free space, Phys. Rev. A 82 (2010) 063825.
[35] K.Y. Bliokh, F.J. Rodríguez-Fortuño, F. Nori, A.V. Zayats, Spin–orbit interactions of light, Nat. Photonics 9 (2015) 796–808.
[36] V.V. Kotylar, A.G. Nalimov, S.S. Stefeev, Inversion of the axial projection of the spin angular momentum in the region of the backward energy flow in sharp focus, Opt. Express 28 (2020) 33830–33840.
[37] L. Marrucci, E. Karimi, S. Slussarenko, B. Piccirillo, E. Santamato, E. Nagali, F. Sciarrino, Spin-to-orbital conversion of the angular momentum of light and its classical and quantum applications, J. Opt. 13 (2011) 064001.
[38] R.C. Devlin, A. Ambrosio, N.A. Rubin, J.P.B. Mueller, F. Capasso, Arbitrary spin-to-orbital angular momentum conversion of light, Science 358 (2017) 896–901.
[39] L.C. Dávila Romero, D.L. Andrews, M. Babiker, A quantum electrodynamics framework for the nonlinear optics of twisted beams, J. Opt. B, Quantum Semiclass. Opt. 4 (2002) S66–S72.
[40] G. Gibson, J. Courtial, M. Padgett, M. Vasnetsov, V. Pas'ko, S. Barnett, S. Franke-Arnold, Free-space information transfer using light beams carrying orbital angular momentum, Opt. Express 12 (2004) 5448–5456.
[41] J. Wang, J.-Y. Yang, I.M. Fazal, N. Ahmed, Y. Yan, H. Huang, Y. Ren, Y. Yue, S. Dolinar, M. Tur, A.E. Willner, Terabit free-space data transmission employing orbital angular momentum multiplexing, Nat. Photonics 6 (2012) 488.
[42] R. Haag, Local Quantum Physics: Fields, Particles, Algebras, Springer, Berlin, Heidelberg, 1996.
[43] A. Stokes, A. Nazir, Implications of gauge-freedom for non-relativistic quantum electrodynamics, arXiv: 2009.10662v1 [quant-ph], 2020.

[44] I. Bialynicki-Birula, Z. Bialynicka-Birula, Why photons cannot be sharply localized, Phys. Rev. A 79 (2009) 032112.
[45] I. Bialynicki-Birula, Z. Bialynicka-Birula, The role of the Riemann-Silberstein vector in classical and quantum theories of electromagnetism, J. Phys. A, Math. Gen. 46 (2013) 053001.
[46] D.L. Andrews, Photon-based and classical descriptions in nanophotonics: a review, J. Nanophotonics 8 (2014) 081599.
[47] S.W. Hell, J. Wichmann, Breaking the diffraction resolution limit by stimulated emission: stimulated-emission-depletion fluorescence microscopy, Opt. Lett. 19 (1994) 780–782.
[48] L. Allen, M.J. Padgett, The Poynting vector in Laguerre-Gaussian beams and the interpretation of their angular momentum density, Opt. Commun. 184 (2000) 67–71.
[49] G.C. Berkhout, M.P. Lavery, J. Courtial, M.W. Beijersbergen, M.J. Padgett, Efficient sorting of orbital angular momentum states of light, Phys. Rev. Lett. 105 (2010) 153601.
[50] M.N. O'Sullivan, M. Mirhosseini, M. Malik, R.W. Boyd, Near-perfect sorting of orbital angular momentum and angular position states of light, Opt. Express 20 (2012) 24444–24449.
[51] S. Franke-Arnold, S.M. Barnett, E. Yao, J. Leach, J. Courtial, M. Padgett, Uncertainty principle for angular position and angular momentum, New J. Phys. 6 (2004) 103.
[52] J. Leach, B. Jack, J. Romero, A.K. Jha, A.M. Yao, S. Franke-Arnold, D.G. Ireland, R.W. Boyd, S.M. Barnett, M.J. Padgett, Quantum correlations in optical angle–orbital angular momentum variables, Science 329 (2010) 662–665.
[53] W. Li, S. Zhao, Manipulating orbital angular momentum entanglement by using the Heisenberg uncertainty principle, Opt. Express 26 (2018) 21725–21735.
[54] D.L. Andrews, K.A. Forbes, Quantum features in the orthogonality of optical modes for structured and plane-wave light, Opt. Lett. 43 (2018) 3249–3252.
[55] M.V. Berry, M.R. Dennis, Quantum cores of optical phase singularities, J. Opt. A 6 (2004) S178–S180.
[56] W. Heitler, The Quantum Theory of Radiation, Dover, Mineola, New York, 1954.
[57] S.M. Barnett, D.T. Pegg, Quantum theory of rotation angles, Phys. Rev. A 41 (1990) 3427–3435.
[58] B. Jack, M.J. Padgett, S. Franke-Arnold, Angular diffraction, New J. Phys. 10 (2008) 103013.
[59] M.D. Williams, D.S. Bradshaw, D.L. Andrews, Quantum issues with structured light, Proc. SPIE 9764 (2016) 976407.
[60] S.J. van Enk, G. Nienhuis, Spin and orbital angular momentum of photons, Europhys. Lett. 25 (1994) 497–501.
[61] S.J. van Enk, G. Nienhuis, Commutation rules and eigenvalues of spin and orbital angular-momentum of radiation-fields, J. Mod. Opt. 41 (1994) 963–977.
[62] S.M. Barnett, L. Allen, Orbital angular momentum and nonparaxial light beams, Opt. Commun. 110 (1994) 670–678.
[63] K.Y. Bliokh, F. Nori, Transverse and longitudinal angular momenta of light, Phys. Rep. 592 (2015) 1–38.
[64] M. Neugebauer, T. Bauer, A. Aiello, P. Banzer, Measuring the transverse spin density of light, Phys. Rev. Lett. 114 (2015) 063901.
[65] L. Marrucci, C. Manzo, D. Paparo, Optical spin-to-orbital angular momentum conversion in inhomogeneous anisotropic media, Phys. Rev. Lett. 96 (2006) 163905.
[66] Y. Zhao, J.S. Edgar, G.D. Jeffries, D. McGloin, D.T. Chiu, Spin-to-orbital angular momentum conversion in a strongly focused optical beam, Phys. Rev. Lett. 99 (2007) 073901.
[67] A. Bekshaev, K.Y. Bliokh, M. Soskin, Internal flows and energy circulation in light beams, J. Opt. 13 (2011) 053001.
[68] A. Afanasev, C.E. Carlson, M. Solyanik, Circular dichroism of twisted photons in non-chiral atomic matter, J. Opt. 19 (2017) 105401.
[69] E. Noether, Invariante variationsprobleme, Nachr. d. König. Gesellsch. d. Wiss. zu Göttingen, Math-phys. Klasse (1918) 235–257.
[70] D. Lipkin, Existence of a new conservation law in electromagnetic theory, J. Math. Phys. 5 (1964) 696–700.

[71] W.I. Fushchich, A.G. Nikitin, The complete set of conservation laws for the electromagnetic field, J. Phys. A, Math. Gen. (1992) L231–L233.
[72] S.C. Anco, J. Pohjanpelto, Classification of local conservation laws of Maxwell's equations, Acta Appl. Math. 69 (2001) 285–327.
[73] S.M. Barnett, On the quantum core of an optical vortex, J. Mod. Opt. 55 (2008) 2279–2292.
[74] M. Babiker, C.R. Bennett, D.L. Andrews, L.C. Dávila Romero, Orbital angular momentum exchange in the interaction of twisted light with molecules, Phys. Rev. Lett. 89 (2002) 143601.
[75] V.E. Lembessis, M. Babiker, Enhanced quadrupole effects for atoms in optical vortices, Phys. Rev. Lett. 110 (2013) 083002.
[76] C.T. Schmiegelow, J. Schulz, H. Kaufmann, T. Ruster, U.G. Poschinger, F. Schmidt-Kaler, Transfer of optical orbital angular momentum to a bound electron, Nat. Commun. 7 (2016) 12998.
[77] M. Babiker, D.L. Andrews, V.E. Lembessis, Atoms in complex twisted light, J. Opt. 21 (2019) 013001.
[78] D.L. Andrews, Optical angular momentum: multipole transitions and photonics, Phys. Rev. A 81 (2010) 033825.
[79] D.L. Andrews, On the conveyance of angular momentum in electronic energy transfer, Phys. Chem. Chem. Phys. 12 (2010) 7409–7417.
[80] D.L. Andrews, Quantum formulation for nanoscale optical and material chirality: symmetry issues, space and time parity, and observables, J. Opt. 20 (2018) 033003.
[81] K.A. Forbes, D.L. Andrews, Spin-orbit interactions and chiroptical effects engaging orbital angular momentum of twisted light in chiral and achiral media, Phys. Rev. A 99 (2019) 023837.
[82] K.A. Forbes, D.L. Andrews, Enhanced optical activity using the orbital angular momentum of structured light, Phys. Rev. Res. 1 (2019) 033080.
[83] K.A. Forbes, Raman optical activity using twisted photons, Phys. Rev. Lett. 122 (2019) 103201.
[84] A. Mair, A. Vaziri, G. Weihs, A. Zeilinger, Entanglement of the orbital angular momentum states of photons, Nature 412 (2001) 313–316.
[85] M. Padgett, J. Courtial, L. Allen, S. Franke-Arnold, S.M. Barnett, Entanglement of orbital angular momentum for the signal and idler beams in parametric down-conversion, J. Mod. Opt. 49 (2002) 777–785.
[86] N. Lal, A. Banerji, A. Biswas, A. Anwar, R.P. Singh, Photon statistics of twisted heralded single photons, J. Mod. Opt. 67 (2020) 126–132.
[87] B.C. Hiesmayr, M.J.A. de Dood, W. Löffler, Observation of four-photon orbital angular momentum entanglement, Phys. Rev. Lett. 116 (2016) 073601.
[88] S. Franke-Arnold, J. Jeffers, Orbital angular momentum in quantum communication and information, in: D.L. Andrews (Ed.), Structured Light and Its Applications: an Introduction to Phase-Structured Beams and Nanoscale Optical Forces, Academic, Amsterdam, Boston, 2008, pp. 271–291.
[89] M. Malik, M. Mirhosseini, M.P. Lavery, J. Leach, M.J. Padgett, R.W. Boyd, Direct measurement of a 27-dimensional orbital-angular-momentum state vector, Nat. Commun. 5 (2014) 3115.
[90] V. Potoček, F.M. Miatto, M. Mirhosseini, O.S. Magaña-Loaiza, A.C. Liapis, D.K.L. Oi, R.W. Boyd, J. Jeffers, Quantum Hilbert hotel, Phys. Rev. Lett. 115 (2015) 160505.
[91] W.K. Wootters, W.H. Zurek, A single quantum cannot be cloned, Nature 299 (1982) 802–803.
[92] L. Sheridan, V. Scarani, Security proof for quantum key distribution using qudit systems, Phys. Rev. A 82 (2010) 030301.
[93] E. Otte, I. Nape, C. Rosales-Guzmán, C. Denz, A. Forbes, B. Ndagano, High-dimensional cryptography with spatial modes of light: tutorial, J. Opt. Soc. Am. B 37 (2020) A309–A323.
[94] M. Mirhosseini, O.S. Magaña-Loaiza, M.N. O'Sullivan, B. Rodenburg, M. Malik, M.P.J. Lavery, M.J. Padgett, D.J. Gauthier, R.W. Boyd, High-dimensional quantum cryptography with twisted light, New J. Phys. 17 (2015) 033033.

CHAPTER 4

Poincaré beams for optical communications

Enrique J. Galvez[a], Behzad Khajavi[a,b], and Brianna M. Holmes[a,c]

[a]Colgate University, Department of Physics and Astronomy, Hamilton, NY, United States [b]University of Houston, Department of Biomedical Engineering, Houston, TX, United States [c]University of Rochester, Institute of Optics, Rochester, NY, United States

Contents

4.1 Introduction 95
4.2 Vortex Poincaré Gaussian beams 96
4.3 Poincaré–Bessel beams 100
4.4 Asymmetric and monstar patterns 102
4.5 Experimental methods 104
4.6 Discussion 104
Acknowledgments 104
References 105

4.1 Introduction

Demands for higher bandwidth in communications have generated interest in using complex light to provide non-binary encoding schemes. Scalar modes have received much attention due to the unbound number of states that can be used in the communication, with particular emphasis being placed on Laguerre-Gaussian modes [1–5]. This type of communication involves encoding information in light beams with a series of spatial modes, with their subsequent detection at the receiving end. Distortions due to atmospheric turbulence are a factor in these possibilities [6–8].

Vector beams that are non-separable superpositions of first-order Laguerre–Gaussian modes and polarization give rise to well-known radial and azimuthal vector beams [9–11]. These modes have also been investigated for communications [12,13]. In this chapter we focus on the potential use of Poincaré modes [14–17], which are also superpositions of polarization and spatial mode, but where the order of the spatial mode is not limited to the first order [18–20]. There is much interest in using these modes for communications [21]. More recently new interesting vector patterns have been investigated [22,23].

Structured Light for Optical Communication
https://doi.org/10.1016/B978-0-12-821510-4.00010-8
Copyright © 2021 Elsevier Inc. All rights reserved.

In this chapter we investigate the ways in which Poincaré modes can be generated and detected for use in communications. We touch on general properties of Poincaré beams, some properties that may be relevant to communications, and various other possibilities, such as the use of Bessel beams and asymmetric monstar patterns.

4.2 Vortex Poincaré Gaussian beams

Poincaré beams feature spatially-variable polarization. The simplest way to understand them is via non-separable superpositions of polarization and spatial mode. If we consider two modes that are orthogonal to each other, μ_1 and μ_2, then, if we superimpose them in conjunction with orthogonal states of polarization, such as right and left circular states, $\hat{e}_R$ and $\hat{e}_L$, respectively, then the non-factorable superposition becomes

$$U = u_1\,\hat{e}_R + u_2\,\hat{e}_L. \tag{4.1}$$

If for the moment we ignore diffraction effects and assume that the propagation direction is along the z-coordinate, we can express the modes as $u_1 = |u_1(x,y)|\exp[i\varphi_1(x,y)]$ and $u_2 = |u_2(x,y)|\exp[i\varphi_2(x,y)]$. Eq. (4.1) can then be expressed as

$$U = \left(|u_1|^2 + |u_2|^2\right)^{-1/2} e^{i\varphi_1}\hat{e}, \tag{4.2}$$

where

$$\hat{e} = \cos\chi\,\hat{e}_R + e^{i\varphi}\sin\chi\,\hat{e}_L \tag{4.3}$$

is the state of polarization at position (x, y), with

$$\chi = \tan^{-1}\frac{|u_2|}{|u_1|} \tag{4.4}$$

determining the ellipticity of the polarization ellipse, and

$$\varphi = \varphi_2 - \varphi_1 \tag{4.5}$$

determining the orientation. If the modes have amplitude and phase differences that vary with position, then the state of polarization $\hat{e}$ varies from point to point.

Let us consider Laguerre–Gaussian modes. When the two modes contain different topological charges, the spatially-variant patterns constitute a mapping of the states on the Poincaré sphere to the transverse mode, and so they are referred to as Poincaré modes [14–16]. In Fig. 4.1(b) and (e) we show two examples of measured Poincaré modes made with Laguerre–Gauss eigenmodes. We draw ellipses corresponding to the state of polarization at the location

Poincaré beams for optical communications 97

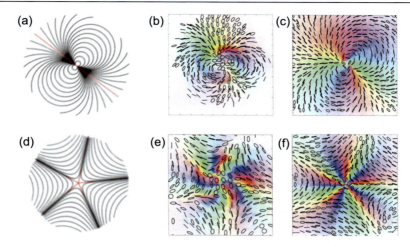

Figure 4.1: *Disclinations in Poincaré beams.* Disclinations (a, d) and polarimetry measurements of non-separable superpositions with topological charges $(\ell_1, \ell_2) = (3, -1)$ for (a–c) producing a pattern with $I_C = 2$; and topological charges $(-1, 2)$, producing a pattern with $I_C = -3/2$ for (d–f). The states of (b, e) used Laguerre–Gauss modes, and (c, f) used plane waves with an angular vortex.

where they are drawn. When the orientation of the semi-major axes of the ellipses is mapped, we get disclination patterns as shown in panes (a) and (d) for the two cases. In panes (c) and (f) we also show the polarization patterns taken by imaging superpositions of collinear plane-wave vortex modes, which produces patterns where all the polarization states are linear. These are produced by the superposition

$$U = f(r) \left(e^{i\ell_1 \phi} \hat{e}_R + e^{i\ell_2 \phi + i\delta} \hat{e}_L \right) \tag{4.6}$$

where δ is a phase difference between the two modes, and $f(r)$ is a slow varying function of the radial coordinate r.

At the center of the disclinations is a C-point polarization singularity [24]. The disclination index I_C quantifies the number of rotations that a tangent to a disclination line describes in a closed path around the C-point. If modes u_1 and u_2 have topological charges ℓ_1 and ℓ_2 (i.e., with $\varphi_1 = \ell_1 \phi$ and $\varphi_2 = \ell_2 \phi$, with ϕ being the azimuthal coordinate), then the index is given by [19]

$$I_C = \frac{\Delta \ell}{2}, \tag{4.7}$$

where $\Delta \ell = \ell_1 - \ell_2$. The polarization-ellipse orientations for the two cases in panes (b, c) and (e, f) can be seen to follow the respective disclinations in panes (a) and (d). The plane-wave mode patterns are much easier to identify.

Table 4.1: List of topological charge of modes that produce Poincaré modes with distinct disclination patterns.

ℓ_1	ℓ_2	$\Delta\ell$	I_C	N
3	−2	5	5/2	3
2	−2	4	2	2
2	−1	3	3/2	1
1	−1	2	1	0[a]
1	0	1	1/2	1
−1	0	−1	−1/2	3
−1	1	−2	−1	4
−2	1	−3	−3/2	5
−2	2	−4	−2	6
−2	3	−5	−5/2	7

[a] Except for the radial mode, where $N = \infty$.

Another distinctive feature of the disclination is that at some angles the orientation of the semi-major axis of the of the polarization ellipse is aligned with the radial direction. A line along this direction is called a radial line [25]. It is a separatrix between angular regions of disclination lines. The latter can have three possible line shapes: elliptic (closed curves that touch the C-point), parabolic (open curves with one endpoint at the C-point), and hyperbolic (open curves with no end points). The number of radial lines is given by [19]

$$N_r = |\Delta\ell - 2|. \tag{4.8}$$

The orientation of the radial lines is given by [19]

$$\phi_r = \frac{2n\pi - \delta}{2 - \Delta\ell}, \tag{4.9}$$

where n is an integer. In the measured patterns of Fig. 4.1 we color-code/shade the images to help identify the orientations. The colors follow the orientation relative to the radial direction, with the radial direction having the lightest shade (yellow-color). Table 4.1 shows a list of possible modes that can produce distinct patterns. The combination of modes that give the same pattern is not unique. Those listed on the table are examples of suitable values. There is no limit to the patterns that we can make. Patterns with increasing absolute value of I_C have higher angular symmetry, as reflected by the number of radial lines.

If the relative phase between the two terms in Eq. (4.3) is changed, the pattern appears to rotate. Thus, when we consider propagation, the pattern based on Laguerre–Gauss modes of different order (in absolute value) will rotate due to the different Gouy phases that the component modes acquire due to diffraction [26]. A particular case of Eq. (4.2) or (4.6) is when one mode has a topological charge of $\ell_1 = +1$ and the other one with $\ell_2 = -1$. This is the setup for the well-known radial and azimuthal vector modes, which has been studied extensively [9]. Laguerre–Gauss eigenmodes have a radial dependence that manifests as a series of rings of alternating phase. Because this complicates the identification of the modes, we will restrict ourselves to the singly-ringed cases in this section. We will come back to this when we discuss Bessel modes.

The position of the radial lines changes with the relative phase δ, and so the change in the position of a radial line with respect to the relative phase is

$$\frac{d\phi_r}{d\delta} = \frac{1}{\Delta\ell - 2}. \tag{4.10}$$

As can be seen in Fig. 4.1, the patterns themselves have distinctive shapes, depending on their C-point index (see also Ref. [27]). Placing a radial or azimuthal polarizer in the path of the beam will reveal the radial lines as a symmetric series of bright and dark radial spokes or spots. If the relative phase between the modes δ is scanned, then the pattern will rotate, with a rotation rate that is inversely proportional to $\Delta\ell - 2$. So a measurement of this rotation could be used as a method of detection. A radial/azimuthal polarizer can consist of a $q = 1/2$ q-plate followed by a linear polarizer [16]. The only challenge with this situation is that the q-plate has to be centered on the beam.

An alternative possibility is to just place a linear polarizer in the path of the Poincaré beam. This results in a symmetric pattern of $\Delta\ell$ spokes or spots. If the transmission axis of the polarizer forms an angle θ with the horizontal, then the orientation of the maxima (i.e., center of the bright spokes/spots) is

$$\phi_m = \frac{2n\pi + \delta - 2\theta}{\Delta\ell}. \tag{4.11}$$

The total number of angular maxima is

$$N_m = |\Delta\ell|. \tag{4.12}$$

One can equally obtain a similar type of equation to identify the minima. If θ is fixed, then

$$\frac{d\phi_m}{d\delta} = \frac{1}{\Delta\ell}. \tag{4.13}$$

In this simpler case the number of spots or their rotation rate is indicative of the Poincaré state.

Another possibility for rotating the pattern is to change the polarizer angle θ, which gives a rotation rate that is also inversely proportional to $\Delta\ell$:

$$\frac{d\phi_m}{d\theta} = -\frac{2}{\Delta\ell}. \tag{4.14}$$

100   Chapter 4

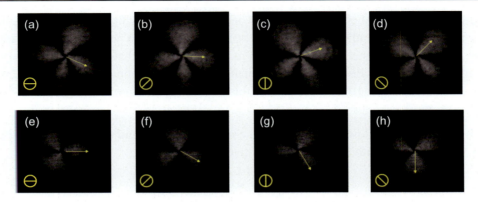

Figure 4.2: *Polarization projections of modes with a linear polarizer.* Polarization projections for the disclination patterns of Fig. 4.1 for the cases: (a–d) where $(\ell_1, \ell_2) = (3, -1)$ involving four maxima/minima; and (e–h) where $(\ell_1, \ell_2) = (-1, 2)$ involving five maxima. The projections were made for polarizer angles that increased by 45 degrees (shown in inserts). Arrows indicate the predicted location of one of the maxima.

Fig. 4.2 shows the pattern for four different polarizer angles. As we rotate the linear polarizer, the pattern is seen to rotate. In this case we opted to show the image of the plane-wave mode as opposed to the Gaussian beam. However, nature can be perverse: the rotation measurement cannot rely on the frequency at which the spokes cross an angular position, in both the radial and the linear polarizer methods, because the number of spokes is inversely proportional to the rotation frequency. Therefore, the detection method has to be based on the angular speed of the rotation of the spokes. A potentially automatic way to impart rotation to the pattern is to shift the electromagnetic frequency of one of the component modes, because δ will depend on $\Delta\omega t$, where $\Delta\omega$ is the difference in angular frequency of the two modes, and t is time. As a consequence, the transmitted Poincaré modes will rotate continuously, as shown previously for the case of scalar modes [28]. The rotations can be clockwise or counter-clockwise depending on the sign of $\Delta\ell$. In Fig. 4.2 we see two cases with different values of I_C (+2 for panes (a–d) and $-3/2$ for panes (e–h)). It can be seen that the sense of rotation of the patterns depends on the sign of the index. The arrows in the picture were computed using Eq. (4.14). They give the expected position of one of the maxima of the pattern for different orientations of the polarizer. The rotation of the pattern per 45-degree rotation of the polarizer is $+22.5°$ for panes (a–d) and $-30°$ for panes (e–h). The patterns follow the expected rotation rates.

4.3 Poincaré–Bessel beams

Another interesting possibility is to use Poincaré modes of Bessel beams due to their diffractionless propagating properties. In the case of Poincaré–Bessel beams, the analytical for-

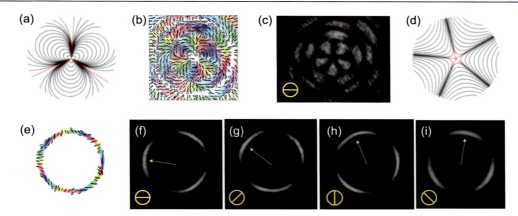

Figure 4.3: *Poincaré patterns of Bessel beams.* (a-c) Case where $(\ell_1, \ell_2) = (+2, -3)$ producing a disclination. Images of this case are: (a) the disclination with index $I_C = +5/2$, (b) the polarimetry of the mode, and (c) the projection of the mode with a horizontal linear polarizer. (d-h) Case where $(\ell_1, \ell_2) = (-2, +1)$. Images for this case are: (d) the disclination with index $I_C = -3/2$, (e) polarimetry of the Fourier transform, and (f-i) polarization projections after a linear polarizer of orientation shown in the insert. Arrows show the predicted angular position of one of the maxima.

malism is similar in structure to Gaussian beams. The Poincaré pattern for Bessel beams is more complicated than for singly-ringed Gaussian modes because of the multi-ringed structure of Bessel beams [29]. In Fig. 4.3 we show the case of a Poincaré–Bessel mode produced by the superposition of topological charges $\ell_1 = +2$ and $\ell_2 = -3$, which has a disclination index $I_C = +5/2$. It entails three radial lines as shown in Fig. 4.3(a) and five angular minima/maxima, as seen in Fig. 4.3(c). However, because of the sign flip between adjacent rings, the maxima and minima interchange positions from one ring to the adjacent one, as can be seen in Fig. 4.3(c). The center is quite clear, but beyond that, the preponderance of staggered maxima and minima can make the identification more challenging. Conversely, if the phase between the relative modes is changed, the pattern as a whole will rotate, so perhaps the abundance of maxima or minima could be advantageous in anchoring the pattern and measuring the rotation frequency by a suitable detection algorithm.

Bessel beams also show an interesting feature: their optical Fourier transform is a ring modulated by the angular phase difference between the two modes. That is, the disclination gets compressed into a ring. Fig. 4.3(d) shows the disclination for the case with topological charges $\ell_1 = -2$ and $\ell_2 = +1$, producing a disclination with an index $I_C = -3/2$. The polarimetry of the optical Fourier transform is shown in Fig. 4.3(e), where the five directions of radial polarization are evident. In Figs. 4.3(f–i) we show the polarization projections of the Fourier transform, which involve a single ring modulated by the angular projection function

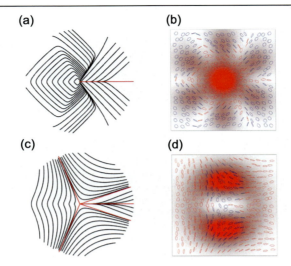

Figure 4.4: *Asymmetric Poincaré patterns.* (a–b) An asymmetric pattern for $(\ell_1, \ell_2) = (3, 0)$, with $\beta = 38.5°$, $\gamma = 180°$ and $\delta = 0$; and (c, d) for $(\ell_1, \ell_2) = (1, 2)$, with $\beta = 40°$, $\gamma = 180°$ and $\delta = 0$. Patterns (a, c) are the disclination patterns and (b, d) are the modeled beam pattern.

that has three symmetric maxima. As can be seen, the maxima rotate with the polarization orientation (shown in the insert). Thus, upon reception of a Bessel beam, a simple optical Fourier transform can simplify the pattern into a modulated ring that can be used to detect the mode.

4.4 Asymmetric and monstar patterns

There have been recent reports on the use of asymmetric space-variant beams for communications [30]. Poincaré beams can also carry asymmetric patterns. In the recent past we investigated these patterns to focus on a particular type of asymmetry called the monstar [19,25,31–34]. If we analyze Eq. (4.1) or (4.6), we note that, if the modes u_a and u_B are Laguerre–Gaussian eigenmodes, in both equations the component modes have a phase that varies linearly with the angular coordinate. Asymmetric modes can be created by a combination of modes that have an asymmetric angular phase variation. We studied the simplest variation: the superposition of modes of same topological charge but different sign. The asymmetric pattern constructed with Laguerre–Gauss modes is given by

$$U = \left(\cos\beta\, LG_0^{\ell_1} + \sin\beta\, LG_0^{-\ell_1} e^{i\gamma}\right) \hat{e}_R + LG_0^{\ell_2} e^{i\delta}\, \hat{e}_L, \qquad (4.15)$$

where LG_0^ℓ represent Laguerre–Gaussian modes with radial index 0 and topological charge ℓ; β is an angular variable that specifies the asymmetry. The closer β is to $\pi/4$ the more asymmetric the pattern will be. γ is a phase that affects the shape of the pattern [33]. Fig. 4.4 shows two examples of asymmetric patterns.

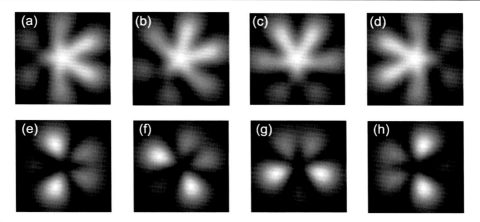

Figure 4.5: *Radial polarization projections.* (a–d) Projections with a radial polarizer for $(\ell_1, \ell_2) = (3, 0)$, with $\beta = 38.5°$, with $(\gamma, \delta) = (180°, 0)$ for (a), $(90°, 45°)$ for (b), $(0, 90°)$ for (c), and $(180°, 180°)$ for (d). (e–h) Projections with a radial polarizer for $(\ell_1, \ell_2) = (1, 2)$, with $\beta = 40°$, with $(\gamma, \delta) = (180°, 0)$ for (a), $(270°, 315°)$ for (b), $(0, 90°)$ for (c), and $(180°, 180°)$ for (d).

Earlier we mentioned the radial lines that are typical of the disclination patterns. When the asymmetry is introduced, the radial lines shift. In some cases, new radial lines appear, creating the pattern known as a monstar. Fig. 4.4(a, b) at $\beta = 38.5°$ is near a monstar region ($38.7° < \beta < 45°$ [27]). The pattern of Fig. 4.4(c, d) is a monstar because it has a different number of radial lines than the symmetric case [19]. If we wish to implement an asymmetric Poincaré mode for communications, the method that we envision would be one where the "characters" in the communication would be the orientation of a pattern with a given asymmetry. The use of a linear polarizer with either an evolving phase with fixed polarizer, or a fixed phase with a rotating polarizer, will not be convenient, because the projected pattern will rotate and change shape, making it confusing to identify. For asymmetric modes, the most convenient projection is one that uses radial or azimuthal polarizers. Such projections will definitely produce an asymmetric object whose orientation can be recognized. Figs. 4.5(a, e) show the calculated radial patterns of intensity that would be obtained from the Poincaré modes of Fig. 4.4.

To rotate the asymmetric/monstar pattern produced via Eq. (4.15) the angles δ and γ need to change in a correlated way. If we want to rotate the pattern an angle $\Delta\phi$, then

$$\Delta\delta = (\Delta\ell - 2)\Delta\phi \tag{4.16}$$

and

$$\Delta\gamma = -2\ell_1\Delta\phi, \tag{4.17}$$

104   Chapter 4

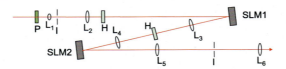

Figure 4.6: *Apparatus for preparing beams.* Schematic of the setup to prepare Poincaré beams, which includes two spatial light modulators (SLM), polarization optics (half-wave plates H and polarizer P), irises (I) and lenses (L).

where $\Delta\ell = \ell_1 - \ell_2$, as before. In Fig. 4.5 we show the rotation of the radial-polarization projection of the mode rotated by 45°, 90° and 180°. The angles to produce the rotations were obtained using Eqs. (4.16) and (4.17).

4.5   Experimental methods

The modes presented in this contribution are best produced via two spatial light modulators (SLM) in an apparatus shown in the schematic of Fig. 4.6. It consists of expanding an incoming laser beam with a polarization that has equal amounts of two linear components (such as vertical and horizontal). Each polarization component is then programmed by a different SLM. The modulators are then imaged onto each other. This way we create an intra-beam superposition. The beam can then be imaged in the far or near field via lenses. Additional polarization and imaging components (not shown) are used to measure the state of the light, and its projections.

4.6   Discussion

In summary, in this article we presented ways in which Poincaré modes can be used in communications. They all exploit the shapes that the modal combinations make when projected by a polarizer. We show that polarization projections with a linear polarizer can be used to detect symmetric modes, exploiting their mode-dependent rotation rates. Radial polarizers can be used to determine the orientation of irregular shapes. They provide alternatives that use polarization as means for encoding and detection. One of the benefits of Poincaré modes is that the beams are made of propagating modes, and so they do not disassemble or distort with propagation. Even in the case of Gaussian modes, pre-dephasing can be used compensate for the Gouy phase shifts present in diffraction. Even for the case of irregular modes, the patterns are stable because they are made of spatial modes with the same propagation dynamics.

Acknowledgments

This work was funded by NSF grants PHY-1506321 and PHY-201672.

References

[1] H. Rubinsztein-Dunlop, A. Forbes, M.V. Berry, M.R. Dennis, D.L. Andrews, M. Mansuripur, C. Denz, C. Alpmann, P. Banzer, T. Bauer, E. Karimi, L. Marrucci, M. Padgett, M. Ritsch-Marte, N.M. Litchinitser, N.P. Bigelow, C. Rosales-Guzmán, A. Belmonte, J.P. Torres, T.W. Neely, M. Baker, R. Gordon, A.B. Stilgoe, J. Romero, A.G. White, R. Fickler, A.E. Willner, G. Xie, B. McMorran, A.M. Weiner, Roadmap on structured light, J. Opt. 19 (2016) 013001.

[2] G. Gibson, J. Courtial, M.J. Padgett, V.P. Vasnetsov, S.M. Barnett, S. Franke-Arnold, Free-space information transfer using light beams carrying orbital angular momentum, Opt. Express 12 (2004) 5448–5456.

[3] J. Wang, J.Y. Yang, I.M. Fazal, N. Ahmed, Y. Yan, H. Huang, Y. Ren, Y. Yue, S. Dolinar, M. Tur, A.E. Willner, Terabit free-space data transmission employing orbital angular momentum multiplexing, Nat. Photonics 340 (2012) 488–496.

[4] M. Krenn, R. Fickler, M. Fink, J. Handsteiner, M. Malik, T. Scheidl, R. Ursin, A. Zeilinger, Communication with spatially modulated light through turbulent air across Vienna, New J. Phys. 16 (2014) 113028.

[5] M. Krenn, J. Handsteiner, M. Fink, R. Fickler, R. Ursin, M. Malik, A. Zeilinger, Twisted light transmission over 143 km, Proc. Natl. Acad. Sci. (2016) 13648–13653.

[6] B. Rodenburg, M.P.J. Lavery, M. Malik, M. O'Sullivan, M. Mirhosseini, D.J. Robertson, M. Padgett, R.W. Boyd, Influence of atmospheric turbulence on states of light carrying orbital angular momentum, Opt. Lett. 37 (2012) 3735–3737.

[7] M. Malik, M. O'Sullivan, B. Rodenburg, M. Mirhosseini, J. Leach, M.P.J. Lavery, M.J. Padgett, R.W. Boyd, Influence of atmospheric turbulence on optical communications using orbital angular momentum for encoding, Opt. Express 20 (2012) 13195–13200.

[8] Y. Ren, Z. Wang, G. Xie, L. Li, A.J. Willner, Y. Cao, Z. Zhao, Y. Yan, N. Ahmed, N. Ashrafi, S. Ashrafi, R. Bock, M. Tur, A.E. Wilner, Atmospheric turbulence mitigation in an OAM-based MIMO free-space optical link using spatial diversity combined with MIMO equalization, Opt. Lett. 41 (2016) 2406–2409.

[9] Q. Zhan, Cylindrical vector beams: from mathematical concepts to applications, Adv. Opt. Photonics 1 (2009) 1–57.

[10] G. Volpe, D. Petrov, Generation of cylindrical vector beams with few-mode fibers excited by Laguerre-Gaussian beams, Opt. Commun. 237 (2004) 89–95.

[11] A. Niv, G. Biener, V. Kleiner, E. Hasman, Rotating vectorial vortices produced by space-variant sub wavelength gratings, Opt. Lett. 30 (2005) 2933–2935.

[12] V. D'Ambrosio, E. Nagali, S. Walborn, A. Aolita, S. Slussarenko, L. Marrucci, F. Sciarrino, Complete experimental toolbox for alignment-free quantum communication, Nat. Commun. 3 (2012) 961.

[13] G. Milione, M.P.J. Lavery, H. Huang, Y. Ren, G. Xie, T.A. Nguyen, E. Karimi, L. Marrucci, D.A. Nolan, R.R. Alfano, A.E. Willner, 4×20 Gbit/s mode division multiplexing over free space using vector modes and a q-plate mode (de)multiplexer, Opt. Lett. 40 (2015) 1980–1983.

[14] A.M. Beckley, T.G. Brown, M.A. Alonso, Full Poincaré beams, Opt. Express 18 (2010) 10777–10785.

[15] E.J. Galvez, S. Khadka, W.H. Schubert, S. Nomoto, Poincaré-beam patterns produced by non-separable superpositions of Laguerre–Gauss and polarization modes of light, Appl. Opt. 51 (2012) 2925–2934.

[16] F. Cardano, E. Karimi, S. Slussarenko, L. Marrucci, C. de Lisio, E. Santamato, Polarization pattern of vector vortex beams generated by q-plates with different topological charges, Appl. Opt. 51 (2012) C1–C6.

[17] V. Kumar, N. Viswanathan, Topological structures in the Poynting vector field: an experimental realization, Opt. Lett. 38 (2013) 3886–3889.

[18] C. Maurer, A. Jesacher, S. Fürhapter, S. Bernet, M. Ritsch-Marte, Tailoring of arbitrary optical vector beams, New J. Phys. 9 (2007) 78.

[19] B. Khajavi, E.J. Galvez, High-order disclinations in space-variant polarization, J. Opt. 18 (2016) 084003.

[20] E. Otte, C. Alpmann, C. Denz, Higher-order polarization singularities in tailored vector beams, J. Opt. 8 (074012) (2016) 1–7.

[21] A. Sit, F. Bouchard, R. Fickler, J. Gagnon-Bischoff, H. Larocque, K. Heshami, D. Elser, C. Peuntinger, K. Günthner, B. Heim, C. Marquardt, G. Leuchs, R.W. Boyd, E. Karimi, High-dimensional intracity quantum cryptography with structured photons, Optica 9 (2017) 1006–1010.

[22] S.N. Khonina, A.V. Ustinov, S.A. Fomchenkov, A.P. Porfirev, Formation of hybrid higher-order cylindrical vector beams using multi-sector phase plates, Sci. Rep. 8 (2018) 14320.
[23] Y. Shen, X. Yang, D. Naidoo, X. Fu, A. Forbes, Structured ray-wave vector vortex beams in multiple degrees of freedom from a laser, Optica 7 (2020) 820–831.
[24] J.F. Nye, Lines of circular polarization in electromagnetic wave fields, Proc. R. Soc. Lond. A 389 (1983) 279–290.
[25] M.V. Berry, J.H. Hannay, Umbilic points on Gaussian random surfaces, J. Phys. A 10 (1977) 1809–1821.
[26] F. Cardano, E. Karimi, L. Marrucci, C. de Lisio, E. Santamato, Generation and dynamics of optical beams with polarization singularities, Opt. Express 21 (2013) 8815–8820.
[27] E.J. Galvez, B. Khajavi, High-order disclinations in the polarization of light, Proc. SPIE 9764 (2016) 97640R.
[28] S. Franke-Arnold, J. Leach, M.J. Padgett, V.E. Lembesis, D. Ellinas, A.J. Wright, J.M. Girkin, P. Ohberg, A.S. Arnold, Optical ferris wheel for ultracold atoms, Opt. Express 15 (2007) 8619–8625.
[29] B.M. Holmes, E.J. Galvez, Poincaré-Bessel beams: structure and propagation, J. Opt. 21 (2019) 104001.
[30] J.E. Holland, I. Moreno, J.A. Davis, M.M. Sanchez-Lopez, D.M. Cottrell, Q-plates with a nonlinear azimuthal distribution of the principal axis: application to encoding binary data, Appl. Opt. 57 (2018) 1005–1010.
[31] V. Kumar, G.M. Philip, N.K. Viswanathan, Formation and morphological transformation of polarization singularities: hunting the monstar, J. Opt. 15 (2013) 044027.
[32] E.J. Galvez, B.L. Rojec, V. Kumar, N.K. Viswanathan, Generation of isolated asymmetric umbilics in light's polarization, Phys. Rev. A 89 (2014) 031801.
[33] E.J. Galvez, B. Khajavi, Monstar disclinations in the polarization of singular optical beams, J. Opt. Soc. Am. A 34 (2017) 568–575.
[34] B.A. Cvarch, B. Khajavi, J.A. Jones, B. Piccirillo, L. Marrucci, E.J. Galvez, Monstar polarization singularities with elliptically-symmetric q-plates, Opt. Express 25 (2017) 14935–14943.

CHAPTER 5

Operators in paraxial quantum optics

Gerard Nienhuis

Huygens-Kamerlingh Onnes Laboratory, Leiden University, Leiden, the Netherlands

Contents

5.1 Introduction 107
5.2 Quantization and conserved quantities of Maxwell field 109
    5.2.1 Discretized plane-wave modes 109
    5.2.2 Continuum of plane-wave modes 110
    5.2.3 Operators for conserved quantities of radiation field 111
5.3 Paraxial quantum fields 114
    5.3.1 Change of variables 114
    5.3.2 Paraxial wave equation 114
    5.3.3 Paraxial limit of quantum field 115
    5.3.4 Discrete transverse modes 116
    5.3.5 Algebra of continuum of bosonic operators 118
5.4 Paraxial modes and harmonic oscillators 119
    5.4.1 Hermite–Gauss modes 120
    5.4.2 Correspondence between paraxial modes and harmonic-oscillator states 121
    5.4.3 Laguerre–Gauss modes 122
5.5 Paraxial energy, momentum and angular momentum 122
5.6 Operator description of Gaussian paraxial modes 124
    5.6.1 Operator description of Hermite–Gauss modes 124
    5.6.2 Operator description of Laguerre–Gauss modes 126
    5.6.3 Elliptical Gaussian modes 128
5.7 Schwinger representation of Laguerre–Gauss modes 130
    5.7.1 The Lie algebra su(2) 130
    5.7.2 The Lie algebra su(1, 1) 132
5.8 Conclusions 134
References 136

5.1 Introduction

Optical experiments most often use light beams, with lasers as source. This is particularly true in experiments in the field of quantum communication and information, where the quantum objects are photons. Such experiments are testing the basis of quantum physics in general [1]. For their description the quantum theory of light is needed. This theory is commonly based on an expansion of the electromagnetic field in modes. A mode is a monochromatic solution of Maxwell's equations, and thereby a purely classical concept, just as the Maxwell field is in general. After quantization, the electromagnetic field is described in terms of quantum operators that act on a Hilbert space spanned by photon number states. A photon can be viewed as an energy quantum of a mode, so that the spatial extension of the photon is determined by the shape of the mode. Nevertheless, it can be attractive to view the radiation field as the wave function of a photon in coordinate representation [2]. Mathematically speaking, modes form the basis of a linear space, which allows superposition. This is a result of the linearity of Maxwell's equations. In a quantum description of radiation, linear operators acting on modes arise next to quantum operators. The relation between these two types of operators is delicate [3], and a proper understanding of the two is worthwhile.

In optical experiments beams of light travel through fibers or free space, connecting optical elements, such as mirrors, lenses, beam splitters, non-linear crystals and detectors. In between these elements the beams can usually be described locally as paraxial modes, with a specific propagation direction. Two standard basis sets of paraxial modes are the Hermite–Gauss modes (HG) and Laguerre–Gauss (LG) modes [4]. Both sets are labeled by two mode indices, which reflects the two-dimensional nature of the transverse profile. The HG modes can be labeled by two indices n_x and n_y, and separate in a product of functions of the Cartesian transverse coordinates x and y. The x- and y-dependence are described by two independent one-dimensional wave equations. The LG modes take their simplest form in polar coordinates, and separate in a product of functions of the radial coordinate R and the azimuthal coordinate ϕ. The corresponding mode indices are denoted as p for the radial index, and m for the azimuthal index. The latter index also describes the orbital angular momentum per photon in the propagation direction (in units of $\hbar$) [5]. The separation between the azimuthal and the radial coordinate does not imply a full independence of the indices p and m, in the sense that the wave equation for the R-dependence still depends on (the square of) the azimuthal index m.

Basis sets of modes, each with two separate mode indices, give rise to a multidimensional state space for single photons, and thereby to entangled states for two or more photons. When LG modes are used for this purpose, it is mostly the azimuthal dependence that is used as the degree of freedom [6]. Also the radial index p has begun to receive attention [7–9]. The full transverse profile can be used for the study of spatial entanglement of photons in space or time [10], and other applications in quantum information [11,12].

In Section 5.2 we briefly summarize the description of the quantized radiation field. We start with a continuum of plane-wave modes. The quantum operators for conserved quantities are expressed in terms of operators acting on mode space. In this way we clarify both the analogy and the difference between mode operators and quantum operators.

In Section 5.3 we derive expressions for the quantum field operators in the paraxial limit. These expressions are based on discrete sets of transverse paraxial mode profiles. The full three-dimensional mode in the paraxial limit contains a carrier wave that has the frequency ω as continuous mode index. The photon creation and annihilation operators are characterized by discrete indices for the mode profile, and the frequency as a continuous variable.

The standard basis sets of paraxial HG and LG modes and their relation are summarized in Section 5.4. As a byproduct we prove a general correspondence between the propagation of paraxial modes and the time evolution of the states of a quantum harmonic oscillator in two dimensions. The quantum operators for conserved quantities in the paraxial limit are discussed in Section 5.5. These operators also generate symmetry transformations of the quantum field.

In this chapter we are also interested in the operators that have the mode indices as eigenvalue. These mode operators characterize the physical significance of the indices. In Section 5.6.1 these operators are given for HG modes. We also present step operators, which change the mode indices of HG modes by one unit. As a result of diffraction, the operators depend in a non-trivial way on the propagation coordinate z. These operators are analogous to the raising and lowering operators for a quantum harmonic oscillator, and their expressions follow directly from the general correspondence discussed in Section 5.4. As demonstrated in Section 5.6.2, the situation is more interesting, but also more delicate for Laguerre–Gauss modes. The operator expression for the azimuthal index m does not depend on z. We also derive the operator expression for the radial index p. This operator does not have a local differential form, in contrast to the other operators for the mode indices. Also the step operators for the index p are specified.

In Section 5.7.1 we show that step operators for the index m arise in a natural way by applying the Schwinger boson representation of the group algebra su(2) for LG modes. Operators for the radial index p arise in an analogous fashion by using the Schwinger boson representation of the algebra su(1, 1). This representation is summarized in Section 5.7.2.

5.2 Quantization and conserved quantities of Maxwell field

5.2.1 Discretized plane-wave modes

The quantum-mechanical description of light is commonly based on an expansion in a basis of discrete orthogonal modes $\mathbf{F}_\lambda(\mathbf{r})$. The mode indices that completely specify the mode are summarized in the single symbol λ. The modes form a complete orthonormal set of complex divergence-free vector solutions of Helmholtz' equation $\nabla^2 \mathbf{F}_\lambda + \omega_\lambda^2 \mathbf{F}_\lambda / c^2 = 0$, with ω_λ the mode frequency. The mode basis spans a linear space, with the modes as elements. The orthonormality relations of the basis are $\int d_3\mathbf{r}\, \mathbf{F}_\lambda^*(\mathbf{r}) \cdot \mathbf{F}_{\lambda'}(\mathbf{r}) = \delta_{\lambda\lambda'}$. Modes and mode space are fully classical concepts. Throughout this chapter bold-face symbols indicate a vector in 3D physical space.

For plane waves in free space the discrete nature of the modes is commonly realized by force, by introducing a cubic quantization volume with sides L, while imposing periodic boundary conditions. The wave vectors $\mathbf{k}$ have components $k_{x,y,z} = 2\pi n_{x,y,z}/L$, with integer values of the indices n_i. For each wave vector $\mathbf{k}$ the polarization direction is orthogonal to $\mathbf{k}$, so that the space of polarizations is two-dimensional. The modes take the form

$$\mathbf{F}_\lambda(\mathbf{r}) = \mathbf{e}_\lambda \exp(i\mathbf{k}_\lambda \cdot \mathbf{r})/L^{3/2}. \tag{5.1}$$

The symbol λ specifies both the wave vector and the polarization vector. At the end of a calculation the limit $L \to \infty$ should be taken.

The quantum field operator for the vector potential $\hat{\mathbf{A}} = \hat{\mathbf{A}}^{(+)} + \hat{\mathbf{A}}^{(-)}$ is defined by its annihilation part [13,14]

$$\hat{\mathbf{A}}^{(+)}(\mathbf{r}) = \sum_\lambda \sqrt{\frac{\hbar}{2\epsilon_0 \omega_\lambda}} \mathbf{F}_\lambda(\mathbf{r}) \hat{a}_\lambda, \tag{5.2}$$

where $\omega_\lambda = ck_\lambda$, and $\hat{a}_\lambda$ is the annihilation operator for a photon in the mode $\mathbf{F}_\lambda$. Symbols with a caret indicate linear operators, acting either on the Hilbert space of quantum states, or on mode space. The operator $\hat{\mathbf{A}}^{(-)}$ is the Hermitian conjugate of $\hat{\mathbf{A}}^{(+)}$, and contains the creation operators $\hat{a}_\lambda^\dagger$ for a photon in the mode $\mathbf{F}_\lambda$. This supports the idea of the modes as a basis of wave functions of a photon [2]. The action of the operators on photon number states of a mode are well known [14]. The nature of $\hat{\mathbf{A}}$ as a quantum operator arises only from the creation and the annihilation operators, which obey the bosonic commutation rules $[\hat{a}_\lambda, \hat{a}_{\lambda'}^\dagger] = \delta_{\lambda\lambda'}$.

5.2.2 Continuum of plane-wave modes

The introduction of the quantization cube to generate discrete wave vectors does not fit well with the geometry of paraxial light beams. For that reason, we first consider the field quantization in the case of a continuum of wave vectors. In free space a plane-wave mode is defined by a wave vector $\mathbf{k}$ and a polarization vector $\mathbf{e}_\nu$ normal to $\mathbf{k}$. The modes take the form [15]

$$\mathbf{F}_\nu(\mathbf{r}; \mathbf{k}) = \mathbf{e}_\nu \exp(i\mathbf{k} \cdot \mathbf{r})/(2\pi)^{3/2}, \tag{5.3}$$

where the wave vector **k** is a continuous variable, and the two unitary polarization vectors $\mathbf{e}_\nu$ are orthogonal to each other and to **k**. The orthonormality property reads

$$\int d_3\mathbf{r}\, \mathbf{F}_\nu^*(\mathbf{r};\mathbf{k}) \cdot \mathbf{F}_{\nu'}(\mathbf{r};\mathbf{k}') = \delta_3(\mathbf{k} - \mathbf{k}')\delta_{\nu\nu'}. \tag{5.4}$$

The expression for the field operator is modified compared with Eqs. (5.2), in that the summation over the wave vectors turns into an integration. The combination (**k**, ν) is the label of the mode. This gives for the annihilation parts of the vector potential, the electric field and the magnetic field

$$\hat{\mathbf{A}}^{(+)}(\mathbf{r}) = \int d_3\mathbf{k} \sum_\nu \sqrt{\frac{\hbar}{2\epsilon_0\omega}} \mathbf{F}_\nu(\mathbf{r};\mathbf{k}) \hat{\alpha}_\nu(\mathbf{k}),$$

$$\hat{\mathbf{E}}^{(+)}(\mathbf{r}) = i \int d_3\mathbf{k} \sum_\nu \sqrt{\frac{\hbar\omega}{2\epsilon_0}} \mathbf{F}_\nu(\mathbf{r};\mathbf{k}) \hat{\alpha}_\nu(\mathbf{k}),$$

$$\hat{\mathbf{B}}^{(+)}(\mathbf{r}) = i \int d_3\mathbf{k} \sum_\nu \sqrt{\frac{\hbar}{2\epsilon_0\omega}} \mathbf{k} \times \mathbf{F}_\nu(\mathbf{r};\mathbf{k}) \hat{\alpha}_\nu(\mathbf{k}). \tag{5.5}$$

The discrete annihilation operators $\hat{a}_\lambda$ are replaced by the annihilation operators $\hat{\alpha}_\nu(\mathbf{k})$ with a continuous label **k** and a discrete binary label ν, which can be taken as $\pm$, for the right and left circularly polarized modes. The bosonic properties of the annihilation and creation operators are expressed by the commutation rule

$$[\hat{\alpha}_\nu(\mathbf{k}), \hat{\alpha}_{\nu'}^\dagger(\mathbf{k}')] = \delta_3(\mathbf{k} - \mathbf{k}')\delta_{\nu\nu'}. \tag{5.6}$$

Note that the right-hand sides of Eqs. (5.4) and (5.6) are identical. This is needed for the bosonic nature of the annihilation and creation operators, and for the validity of the expressions of the operators (5.5). These commutation rules ensure that the commutators of the field operators with continuous modes are the same as with discrete modes, in the limit of an infinite quantization box. As a result, the field dynamics is the same in both descriptions.

We shall denote bosonic operators with a continuous label as $\hat{\alpha}$. These operators are not dimensionless, while operators denoted as $\hat{a}$ are.

5.2.3 Operators for conserved quantities of radiation field

Before discussing quantum operators in the paraxial limit, we briefly recall the structure of quantum operators for conserved quantities of a radiation field. Recently we have shown that quantum operators $\hat{Q}_{qu}$ for a conserved quantity of the Maxwell field can be expressed in terms of matrix elements of a corresponding operator $\hat{Q}_{\text{mod}}$ acting on modes [16]. When the field operators are expanded as in Section 5.2.1 on a discrete basis of modes $\mathbf{F}_\lambda(\mathbf{r})$, the quantum operator has the structure

$$\hat{Q}_{\text{qu}} = \frac{1}{2} \sum_{\lambda\lambda'} \left[\hat{a}_{\lambda'} \hat{a}_\lambda^\dagger + \hat{a}_\lambda^\dagger \hat{a}_{\lambda'} \right] \int d_3\mathbf{r}\, \mathbf{F}_\lambda^* \cdot \hat{Q}_{\text{mod}} \mathbf{F}_{\lambda'}. \tag{5.7}$$

The mode operator $\hat{Q}_{\text{mod}}$ is reminiscent of the analogous operator acting on the wave function for a single quantum particle. The structure of the quantum operator $\hat{Q}_{\text{qu}}$ is typical for quantum field theory of a system of non-interacting bosonic particles.

The quantum operators for the total energy and the total momentum of the Maxwell field are

$$\hat{U}_{\text{qu}} = \frac{1}{2} \int d_3\mathbf{r}\, \epsilon_0 \left(\hat{\mathbf{E}}^2 + c^2 \hat{\mathbf{B}}^2 \right), \quad \hat{\mathbf{P}}_{\text{qu}} = \int d_3\mathbf{r}\, \epsilon_0 \hat{\mathbf{E}} \times \hat{\mathbf{B}}. \tag{5.8}$$

The squares $\hat{\mathbf{E}}^2$ and $\hat{\mathbf{B}}^2$ of the vector operators $\hat{\mathbf{E}}$ and $\hat{\mathbf{B}}$ are scalar operators, while the cross product $\hat{\mathbf{E}} \times \hat{\mathbf{B}}$ is a vector operator. Starting from Eq. (5.2) one arrives at the mode expansions for the quantum operators [16]

$$\hat{U}_{\text{qu}} = \frac{1}{2} \sum_\lambda \hbar\omega_\lambda \left[\hat{a}_\lambda \hat{a}_\lambda^\dagger + \hat{a}_\lambda^\dagger \hat{a}_\lambda \right],$$

$$\hat{\mathbf{P}}_{\text{qu}} = \frac{\hbar}{2i} \sum_{\lambda\lambda'} \left[\hat{a}_{\lambda'} \hat{a}_\lambda^\dagger + \hat{a}_\lambda^\dagger \hat{a}_{\lambda'} \right] \int d_3\mathbf{r}\, \mathbf{F}_\lambda^* \cdot (\boldsymbol{\partial}_\mathbf{r}) \mathbf{F}_{\lambda'}. \tag{5.9}$$

In the second equation (5.9) the dot denotes the product of the two vector modes. The three components of the vector operator of the momentum correspond to the three components of the gradient operator $\boldsymbol{\partial}_\mathbf{r} = (\partial_x, \partial_y, \partial_z)$.

In the zero-photon state the ordering $\hat{a}_\lambda \hat{a}_\lambda^\dagger$ still gives a finite contribution 1. This leads to a diverging energy contribution, which represents vacuum fluctuations. In the context of the present contribution this constant but infinite term can be safely ignored. These expressions (5.9) have the structure of Eq. (5.7), with the momentum mode operator $\hat{\mathbf{P}}_{\text{mod}} = (\hbar/i)\boldsymbol{\partial}_\mathbf{r}$, while all modes $\mathbf{F}_\lambda$ are eigenvector of the energy mode operator $\hat{U}_{\text{mod}}$, with eigenvalue the energy quantum of light $\hbar\omega_\lambda$.

For the angular momentum the situation is more subtle. The term $\hat{\mathbf{p}}_{\text{qu}} = \epsilon_0 \hat{\mathbf{E}} \times \hat{\mathbf{B}}$ may be viewed as the operator for the momentum density, and the quantum operator $\hat{\mathbf{j}}_{\text{qu}} = \mathbf{r} \times \hat{\mathbf{p}}_{\text{qu}}$ represents the density of angular momentum. Therefore $\hat{\mathbf{J}}_{\text{qu}} = \int d_3\mathbf{r}\, \hat{\mathbf{j}}_{\text{qu}}$ is the operator for the total angular momentum. It has been shown that after integration by parts, this operator can be separated into two contributions, so that $\hat{\mathbf{J}}_{qu} = \hat{\mathbf{L}}_{qu} + \hat{\mathbf{S}}_{qu}$. The separate terms are given by the identities [17]

$$\hat{\mathbf{L}}_{qu} = \frac{\hbar}{2i} \sum_{\lambda\lambda'} \left[\hat{a}_{\lambda'} \hat{a}_\lambda^\dagger + \hat{a}_\lambda^\dagger \hat{a}_{\lambda'} \right] \int d_3 \mathbf{r}\, \mathbf{F}_\lambda^* \cdot (\mathbf{r} \times \partial_\mathbf{r}) \mathbf{F}_{\lambda'},$$

$$\hat{\mathbf{S}}_{qu} = \frac{\hbar}{2i} \sum_{\lambda\lambda'} \left[\hat{a}_{\lambda'} \hat{a}_\lambda^\dagger + \hat{a}_\lambda^\dagger \hat{a}_{\lambda'} \right] \int d_3 \mathbf{r}\, [\mathbf{F}_\lambda^* \times \mathbf{F}_{\lambda'}]. \tag{5.10}$$

One may consider this result as a separation of the angular momentum into an orbital part $\hat{\mathbf{L}}_{qu}$ and a spin part $\hat{\mathbf{S}}_{qu}$.

A comparison with Eq. (5.7) defines the corresponding mode operators. This means that $\hat{\mathbf{L}}_{mod} = (\hbar/i)(\mathbf{r} \times \partial_\mathbf{r})$. The properties of this operator are well known from elementary quantum mechanics. The mode operator $\hat{\mathbf{S}}_{mode}$ is a vector of three 3×3 matrices that act exclusively upon the vector nature of the modes, so that

$$\mathbf{F}_\lambda^* \cdot \hat{\mathbf{S}}_{mod} \cdot \mathbf{F}_{\lambda'} = \frac{\hbar}{i} \mathbf{F}_\lambda^* \times \mathbf{F}_{\lambda'}. \tag{5.11}$$

The components of the mode operators $\hat{\mathbf{L}}_{mod}$, $\hat{\mathbf{S}}_{mod}$ and their sum $\hat{\mathbf{J}}_{mod}$ obey the standard commutation rules for angular momentum. When acting on a mode, the operator $\hat{\mathbf{S}}_{mod}$ generates rotations of the vector nature only, while leaving the $\mathbf{r}$-dependence unchanged. On the other hand, the operator $\hat{\mathbf{L}}_{mod}$ generates rotations of the position dependence of a vector field, without modifying the direction of the vector. When a partial rotation of only the vector nature or only the $\mathbf{r}$-dependence is applied to a mode, the resulting field in general does not have a vanishing divergence, and therefore it is not an element of mode space. The rotations generated by the separate mode operators carries a mode outside mode space. In contrast, when the same rotation is carried out on both the position dependence and the vector direction a mode is transformed into another mode. Such full rotations are generated by the mode operator $\hat{\mathbf{J}}_{mod}$.

The properties of the separate quantum operators $\hat{\mathbf{L}}_{qu}$ and $\hat{\mathbf{S}}_{qu}$ are different. These operators act within the Fock space of states of the quantized Maxwell field. However, they do not separately obey the commutation rules for angular momentum. For instance, the three components of $\hat{\mathbf{S}}_{qu}$ have vanishing commutators [17–19]. Components of the sum $\hat{\mathbf{J}}_{qu} = \hat{\mathbf{L}}_{qu} + \hat{\mathbf{S}}_{qu}$ do obey the commutation rules of an angular momentum.

The operators for energy, momentum and angular momentum generate symmetry transformations of the electromagnetic field. These transformations can be expressed in terms of the quantum operators that act on the Fock space of photon states. As a result of the general relation (5.7) between quantum operators and mode operators, the transformations can also be expressed in terms of the mode operators that act on mode space. From the general relation (5.7) one derives

$$\exp(i\kappa \hat{Q}_{qu})\hat{\mathbf{A}}^+ \exp(-i\kappa \hat{Q}_{qu}) = \exp(-i\kappa \hat{Q}_{mod})\hat{\mathbf{A}}^+ . \qquad (5.12)$$

The operators (5.5) for the electric and the magnetic field transform in the same way. The energy operators $\hat{U}$ act as a Hamiltonian, and generates time evolution in the Heisenberg picture. The momentum operators $\hat{\mathbf{P}}$ generate translations, and the angular momentum operators generate rotations [16].

5.3 Paraxial quantum fields

5.3.1 Change of variables

We replace the component k_z by the frequency $\omega = ck$ as integration variable in Eqs. (5.5). The projection of the wave vector $\mathbf{k}$ on the xy-plane is denoted as $\mathbf{K} = (k_x, k_y)$, so that the wave vector is determined by the pair $(\mathbf{K}, \omega)$. We also renormalize the creation and annihilation operators, in such a way that they obey the commutation rules

$$[\hat{\alpha}_\nu(\mathbf{K},\omega), \hat{\alpha}^\dagger_{\nu'}(\mathbf{K}',\omega')] = \delta_2(\mathbf{K}-\mathbf{K}')\delta(\omega-\omega')\delta_{\nu\nu'} . \qquad (5.13)$$

From the relation $\omega^2 = c^2(\mathbf{K}^2 + k_z^2)$ one finds that $d\omega/dk_z = c\cos\theta$ with θ the angle between the wave vector $\mathbf{k}$ and the z-axis. This implies that the annihilation operators in the new variables $(\mathbf{K}, \omega)$ are given by the relation $\hat{\alpha}_\nu(\mathbf{k}) = \hat{\alpha}_\nu(\mathbf{K}, \omega)\sqrt{c|\cos\theta|}$. The operator for the vector potential in the new variables takes the form

$$\hat{\mathbf{A}}^{(+)}(\mathbf{r}) = \frac{1}{(2\pi)^{3/2}} \int d_2\mathbf{K} d\omega \frac{1}{\sqrt{|\cos\theta|}} \sqrt{\frac{\hbar}{2\epsilon_0\omega c}} \exp(i\mathbf{K}\cdot\mathbf{R} + i\omega z\cos\theta/c) \sum_\nu \hat{\alpha}_\nu(\mathbf{K},\omega)\mathbf{e}_\nu . \qquad (5.14)$$

This expression is still equivalent to Eqs. (5.5), and no approximation has been made. A similar change of variables is also applied in Ref. [20].

5.3.2 Paraxial wave equation

For traveling light beams, we take the positive z-axis in the propagation direction. Planes normal to the axis are termed transverse planes. We wish to express the field operators in terms of basis sets of paraxial modes. These modes are based on exact solutions of the scalar paraxial wave equation [4,21]

$$\partial_z u = \frac{ic}{2\omega}\partial^2_{\mathbf{R}} u . \qquad (5.15)$$

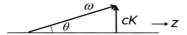

Figure 5.1: Frequency ω replaces k_z as variable. In the right-angled triangle the hypotenuse has a length equal to the frequency ω, and the smallest side has the length cK. For a fixed value of the frequency, the value of k_z decreases for increasing values of K.

For each value of z, solutions $u(\mathbf{R}; z, \omega)$ of this equation determine the transverse profile of a light beam with frequency ω. They are often called transverse paraxial modes for short. A full vector mode in three dimensions is obtained when we multiply a solution of (5.15) by a uniform polarization vector $\mathbf{e}_\nu$ in the transverse (x–y) plane, and by the normalized plane carrier wave

$$w(z, \omega) = \exp(i\omega z/c)/\sqrt{2\pi c}. \tag{5.16}$$

The result is an approximate vector solution to the Helmholtz equation. This can be viewed as the lowest-order term of an expansion of modes in the small opening angle θ [22]. The field u varies only slowly with z, so that the second derivative with respect to z can be ignored. Moreover the field varies slowly with the frequency ω, which also figures in Eq. (5.15). The scalar paraxial wave $u(\mathbf{R}; z, \omega)$ serves as an envelope function.

It is noteworthy that the paraxial wave equation has the same structure as the Schrödinger equation for a free quantum particle in two dimensions, where the propagation coordinate z plays the role of time. This suggests to introduce the Dirac notation $\langle \mathbf{R}|u(z, \omega)\rangle \equiv u(\mathbf{R}; z, \omega)$, in analogy to the notation $\langle \mathbf{r}|\psi\rangle \equiv \psi(\mathbf{r})$ for the wave function of a free quantum particle. Note that the ket vector $|u(z, \omega)\rangle$ still depends on z and ω as parameters. Since Eq. (5.15) contains only a first derivative in the propagation variable z, the transverse profile $\langle \mathbf{R}|u(z, \omega)\rangle$ for all values of z is fully determined by the transverse profile $\langle \mathbf{R}|u(0, \omega)\rangle$ in a single reference plane, which we take as the plane $z = 0$.

5.3.3 Paraxial limit of quantum field

The paraxial limit of field operators is reached when only paraxial modes contain an appreciable number of photons. In a paraxial mode the angle θ is small for the modes containing photons. In that case the factor $\sqrt{\cos\theta}$ in the denominator in Eq. (5.14) can be ignored. In the exponent of the same equation it is justified to approximate $\cos\theta \approx 1 - \theta^2/2$ with $\theta \approx \omega/(cK)$. This determines the z-component of the wave vector $\mathbf{k}$, as is illustrated in Fig. 5.1.

After these substitutions into Eq. (5.14) the quantum operator $\hat{\mathbf{A}}^{(+)}$ can be expressed as

$$\hat{\mathbf{A}}^{(+)}(\mathbf{r}) = \int d\omega \sqrt{\frac{\hbar}{2\epsilon_0 \omega}} \sum_\nu \int d_2\mathbf{K}\, \mathbf{F}_\nu(\mathbf{R}, z; \mathbf{K}, \omega) \hat{\alpha}_\nu(\mathbf{K}, \omega), \tag{5.17}$$

in terms of the full three-dimensional mode

$$\mathbf{F}_\nu(\mathbf{R}, z; \mathbf{K}, \omega) = w(z, \omega)\langle\mathbf{R}|u(z, \mathbf{K}, \omega)\rangle\mathbf{e}_\nu. \tag{5.18}$$

This mode function is the product of the carrier wave (5.16), the uniform polarization vector $\mathbf{e}_\nu$, and the scalar transverse mode

$$\langle\mathbf{R}|u(z, \mathbf{K}, \omega)\rangle = \frac{1}{2\pi}\exp\left(i\mathbf{K}\cdot\mathbf{R} - iz\frac{cK^2}{2\omega}\right). \tag{5.19}$$

Eq. (5.17) for the vector potential has the same general structure as in Eq. (5.2). An extensive description of quantum paraxial optics is also given in Ref. [23].

The transverse mode (5.19) is an exact solution of the paraxial wave equation (5.15), and for all values of the frequency ω and the propagation coordinate z it obeys the orthonormalization relation

$$\langle u(z, \mathbf{K}, \omega)|u(z, \mathbf{K}', \omega)\rangle = \delta_2(\mathbf{K} - \mathbf{K}'). \tag{5.20}$$

The full modes (5.18) are approximate solutions of the Helmholtz equation. Their orthonormality relations

$$\int dz \int d_2\mathbf{R}\, \mathbf{F}_\nu^*(\mathbf{R}, z; \mathbf{K}, \omega) \cdot \mathbf{F}_{\nu'}(\mathbf{R}, z; \mathbf{K}', \omega') = \delta_2(\mathbf{K} - \mathbf{K}')\delta(\omega - \omega')\delta_{\nu\nu'} \tag{5.21}$$

hold in the paraxial limit. As shown in Eq. (5.13), this coincides with the commutator $[\hat{\alpha}_\nu(\mathbf{K}, \omega), \hat{\alpha}_{\nu'}^\dagger(\mathbf{K}', \omega')]$.

5.3.4 Discrete transverse modes

Since linear combinations of modes with the same frequency produce modes, there is an infinity of possible choices of a basis set of transverse paraxial modes. While modes $|u(z, \mathbf{K}, \omega)\rangle$ are labeled by the continuous variable $\mathbf{K}$, the well-known basis sets of Hermite–Gauss and Laguerre–Gauss modes are discrete [4]. They are labeled by pairs of natural numbers, with a specific physical significance. These will be discussed later in this chapter. For now, we label paraxial modes in a discrete basis set by the single symbol μ. Since the propagation as described by the paraxial wave equation is unitary, both completeness and orthogonality hold for any transverse plane, so that we can write

$$\langle u_\mu(z, \omega)|u_{\mu'}(z, \omega)\rangle \equiv \int d_2\mathbf{R}\, u_\mu^*(\mathbf{R}; z, \omega) u_{\mu'}(\mathbf{R}; z, \omega) = \delta_{\mu\mu'}. \tag{5.22}$$

These orthonormality relations hold exactly for all values of z and ω. As a result of the completeness, the state of a photon can be perfectly localized in a transverse plane. Diffraction during propagation then causes spreading in other transverse planes.

The three-dimensional modes are labeled by the frequency ω, the paraxial mode index μ and the polarization index ν. They are described by the product of a carrier wave, a paraxial mode and a polarization vector

$$\mathbf{F}_{\mu,\nu}(\mathbf{R}, z; \omega) = w(z, \omega)\langle \mathbf{R}|u_\mu(z, \omega)\rangle \mathbf{e}_\nu. \qquad (5.23)$$

In the paraxial limit they obey the orthonormality relations in three dimensions

$$\int dz \int d_2\mathbf{R}\, \mathbf{F}^*_{\mu,\nu}(\mathbf{R}, z; \omega) \cdot \mathbf{F}_{\mu',\nu'}(\mathbf{R}, z; \omega') = \delta_{\mu\mu'}\delta_{\nu\nu'}\delta(\omega - \omega'). \qquad (5.24)$$

The modes (5.23) have two discrete indices ν and μ, and a continuous index ω. The expansion of the paraxial field operator $\hat{\mathbf{A}}^{(+)}$ in these modes is expressed in analogy to Eq. (5.17) in the form

$$\hat{\mathbf{A}}^{(+)}(\mathbf{r}) = \int d\omega \sqrt{\frac{\hbar}{2\epsilon_0 \omega}} \sum_\nu \sum_\mu \mathbf{F}_{\mu,\nu}(\mathbf{R}, z; \omega)\hat{\alpha}_{\mu,\nu}(\omega). \qquad (5.25)$$

The commutator $[\hat{\alpha}_{\mu',\nu'}(\omega'), \hat{\alpha}^\dagger_{\mu,\nu}(\omega)]$ between creation and annihilation operators is identical to the right-hand side of Eq. (5.24).

The operators for the electric and the magnetic field in the paraxial limit follow directly from Eq. (5.25) for the vector potential in the same limit. The electric-field operator $\hat{\mathbf{E}}$ is minus the time derivative of the vector potential (in the Heisenberg picture). The resulting expression for its annihilation part is

$$\hat{\mathbf{E}}^{(+)}(\mathbf{r}) = i \int d\omega \sqrt{\frac{\hbar\omega}{2\epsilon_0}} \sum_\nu \sum_\mu \mathbf{F}_{\mu,\nu}(\mathbf{R}, z; \omega)\hat{\alpha}_{\mu,\nu}(\omega). \qquad (5.26)$$

The magnetic field operator $\hat{\mathbf{B}}$ is the curl of the vector potential operator. In the paraxial limit the polarization vector $\mathbf{e}_\nu$ lies in the xy plane, while the wave vector with length ω/c is directed along the z-axis. For its annihilation part this gives the expression

$$\hat{\mathbf{B}}^{(+)}(\mathbf{r}) = \frac{i}{c} \int d\omega \sqrt{\frac{\hbar\omega}{2\epsilon_0}} \sum_\nu \sum_\mu \mathbf{e}_z \times \mathbf{F}_{\mu,\nu}(\mathbf{R}, z; \omega)\hat{\alpha}_{\mu,\nu}(\omega). \qquad (5.27)$$

The mode $\mathbf{F}_{\mu,\nu}$ has the uniform polarization $\mathbf{e}_\nu$, so that each summand in this equation (5.27) has a polarization vector $\mathbf{e}_z \times \mathbf{e}_\nu$, which also lies in the xy-plane. It is the electric polarization vector rotated over an angle $\pi/2$ around the propagation direction. In the remainder of this chapter we shall use the expressions (5.25)–(5.27) for the quantum field operators in the paraxial limit.

118 Chapter 5

Eqs. (5.17) and (5.25) give two different expansions of the same operator $\hat{\mathbf{A}}^{(+)}(\mathbf{r})$. Inspection of the expressions (5.18) and (5.23) of the two basis sets of modes show that they differ only in the basis of transverse paraxial modes. A comparison of the two expansions reveals the identity

$$\int d_2\mathbf{K}\, |u(z,\mathbf{K},\omega)\rangle \hat{\alpha}_\nu(\mathbf{K},\omega) = \sum_\mu |u_\mu(z,\omega)\rangle \hat{\alpha}_{\mu,\nu}(\omega). \tag{5.28}$$

By taking inner products of this identity with the bra vectors $\langle u(z,\mathbf{K}',\omega)|$ or $\langle u_{\mu'\nu},\omega|$, while using the orthonormality relations (5.20) and (5.22), one finds the transformation rules between the two sets of annihilation operators $\hat{\alpha}_\nu(\mathbf{K},\omega)$ and $\hat{\alpha}_{\mu,\nu}(\omega)$.

5.3.5 Algebra of continuum of bosonic operators

The three-dimensional modes (5.23) in the paraxial limit are specified by the discrete indices ν for the polarization and μ for the transverse paraxial mode, and the frequency ω as a continuous variable. The relation between the commutation rules for continuous bosonic operators and the orthogonality of modes is delicate [24]. In this section we consider the algebra of the annihilation and creation operators for a continuum of modes, for fixed values of the discrete indices μ and ν. We omit these indices, and simply denote the operators as $\hat{\alpha}(\omega)$ and $\hat{\alpha}^\dagger(\omega)$ with the commutation rules

$$\left[\hat{\alpha}(\omega'),\hat{\alpha}^\dagger(\omega)\right] = \delta(\omega-\omega'), \quad \left[\hat{\alpha}(\omega'),\hat{\alpha}(\omega)\right] = 0. \tag{5.29}$$

Just as in the simpler case of a single bosonic pair of operators $\hat{a}$ and $\hat{a}^\dagger$, the algebra of these operators and their action on number states is fully determined by their commutation rules. First we consider the operator

$$\hat{N} = \int d\omega\, \hat{\alpha}^\dagger(\omega)\hat{\alpha}(\omega). \tag{5.30}$$

It is obviously Hermitian and has real non-negative eigenvalues. From Eq. (5.29) we find the commutation relations

$$\left[\hat{N},\hat{\alpha}(\omega)\right] = -\hat{\alpha}(\omega), \quad \left[\hat{N},\hat{\alpha}^\dagger(\omega)\right] = \hat{\alpha}^\dagger(\omega). \tag{5.31}$$

When the operator equalities (5.31) act on an eigenstate of $\hat{N}$ with eigenvalue N, it is obvious that the operators $\hat{\alpha}(\omega)$ and $\hat{\alpha}^\dagger(\omega)$ applied to this eigenstate produce other eigenstates of $\hat{N}$ with eigenvalue $N-1$ and $N+1$, respectively. The eigenvalues form a ladder, with $\hat{\alpha}^\dagger(\omega)$ as raising operators, and $\hat{\alpha}(\omega)$ as lowering operators for all values of ω. Each step in the ladder can be viewed as a photon. Since eigenvalues of $\hat{N}$ cannot be negative, the eigenvalues must be integers, with minimum value $N = 0$. The operator $\hat{N}$ has the total number of photons as eigenvalues.

When $|0\rangle$ is the vacuum state with no photons, $\hat{\alpha}(\omega)|0\rangle = 0$ for all values of ω. This state can be normalized as $\langle 0|0\rangle = 1$. One-photon states arise when a raising operator acts on the vacuum state. Then $\hat{\alpha}^\dagger(\omega)|0\rangle \equiv |\omega\rangle$ is the state of a single photon in the mode with frequency ω. It follows from Eq. (5.29) that these states are normalized as $\langle \omega'|\omega\rangle = \delta(\omega - \omega')$. A basis of states with N photons consists of the fully symmetrized states of identical particles. This leads to the expression

$$|\omega_1, \omega_2, \ldots, \omega_N\rangle = \frac{1}{\sqrt{N!}} \hat{\alpha}^\dagger(\omega_1)\hat{\alpha}^\dagger(\omega_2)\ldots\hat{\alpha}^\dagger(\omega_N)|0\rangle, \tag{5.32}$$

which implies that $\hat{\alpha}^\dagger(\omega)|\omega_1, \omega_2, \ldots, \omega_N\rangle = \sqrt{N+1}|\omega, \omega_1, \omega_2, \ldots, \omega_N\rangle$. It follows from the commutation rule (5.29) that the state (5.32) is invariant for a permutation of the frequencies ω_i.

A coherent state for a continuum of modes is specified by a complex function $f(\omega)$ of the continuous mode index. A general expression of coherent states is then

$$|\Psi\rangle = \frac{1}{Z} \exp\left(\int d\omega f(\omega)\hat{\alpha}^\dagger(\omega)\right)|0\rangle, \tag{5.33}$$

with the normalization factor $Z = \exp(\int d\omega |f(\omega)|^2/2)$. This state is normalized, so that $\langle \Psi|\Psi\rangle = 1$. It is an eigenstate of $\hat{\alpha}(\omega)$ with eigenvalue $f(\omega)$, and the expectation value of the number operator is equal to $\langle \Psi|\hat{N}|\Psi\rangle = \int d\omega |f(\omega)|^2$.

When we no longer suppress the mode indices μ and ν, the single-photon states take the form $|\omega\rangle_{\mu,\nu} = \hat{\alpha}^\dagger_{\mu,\nu}(\omega)|0\rangle$. The operator for the number of photons in the paraxial mode $|u_\mu\rangle$ with polarization $\mathbf{e}_\nu$ is given by $\hat{N}_{\mu,\nu} = \int d\omega\, \hat{\alpha}^\dagger_{\mu,\nu}(\omega)\hat{\alpha}_{\mu,\nu}(\omega)$. When the quantum field for the paraxial mode described by the pair of indices (μ, ν) is in the coherent state (5.33), the contribution of this paraxial mode to the expectation value of the quantum fields (5.25)–(5.27) is found by replacing the operator $\hat{\alpha}_{\mu,\nu}(\omega)$ by the function $f(\omega)$.

5.4 Paraxial modes and harmonic oscillators

The paraxial mode operators that we considered so far represent conserved quantities of light modes. Modes are reminiscent of stationary states of quantum systems. Quantum states are commonly labeled by quantum numbers, which play a similar role as mode indices that characterize a mode. We will be interested in mode operators which have mode indices as eigenvalues, or which change mode indices by one unit. We start with the most common paraxial modes, which are known as the Hermite–Gauss (HG) and the Laguerre–Gauss (LG) modes [4,21,25].

5.4.1 Hermite–Gauss modes

The HG modes can be expressed in the condensed form [26]

$$u_{n_x,n_y}(\mathbf{R}; z, \omega) = \frac{1}{\gamma} \psi_{n_x}(\xi) \psi_{n_y}(\eta) e^{-i\chi(n_x+n_y+1)} \exp\left(\frac{ikR^2}{2q}\right), \qquad (5.34)$$

where $k = \omega/c$ is the wave number. Here the dimensionless transverse coordinates $\xi = x/\gamma$ and $\eta = y/\gamma$ are scaled by the factor γ, which can be viewed as the width of the mode profile in the transverse plane. The phase χ is the Gouy phase, and q is the radius of curvature of the wave fronts. These three mode parameters γ, χ and q vary with the propagation coordinate z. They are given by

$$\gamma = \sqrt{\frac{b^2+z^2}{kb}}, \quad q = \frac{b^2+z^2}{z}, \quad \tan\chi = \frac{z}{b}, \qquad (5.35)$$

where b is the diffraction length (or the Rayleigh range), which is a measure of the length of the focal region. The Gouy phase χ increases by an amount π from $z = -\infty$ to ∞. At the focal plane $z = 0$ the width attains its minimal value $\gamma(0) = \sqrt{b/k} \equiv \gamma_0$, which is a measure of the beam waist. The opening angle of the beam is of the order of $\theta = \sqrt{1/kb}$, which must be small in order to validate the paraxial limit.

The functions $\psi_{n_x}(\xi)$ for $n_x = 0, 1, \ldots$ form an orthonormal basis of eigenfunctions with eigenvalue $n_x + 1/2$ of the dimensionless Hamiltonian

$$\hat{H}_\xi = \frac{1}{2}\left(-\partial_\xi^2 + \xi^2\right) \qquad (5.36)$$

for the quantum harmonic oscillator (HO) in one dimension [27]. Their explicit expressions are

$$\psi_{n_x}(\xi) = \frac{1}{\sqrt{2^{n_x} n_x! \sqrt{\pi}}} \exp(-\xi^2/2) H_{n_x}(\xi), \qquad (5.37)$$

with H_{n_x} being the Hermite polynomials. Likewise, the functions $\psi_{n_y}(\eta)$ are the real normalized eigenfunctions with eigenvalue $n_y + 1/2$ of the dimensionless Hamiltonian $\hat{H}_\eta$.

It is important to notice that the modes (5.34) are exact solutions of the paraxial wave equation (5.15), which in turn is an approximation of the Helmholtz equation. For each transverse plane the modes form a complete orthonormal basis, which implies that an arbitrary solution of the paraxial wave equation can be expanded on this basis. In the Dirac notation of Section 5.3.2, the orthonormality relation takes the form

$$\langle u_{n_x,n_y}(z,\omega) | u_{n'_x,n'_y}(z,\omega) \rangle \equiv \int d^2\mathbf{R}\, u^*_{n_x,n_y}(\mathbf{R}; z, \omega) u_{n'_x,n'_y}(\mathbf{R}; z, \omega) = \delta_{n_x,n'_x} \delta_{n_y,n'_y}, \qquad (5.38)$$

for each value of z and ω.

5.4.2 Correspondence between paraxial modes and harmonic-oscillator states

The obvious similarity of the HG modes (5.34) to the eigenstates of a quantum-mechanical harmonic oscillator in a plane allows us to derive a correspondence between the time-dependent wave functions of the oscillator and paraxial modes. One notices that Eq. (5.34) contains the products $\psi_{n_x}(\xi)\psi_{n_y}(\eta)$, which are eigenfunctions of the Hamiltonian

$$\hat{H} = \hat{H}_\xi + \hat{H}_\eta, \tag{5.39}$$

with eigenvalues $n_x + n_y + 1$. The right-hand side of Eq. (5.34) contains the curvature term as a common factor, which does not depend on the mode indices. Omitting this factor leaves us with the normalized wave functions for the stationary states of a plane HO

$$\Psi_{n_x,n_y}(\xi, \eta; \chi) = \psi_{n_x}(\xi)\psi_{n_y}(\eta)e^{-i\chi(n_x+n_y+1)}, \tag{5.40}$$

with position variables ξ and η, and where χ plays the role of time. A full cycle of the oscillator is represented by an interval $\Delta\chi = 2\pi$. These stationary states (5.40) form a complete set of normalized solutions of the Schrödinger equation

$$\partial_\chi \Psi(\xi, \eta; \chi) = -i\hat{H}\Psi(\xi, \eta; \chi). \tag{5.41}$$

The general linear relation between HG modes (5.34) and HO eigenstates (5.40) leads to a correspondence between solutions $u(\mathbf{R}; z, \omega)$ of the paraxial wave equation and HO wave functions $\Psi(\xi, \eta; \chi)$, in the form [28]

$$u(\mathbf{R}; z, \omega) = \frac{1}{\gamma}\Psi(\xi, \eta; \chi)\exp\left(\frac{ikR^2}{2q}\right). \tag{5.42}$$

The expansion of the paraxial mode u on the basis of HG modes (5.34) has the same coefficients as the expansion of the HO wave function Ψ on the basis of the stationary states (5.40).

The exact correspondence (5.42) between solutions u of the paraxial wave equation and HO wave functions Ψ works both ways. For a given solution Ψ of the time-dependent Schrödinger equation for the HO we find a paraxial mode u, by choosing a value for the Rayleigh range b. This determines the z-dependent mode parameters γ, q and χ. Conversely, when we substitute an arbitrary solution of the paraxial wave equation (5.15) in the left-hand side of Eq. (5.42), the correspondence produces a solution $\Psi(\xi, \eta; \chi)$ of the Schrödinger equation (5.41). The overlap of two modes is the same as the overlap of the two corresponding wave functions. In particular, a normalized mode u corresponds to a normalized wave function Ψ. Since the Gouy phase χ increases by an amount π between $z = -\infty$ and ∞, any paraxial mode u is compressed to half a cycle of the oscillator.

Eigenstates of the HO with the same value of $n = n_x + n_y$ are degenerate, so that linear superpositions of these states are stationary. Their dependence on χ is given by the phase factor $\exp(-i\chi(n+1))$. This means that the corresponding paraxial profile is self-similar under propagation, apart from this phase factor and the scaling factor $\gamma(z)$.

5.4.3 Laguerre–Gauss modes

From elementary quantum mechanics [27] it is known that stationary states of a plane HO that separate in polar coordinates ρ and ϕ with $\xi + i\eta = \rho \exp(i\phi)$ are given by the expressions

$$\Psi_{p,m}(\rho, \phi, \chi) = \sqrt{\frac{p!}{\pi(p+|m|)!}} e^{im\phi} \rho^{|m|} e^{-\rho^2/2} L_p^{|m|}(\rho^2) e^{-i(2p+|m|+1)\chi}. \tag{5.43}$$

Here $L_p^{|m|}$ is the generalized Laguerre polynomial of order p and degree $|m|$, with p a natural number, and m an integer. For a plane oscillator these states are eigenstates of the Hamiltonian (5.39) with eigenvalue $2p + |m| + 1$, and of angular momentum with eigenvalue m. Obviously, the state (5.43) is a linear combination of the eigenstates (5.40) for which the quantum numbers obey the relation $n_x + n_y = 2p + |m|$.

In polar coordinates we denote $x + iy = R\exp(i\phi)$, so that the z-component of the mode operator for orbital angular momentum takes the form $\hat{L}_{\text{mod},z} = (\hbar/i)\partial_\phi$. It is convenient to use as basis for the paraxial modes the Laguerre–Gauss modes, which correspond to the HO states (5.43), according to the expression

$$u_{p,m}(R, \phi; z) = \frac{1}{\gamma} \Psi_{p,m}(\rho, \phi; \chi) \exp\left(\frac{ikR^2}{2q}\right). \tag{5.44}$$

They are characterized by their azimuthal dependence $\exp(im\phi)$ [4], so that they are eigenmodes of the mode operator $\hat{L}_{\text{mod},z}$ with eigenvalue $\hbar m$. This confirms that the Laguerre–Gauss modes carry an orbital angular momentum $\hbar m$ per photon in the propagation direction, as has been shown some time ago [5].

5.5 Paraxial energy, momentum and angular momentum

Expressions are obtained for the quantum operators for energy, linear momentum and angular momentum in the paraxial limit when we replace the discrete basis of modes in Eqs. (5.9) and (5.10) by the paraxial modes (5.23). Since we only take into account paraxial modes, the vacuum fluctuations of the electromagnetic field are not properly described, and we simply ignore them. For the energy quantum operator in the paraxial limit this leads to the simple expression

$$\hat{U}_{\mathrm{qu}} = \int d\omega \sum_{\nu} \sum_{\mu} \hbar\omega \hat{\alpha}^{\dagger}_{\mu,\nu}(\omega)\hat{\alpha}_{\mu,\nu}(\omega). \qquad (5.45)$$

The time evolution of the fields is described by treating the energy operator as a Hamiltonian. Then by using the bosonic commutation rule we find for the annihilation operator the time evolution

$$e^{i\hat{U}_{\mathrm{qu}}t/\hbar}\hat{\alpha}_{\mu,\nu}(\omega)e^{-i\hat{U}_{\mathrm{qu}}t/\hbar} = \hat{\alpha}_{\mu,\nu}(\omega)\exp(-i\omega t). \qquad (5.46)$$

This also determines the time evolution of the quantum operators for the electric and the magnetic field Eqs. (5.26) and (5.27).

The mode operator for the momentum is $(\hbar/i)$ times the gradient operator $\partial_{\mathbf{r}}$. In the paraxial limit it is natural to separate the momentum quantum operator $\hat{\mathbf{P}}_{\mathrm{qu}}$ in the two-dimensional projection $\hat{\mathbf{P}}_{\mathrm{qu},xy}$ and the z-component $\hat{P}_{\mathrm{qu},z}$. The component in the propagation direction is determined by the z-derivative of the modes. Since the z-dependence of the transverse mode $\langle \mathbf{R}|u_{\mu}(z,\omega)\rangle$ is slow, the dominant order is determined by the derivative of the carrier wave $w(z)$, which contributes the factor $\hbar\omega/c$. This gives the result $\hat{P}_{z,\mathrm{qu}} = \hat{U}_{\mathrm{qu}}/c$, which simply states that for a paraxial light beam the ratio of energy and momentum is equal to the velocity of light. The component of the momentum quantum operator in the xy-plane is found as

$$\hat{\mathbf{P}}_{\mathrm{qu},xy} = \frac{\hbar}{i}\int d\omega \sum_{\nu}\sum_{\mu,\mu'} \hat{\alpha}^{\dagger}_{\mu,\nu}(\omega)\hat{\alpha}_{\mu',\nu}(\omega)\langle u_{\mu}(z,\omega)|\partial_{\mathbf{R}}|u_{\mu'}(z,\omega)\rangle. \qquad (5.47)$$

These operators generate translations in the propagation direction and in the transverse plane. For paraxial beams the z-component of the momentum is large compared with the component in the transverse plane.

For structured paraxial beams the interesting part of the angular momentum is the component in the propagation direction. It is convenient to consider the basis of circularly polarized LG modes, so that the scalar mode index μ is replaced by the pair of LG indices (p,m). The expressions (5.10) show that $\hat{L}_{\mathrm{mod},z} = (\hbar/i)\partial_{\phi}$, which has the LG modes as eigenmodes with eigenvalue $\hbar m$. The two circular polarization vectors $\mathbf{e}_{\pm} = (\mathbf{e}_{x} \pm i\mathbf{e}_{y})/\sqrt{2}$ are eigenvectors of $\hat{S}_{\mathrm{mod},z}$, with eigenvalues $\pm\hbar$. This gives for the quantum operators the expressions

$$\begin{aligned}\hat{L}_{\mathrm{qu},z} &= \hbar \int d\omega \sum_{\nu}\sum_{p,m} m\hat{\alpha}^{\dagger}_{pm,\nu}(\omega)\hat{\alpha}_{pm,\nu}(\omega), \\ \hat{S}_{\mathrm{qu},z} &= \hbar \int d\omega \sum_{p,m}\left[\hat{\alpha}^{\dagger}_{pm,+}(\omega)\hat{\alpha}_{pm,+}(\omega) - \hat{\alpha}^{\dagger}_{pm,-}(\omega)\hat{\alpha}_{pm,-}(\omega)\right]. \end{aligned} \qquad (5.48)$$

These operators generate rotations about the z-axis. Their effects on the operators $\hat{\alpha}_{pm,+}(\omega)$ are expressed by the relations

$$e^{i\hat{L}_{\text{qu},z}\psi/\hbar}\hat{\alpha}_{pm,+}(\omega)e^{-i\hat{L}_{\text{qu},z}\psi/\hbar} = \hat{\alpha}_{pm,+}(\omega)e^{-i\psi m},$$
$$e^{i\hat{S}_{\text{qu},z}\psi/\hbar}\hat{\alpha}_{pm,+}(\omega)e^{-i\hat{S}_{\text{qu},z}\psi/\hbar} = \hat{\alpha}_{pm,+}(\omega)e^{\mp i\psi}. \tag{5.49}$$

The transformations (5.46), (5.48) and (5.49) also determine the transformations of the quantum operators in Eqs. (5.25)–(5.27). These are examples of the symmetry transformation (5.12), with the proper mode operator.

5.6 Operator description of Gaussian paraxial modes

5.6.1 Operator description of Hermite–Gauss modes

As is well known, the eigenvalue spectrum of a quantum harmonic oscillator can be derived by algebraic means [27]. From the correspondence (5.42) it follows that the expressions (5.34) for the HG modes can likewise be obtained by bosonic operator algebra. To show this we introduce the (lowering and raising) ladder operators for the HO,

$$\hat{B}_\xi(\chi) = \frac{1}{\sqrt{2}}\left(\xi + \partial_\xi\right)e^{i\chi}, \quad \hat{B}_\xi^\dagger(\chi) = \frac{1}{\sqrt{2}}\left(\xi - \partial_\xi\right)e^{-i\chi}, \tag{5.50}$$

for all values of χ. They obey the bosonic commutation rules $[\hat{B}_\xi, \hat{B}_\xi^\dagger] = 1$. From these commutation rules it follows that the operator $\hat{B}_\xi^\dagger \hat{B}_\xi$ has the natural numbers $n_x = 0, 1, 2, \ldots$ as eigenvalues. The eigenstates are given in Eq. (5.40). Neighboring stationary states are connected in the usual way:

$$\hat{B}_\xi \Psi_{n_x,n_y} = \sqrt{n_x}\Psi_{n_x-1,n_y}, \quad \hat{B}_\xi^\dagger \Psi_{n_x,n_y} = \sqrt{n_x+1}\Psi_{n_x+1,n_y}. \tag{5.51}$$

Similar expressions hold for $\hat{B}_\eta(\chi)$ and its Hermitian conjugate.

The normalized stationary states Ψ_{n_x,n_y} of the HO can be reached from the ground state by repeated application of the raising operators $\hat{B}_\xi^\dagger$ and $\hat{B}_\eta^\dagger$, so that

$$\Psi_{n_x,n_y}(\xi,\eta;\chi) = \frac{1}{\sqrt{n_x!n_y!}}\left(\hat{B}_\xi^\dagger(\chi)\right)^{n_x}\left(\hat{B}_\eta^\dagger(\chi)\right)^{n_y}\Psi_{0,0}(\xi,\eta;\chi). \tag{5.52}$$

The ground state $\Psi_{0,0}$ of the HO is eigenstate of the lowering operators $\hat{B}_\xi$ and $\hat{B}_\eta$, with eigenvalue 0. The ladder operators transform any solution $\Psi(\xi,\eta;\chi)$ of the Schrödinger equation (5.41) into another solution.

The correspondence (5.42) between paraxial modes and HO wave functions implies the existence of similar ladder operators $\hat{A}_x(z)$ and $\hat{A}_x^\dagger(z)$ for the modes. The lowering operator $\hat{A}_x$ is defined by the requirement that when the HO state Ψ and the paraxial beam u correspond to each other in the sense of Eq. (5.42), also the HO state $\hat{B}_\xi \Psi$ and the paraxial mode $\hat{A}_x u$ correspond to each other. This requirement gives the connection between the ladder operators

$$\hat{A}_x(z) = \exp\left(\frac{ikR^2}{2q}\right) \hat{B}_\xi(\chi) \exp\left(\frac{-ikR^2}{2q}\right). \tag{5.53}$$

When one uses the transformation between the variables (ξ, η, χ) and (x, y, z), while using the properties (5.35), one arrives at the result for the ladder operators

$$\hat{A}_x(z) = \frac{1}{\sqrt{2bk}}[kx + (b+iz)\partial_x], \quad \hat{A}_x^\dagger(z) = \frac{1}{\sqrt{2bk}}[kx - (b-iz)\partial_x]. \tag{5.54}$$

For each value of z, these operators obey the same bosonic commutation rules as $\hat{B}_\xi$ and $\hat{B}_\xi^\dagger$. They connect HG modes of different order n_x, in analogy to Eq. (5.51). Similar expressions hold for the lowering operator $\hat{A}_y$ and its Hermitian conjugate, which change the index n_y by one unit. These expressions were derived before in a different manner [26].

The paraxial modes that correspond to the stationary HO states (5.40) are the HG modes (5.34). The ladder operators (5.54) acting on the HG modes have the same effect as the ladder operators (5.50) on the stationary states. For each value of z, the HG modes are eigenmodes of the Cartesian mode-number operators $\hat{n}_x \equiv \hat{A}_x^\dagger(z)\hat{A}_x(z)$, $\hat{n}_y \equiv \hat{A}_y^\dagger(z)\hat{A}_y(z)$, with eigenvalues n_x and n_y. These modes are obtained from the fundamental mode $u_{0,0}$ by repeated application of the raising operators $\hat{A}_x^\dagger(z)$ and $\hat{A}_y^\dagger(z)$, in analogy to Eq. (5.52). The explicit form of $\hat{n}_x$ is

$$\hat{n}_x = -\frac{1}{2} + \frac{iz}{b}\left(\frac{1}{2} + x\partial_x\right) + \frac{k}{2b}x^2 - \frac{b^2+z^2}{2b}\partial_x^2. \tag{5.55}$$

For $z = 0$, where $\xi = x/\gamma_0$, this operator reduces to $\hat{H}_\xi - 1/2$, as it should. A similar form holds for $\hat{n}_y$. The operators $\hat{n}_x$ and $\hat{n}_y$ act in the Hilbert space of paraxial modes, and commute with each other. When acting on HG modes, they return the mode index n_x or n_y. The ladder operator $\hat{A}_x^\dagger(z)$ acting on an HG mode raises the mode index n_x by one, while the operator $\hat{A}_x(z)$ lowers the mode index by one. In other words, the ladder operators (5.54) are step operators for the mode index n_x.

A common feature of the mode index operators and the ladder operators is that, when acting on an arbitrary solution of the paraxial wave equation, they produce another solution. In this light, the z-dependence of these operators is remarkably simple, given the subtle z-dependence of the beam parameters γ, q and χ. It is important to notice that these operators do not refer to the number of photons or their annihilation or creation. They just connect modes of different order.

5.6.2 Operator description of Laguerre–Gauss modes

As shown in Section 5.4.3, LG modes are labeled by two mode indices, the azimuthal mode index m and the radial mode index p. They form an alternative basis of paraxial modes. We are interested in expressions for the operators that return the mode indices, just as the operators $\hat{n}_x$ and $\hat{n}_y$ do for HG modes.

We introduce bosonic ladder operators with a circular nature

$$\hat{B}_\pm = \frac{1}{\sqrt{2}}(\hat{B}_\xi \mp i\hat{B}_\eta), \quad \hat{B}^\dagger_\pm = \frac{1}{\sqrt{2}}(\hat{B}^\dagger_\xi \pm i\hat{B}^\dagger_\eta). \tag{5.56}$$

Expressed in polar coordinates, these operators have the explicit analytical form

$$\hat{B}_\pm(\chi) = \frac{1}{2}e^{i\chi \mp i\phi}\left(\rho + \partial_\rho \mp \frac{i}{\rho}\partial_\phi\right), \quad \hat{B}^\dagger_\pm(\chi) = \frac{1}{2}e^{-i\chi \pm i\phi}\left(\rho - \partial_\rho \mp \frac{i}{\rho}\partial_\phi\right). \tag{5.57}$$

Starting from the fundamental mode $\Psi_{0,0}$, the circular raising operators generate stationary states of the HO with a circular nature. In analogy to the linear stationary states (5.52) of the HO, we denote the basis set of normalized circular stationary states in the form

$$\Psi_{n_+,n_-} = \frac{1}{\sqrt{n_+!n_-!}}\left(\hat{B}^\dagger_+(\chi)\right)^{n_+}\left(\hat{B}^\dagger_-(\chi)\right)^{n_-}\Psi_{0,0}, \tag{5.58}$$

with $n_\pm$ natural numbers. It is obvious from the expressions (5.56) for $\hat{B}^\dagger_\pm$ that this stationary state is a linear combination of the states (5.52) with the same value $n_x + n_y = n = n_+ + n_-$. This is the degree of excitation of the state, while the value $n+1$ is the degree of degeneracy.

When the pair of quantum numbers (n_+, n_-) is translated to the pair (p, m) by the relations

$$m = n_+ - n_-, \quad p = (n - |m|)/2 = \min(n_+, n_-), \tag{5.59}$$

the HO state (5.58) is identical to the state (5.43). Since these HO states are stationary, the corresponding Laguerre–Gauss modes paraxial modes are self-similar under propagation.

The operators that have the mode indices n_+ and n_- as eigenvalue are now directly obtained. We transform the circular HO ladder operators $\hat{B}_\pm$ and $\hat{B}^\dagger_\pm$ into mode operators, in analogy to Eq. (5.53). In polar coordinates this leads to the expressions

$$\hat{A}_\pm(z) = \frac{1}{2\sqrt{bk}}e^{\mp i\phi}\left[kR + (b+iz)\left(\partial_R \mp \frac{i}{R}\partial_\phi\right)\right],$$
$$\hat{A}^\dagger_\pm(z) = \frac{1}{2\sqrt{bk}}e^{\pm i\phi}\left[kR - (b-iz)\left(\partial_R \pm \frac{i}{R}\partial_\phi\right)\right], \tag{5.60}$$

Operators in paraxial quantum optics 127

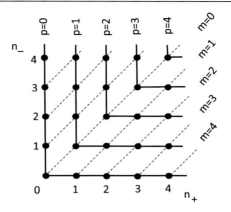

Figure 5.2: Illustration of transformation between mode indices (n_+, n_-) and (p, m). The LG modes are naturally arranged as points in a square lattice, with the mode numbers n_+ and n_- indicated along the horizontal and the vertical axis. The lines connecting the points with a fixed value of m are the dashed diagonal lines, whereas the lines for a fixed value of p are the drawn lines.

for ladder operators acting on modes. Mode-number operators for the LG indices n_+ and n_- are defined by the expressions $\hat{n}_+ \equiv \hat{A}_+^\dagger \hat{A}_+$ and $\hat{n}_- \equiv \hat{A}_-^\dagger \hat{A}_-$. The sum of these two commuting number operators obeys the identity

$$\hat{n}_x + \hat{n}_y = \hat{n}_+ + \hat{n}_- \equiv \hat{n}. \tag{5.61}$$

Using Eq. (5.55) and the analogous expression for $\hat{n}_y$ one finds in polar coordinates

$$\hat{n} = -1 + \frac{iz}{b}(1 + R\partial_R) + \frac{k}{2b}R^2 - \frac{b^2 + z^2}{2kb}\left(\partial_R^2 + \frac{1}{R}\partial_R + \frac{1}{R^2}\partial_\phi^2\right). \tag{5.62}$$

The operator $\hat{m}$ that has the mode index m as eigenvalue is the same as $\hat{L}_{\text{mod},z}/\hbar$, so that

$$\hat{m} = -i\partial_\phi. \tag{5.63}$$

The combination of Eqs. (5.62) and (5.63) gives analytical expressions for the operators

$$\hat{n}_\pm = (\hat{n} \pm \hat{m})/2, \tag{5.64}$$

which have $n_\pm$ as eigenvalues.

We find that the LG modes $u_{p,m}$ are eigenmodes of the operator $\hat{m}$ with eigenvalue m, of the operators $\hat{n}_\pm$ with eigenvalues $p + (|m| \pm m)/2$, and of the operator $\hat{n}$ with eigenvalue $2p + |m|$. In Fig. 5.2 the transformation between these two pairs of mode indices is illustrated.

The radial mode index given by the natural number $p = (n - |m|)/2$ is equal to the number of dark rings in the mode profile [8]. The operator that has this index $\hat{p}$ as eigenvalue can be written as

$$\hat{p} = \frac{1}{2}\left(\hat{n} - \hat{m}_{\text{abs}}\right). \tag{5.65}$$

An explicit differential form for $\hat{n}$ is given in Eq. (5.62). The operator $\hat{m}_{\text{abs}}$ only acts on the ϕ-dependence of a mode function $u(R, \phi, z)$, and it is specified by the property that it has the functions $\exp(im\phi)$ as eigenfunctions, with eigenvalue $|m|$. This operator cannot be simply expressed in a differential form, when it acts on an arbitrary (periodic) function $g(\phi)$. In integral form it is specified by the expression

$$[\hat{m}_{\text{abs}} g](\phi) = \int_0^{2\pi} d\phi' \, M_{\text{abs}}(\phi, \phi') g(\phi'), \tag{5.66}$$

in terms of the integration kernel

$$M_{\text{abs}}(\phi, \phi') = \frac{1}{\pi} \sum_{m=1}^{\infty} m \cos[m(\phi - \phi')]. \tag{5.67}$$

Only in the special case that it acts upon a superposition of LG modes $u_{p,m}$ with only positive (negative) values of m can one replace the operator $\hat{m}_{\text{abs}}$ by the differential operator $-i\partial_\phi$ ($i\partial_\phi$). An expression for the operator $\hat{p}$ follows after substitution of the expressions (5.62) and (5.66) in Eq. (5.65). Expressions for the operator $\hat{p}$ acting in the focal plane of superpositions of LG beams with only positive values of m have been obtained by analytical techniques in Refs. [7–9]. Our algebraic method produces in a compact way operator expressions that are valid for all values of z.

5.6.3 Elliptical Gaussian modes

The expressions (5.56) show that the relation between the circular raising operators $\hat{B}_\pm^\dagger$ and the Cartesian operators $\hat{B}_\xi^\dagger$ and $\hat{B}_\eta^\dagger$ for the HO is analogous to the relation $\mathbf{e}_\pm = (\mathbf{e}_x \pm i\mathbf{e}_y)/\sqrt{2}$ between the circular and linear polarization vectors. Furthermore it is well known that all possible unitary polarization vectors in the xy-plane can be mapped on a unit sphere, which is termed the Poincaré sphere. The polarization vector corresponding to the polar angle θ and azimuthal angle ϕ is given by the expression

$$\mathbf{e}(\theta, \phi) = e^{-i\phi/2} \cos\left(\frac{\theta}{2}\right) \mathbf{e}_+ + e^{i\phi/2} \sin\left(\frac{\theta}{2}\right) \mathbf{e}_-. \tag{5.68}$$

At the poles ($\theta = 0$ or $\theta = \pi/2$) the polarization is circular, and points at the equator ($\theta = \pi/4$) correspond to linear polarization, with a direction that varies with the azimuthal angle ϕ.

Points between the poles and the equator represent elliptical polarization. Opposite points on the sphere correspond to two orthogonal polarizations.

The unit sphere also serves to map the states of a spin 1/2, when we substitute in Eq. (5.68) the states 'spin up' for $\mathbf{e}_+$, and 'spin down' for $\mathbf{e}_-$. The point (θ, ϕ) on the sphere specifies the direction of the spin in the resulting state. The sphere representing the spin direction is termed the Bloch sphere.

Another analogy of the spherical representation of a two-dimensional state space has been mentioned in Ref. [29]. Here the two basis states represented by the poles of the sphere are the LG modes with radial mode number $p = 0$ and opposite indices $m = \pm 1$. Other points on the sphere represent linear combinations of these basis modes. Points at the equator correspond to HG modes with a varying orientation of the axes. Analogous experiments are performed with states of polarization and with the modes that trace out the same trajectory on the Poincaré sphere and on the two-mode sphere. This leads to the same geometric phase effect for spin and for angular momentum of photons. Spherical representations have also been studied in the case that the two basis states are paraxial modes with non-uniform polarization, and a vectorial vortex on the beam axis [30].

The spherical representation can also be applied to pairs of opposite raising operators, defined by

$$\hat{A}_1^\dagger(\theta, \phi) = e^{-i\phi/2} \cos\left(\frac{\theta}{2}\right) \hat{A}_+^\dagger + e^{i\phi/2} \sin\left(\frac{\theta}{2}\right) \hat{A}_-^\dagger,$$
$$\hat{A}_2^\dagger(\theta, \phi) = -e^{-i\phi/2} \sin\left(\frac{\theta}{2}\right) \hat{A}_+^\dagger + e^{i\phi/2} \cos\left(\frac{\theta}{2}\right) \hat{A}_-^\dagger. \tag{5.69}$$

These pairs are a unitary transformation of the raising operators (5.60) for LG modes. Since the ladder operators $\hat{A}_1$, $\hat{A}_1^\dagger$, $\hat{A}_2$ and $\hat{A}_2^\dagger$ obey the bosonic commutation rules for each point (θ, ϕ) on the sphere, they generate a basis of modes in algebraic form. The expression in Dirac notation reads

$$|u_{n_1, n_2}\rangle = \frac{1}{\sqrt{n_1! n_2!}} \left(\hat{A}_1^\dagger\right)^{n_1} \left(\hat{A}_2^\dagger\right)^{n_2} |u_{0,0}\rangle. \tag{5.70}$$

Recall that both the modes $|u_{n_1, n_2}\rangle$ and the raising operators depend on z. These modes correspond to stationary states of the HO, so that they are self-similar under propagation.

This notation is also convenient to obtain the expansion of these modes on the basis of either the LG modes or the HG modes. The expectation value of the mode operator for orbital angular momentum in these modes is found to be

$$\langle u_{n_1, n_2} | \hat{L}_{\text{mod}, z} | u_{n_1, n_2} \rangle = \hbar (n_1 - n_2) \cos\theta. \tag{5.71}$$

As pointed out already in Ref. [26], these basis sets of modes are intermediate between the LG modes on the poles and the HG modes on the equator. This justifies one to introduce the name Hermite–Laguerre sphere. The basis sets have an elliptical nature, analogous to the elliptical polarization (5.68). Other formulations of these generalized Gaussian modes are given by analytical techniques in elliptical coordinates, under the name Ince–Gaussian beams [31,32]. A more extensive analysis in quantum terms has been given in Ref. [33].

We emphasize that all operators derived in Sections 5.6 have the property that acting on arbitrary solutions of the paraxial equation they produce another solution.

5.7 Schwinger representation of Laguerre–Gauss modes

5.7.1 The Lie algebra su(2)

Among physicists the group SU(2) is probably the best-known Lie group. The group elements are the unitary matrices in two dimensions with determinant 1. They arise after exponentiation of linear combinations of the three matrices $\hat{S}_i$ ($i = 1, 2, 3$) that describe the components of a spin $1/2$. These matrices are equal to half the Pauli matrices. The group of rotation matrices of spins of any (integer or half-integer) value are representations of SU(2). They are spanned by an algebra su(2) of three Hermitian operators $\hat{J}_i$, which form the angular momentum vector $\hat{\mathbf{J}} = (\hat{J}_1, \hat{J}_2, \hat{J}_3)$, and which obey the same commutation rules as the spin matrices $\hat{S}_i$. When we reorder the three operators by the expressions $\hat{J}_\pm = \hat{J}_1 \pm i\hat{J}_2$, and $\hat{J}_0 = \hat{J}_3$, the commutation rules that define the algebra can be summarized in the form

$$[\hat{J}_0, \hat{J}_\pm] = \pm \hat{J}_\pm, \quad [\hat{J}_+, \hat{J}_-] = 2\hat{J}_0. \tag{5.72}$$

The operator $\hat{\mathbf{J}}^2 \equiv \hat{J}_0^2 + (\hat{J}_+\hat{J}_- + \hat{J}_-\hat{J}_+)/2$ is a Casimir operator, which implies that it commutes with the three operators $\hat{J}_\pm$ and $\hat{J}_0$. As is derived in any textbook on quantum mechanics, it follows from the commutation rules (5.72) that the common eigenstates of $\hat{\mathbf{J}}^2$ and $\hat{J}_0$ can be arranged in ladders of $2J + 1$ states $|JM\rangle$. Each state in a ladder has the same eigenvalue $J(J + 1)$ of $\hat{\mathbf{J}}^2$, and an eigenvalue M of $\hat{J}_0$ that can take the values $M = -J$, $M = -J + 1, \ldots, M = J$. The quantum number J can only take non-negative integer or half-integer values $J = 0, 1/2, 1, \ldots$, so that the number $2J + 1$ of dimensions of the resulting representation of the Lie group SU(2) can have the value of any natural number. The operators $\hat{J}_\pm$ act as step operators, with the effect of raising or lowering M by one unit. The action of the operators $\hat{J}_\pm$ and $\hat{J}_0$ on the states $|JM\rangle$ is specified by the equalities

$$\hat{J}_\pm |JM\rangle = \sqrt{J(J+1) - M(M \pm 1)}|JM \pm 1\rangle, \quad \hat{J}_0|JM\rangle = M|JM\rangle. \tag{5.73}$$

Operators in paraxial quantum optics 131

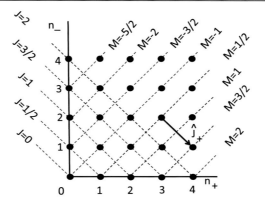

Figure 5.3: Illustration of transformation between mode indices (n_+, n_-) and (J, M).

The Schwinger boson representation of the algebra su(2) is realized by pairs of bosonic operators $\hat{A}_+$, $\hat{A}_-$, $\hat{A}_+^\dagger$, $\hat{A}_+^\dagger$. The su(2) commutation rules (5.72) for the operators

$$\hat{J}_+ \equiv \hat{A}_+^\dagger \hat{A}_-, \quad \hat{J}_- \equiv \hat{A}_-^\dagger \hat{A}_+, \quad \hat{J}_0 \equiv (\hat{A}_+^\dagger \hat{A}_+ - \hat{A}_-^\dagger \hat{A}_-)/2 \tag{5.74}$$

follow immediately from the bosonic commutation rules [34]. The operators $\hat{J}_\pm$ and $\hat{J}_0$ then act on the space of paraxial modes. This space is spanned by the basis of LG modes. In terms of the mode indices n_+ and n_-, the eigenvalues of $\hat{J}_0$ are $M = (n_+ - n_-)/2 = m/2$, which runs from $-n/2$ to $n/2$, when n_+ runs from 0 to n. The number of states in the ladder is $n+1$, so that $J = n/2$. The boson Schwinger construction of the group SU(2) contains every unitary irreducible representation exactly once, so that each LG mode is uniquely labeled by the quantum numbers J and M. The transformation between the mode indices (n_+, n_-) and the quantum numbers (J, M) is shown in Fig. 5.3.

The transformation of the indices (p, m) to the quantum numbers (J, M) takes the simple form

$$p = J - |M|, \quad m = 2M. \tag{5.75}$$

Eq. (5.63) for the mode index operator $\hat{m}$ also determines the form of the operator $\hat{J}_0 = \hat{m}/2$.

Analytical forms of the step operators $\hat{J}_\pm$ follow directly from the defining relations (5.74). By using Eqs. (5.60) we obtain the result

$$\hat{J}_\pm = \frac{1}{4bk} e^{\pm 2i\phi}$$
$$\times \left[k^2 R^2 + 2ikz(R\partial_R \pm i\partial_\phi) + \frac{b^2 + z^2}{R^2} \left(-R^2 \partial_R^2 + \partial_\phi^2 + R\partial_R \pm 2i(1 - R\partial_R)\partial_\phi \right) \right]. \tag{5.76}$$

These rather complicated expressions have been derived without tedious calculations. These operators depend on z, just as the ladder operators (5.60). Their algebraic properties are identical for all values of z.

One might hope that the step operators $\hat{J}_\pm$ change only the mode index m, while leaving the radial index p unchanged. However, this is not true. After applying the transformation (5.75) to the equalities (5.73), we find that the action of $\hat{J}_+$ on the LG modes $u_{p,m}$ takes the form

$$\hat{J}_+ u_{p,-1} = u_{p,1}(p+1),$$
$$\hat{J}_+ u_{p,m} = u_{p-1,m+2}\sqrt{p(p+m+1)} \quad \text{for} \quad m \geq 0,$$
$$\hat{J}_+ u_{p,m} = u_{p+1,m+2}\sqrt{(p+1)(p-m)} \quad \text{for} \quad m \leq -2. \tag{5.77}$$

The operator $\hat{J}_+$ raises the value of m by 2, as expected, but it leaves the value of p unchanged only when $m = -1$. For other negative values of m it raises the index p by 1, while for positive m-values it lowers p by 1. This feature of the raising step operator can be traced back to the fact that in the LG modes the separation between the polar coordinates R and ϕ is not simply labeled by the mode indices p and m. The R-dependent part of the mode varies with the index p as well as with (the absolute value of) the index m. The modes with index $m = -1$ are special in that the raising operator $\hat{J}_+$ does not change the absolute value of m. The action of $\hat{J}_+$ can also be understood by inspection of Figs. 5.2 and 5.3.

5.7.2 The Lie algebra su(1, 1)

The Lie group SU(1, 1) has many applications in quantum optics [35–38]. The corresponding algebra su(1, 1) is spanned by three operators $\hat{K}_i$, with $i = 1, 2, 3$, which can also be represented by the three operators $\hat{K}_\pm \equiv \hat{K}_1 \pm i\hat{K}_2$, $\hat{K}_0 \equiv \hat{K}_3$. The algebra is defined by the commutation rules

$$[\hat{K}_0, \hat{K}_\pm] = \pm \hat{K}_\pm, \quad [\hat{K}_+, \hat{K}_-] = -2\hat{K}_0, \tag{5.78}$$

that differ only slightly from the rules (5.72) for su(2). The Casimir operator, which commutes with all three operators, is $\hat{\mathbf{K}}^2 \equiv \hat{K}_0^2 - (\hat{K}_+\hat{K}_- + \hat{K}_-\hat{K}_+)/2$.

The elements of the group SU(2) are unitary by definition, but the elements of SU(1, 1) are not. Unitary representations of SU(1, 1) are realized by any algebra spanned by three operators $\hat{K}_i$ that obey the commutation rules (5.78), and are Hermitian. Since the Hermitian operators $\hat{K}_0$ and $\hat{\mathbf{K}}^2$ commute, a basis of common eigenstates must exist. Again, the first commutation rule (5.78) ensures that $\hat{K}_\pm$ are step operators, in the sense that they change the eigenvalue of $\hat{K}_0$ by ± 1. Moreover, they leave the eigenvalue of $\hat{\mathbf{K}}^2$ unchanged. By starting from the commutation rules (5.78) one can prove that there are no finite ladders of eigenstates of $\hat{K}_0$. As it will turn out, for the description of LG modes the relevant irreducible representations have half-infinite ladders, where the eigenvalues Q of $\hat{K}_0$ have a lower bound $Q = K$, and no upper bound. This means that $\hat{K}_-$ gives zero when acting on this bottom state.

In analogy to the derivation of the possible representations of SU(2) one can derive from the commutation rules (5.78) that for each positive real value of the bottom eigenvalue K there is a half-infinite ladder of common eigenstates $|K\,Q\rangle$ of $\hat{K}_0$ and $\mathbf{K}^2$. These states all have the same eigenvalue $K(K-1)$ of the Casimir operator, whereas the eigenvalues of $\hat{K}_0$ are $Q = K, Q = K+1, Q = K+2, \ldots$. The action of the operators $\hat{K}_\pm$ on the states $|K\,Q\rangle$ can be specified by the expressions

$$\hat{K}_\pm |K\,Q\rangle = \sqrt{Q(Q \pm 1) - K(K-1)} |K\,Q \pm 1\rangle, \quad \hat{K}_0 |K\,Q\rangle = Q|K\,Q\rangle. \tag{5.79}$$

They give a Hermitian realization of the algebra su(1, 1), and thereby an irreducible unitary representation of the group SU(1, 1).

A Hermitian Schwinger boson representation of the algebra su(1, 1) in terms of two bosonic ladder operators is given by the definitions [34]

$$\hat{K}_+ \equiv \hat{A}_+^\dagger \hat{A}_-^\dagger, \quad \hat{K}_- \equiv \hat{A}_- \hat{A}_+, \quad \hat{K}_0 \equiv (\hat{A}_+^\dagger \hat{A}_+ + \hat{A}_-^\dagger \hat{A}_- + 1)/2. \tag{5.80}$$

They indeed obey the commutation rules (5.78). The LG modes with mode indices (n_+, n_-) are obviously eigenstate of $\hat{K}_0$, with eigenvalue $Q = (n+1)/2$. They are also eigenstates of the Casimir operator $\mathbf{K}^2$ with eigenvalue $(m^2 - 1)/4 = K(K-1)$, with $K = (|m|+1)/2$ the minimal value of the eigenvalue Q of $\hat{K}_0$. For each non-zero value of m there are two ladders with the same value of K, so that the quantum numbers K and Q determine the radial mode index m apart from its sign. The step operators $\hat{K}_\pm$ do not change the value of m. The modes with indices n_+ and n_- for a fixed value of $m = n_+ - n_-$ form a half-infinite ladder of states $|K\,Q\rangle$. The value of Q runs upwards from its minimal value K with unit steps. Each ladder is characterized by a single value of K, which can be half any natural number. Each of the modes with indices (n_+, n_-) is a rung in one of these ladders. The position of the ladders within the grid of mode indices (n_+, n_-) is illustrated in Fig. 5.4.

Differential forms for the step operators $\hat{K}_\pm$ are obtained after substitution of the expressions (5.60) for the ladder operators in Eqs. (5.80). The result is

$$\hat{K}_\pm = \frac{1}{4bk} \left[k^2 R^2 + 2k(iz \mp b)(1 + R\partial_R) + \frac{(iz \mp b)^2}{R^2} \left(R^2 \partial_R^2 + \partial_\phi^2 + R\partial_R \right) \right] \tag{5.81}$$

Notice again that these operators depend on the propagation coordinate z. These expressions are not very illuminating in themselves.

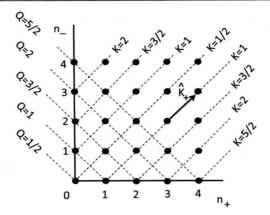

Figure 5.4: Illustration of transformation between mode indices (n_+, n_-) and (K, Q).

Their action on the LG modes $u_{p,m}$ follows from Eqs. (5.79), combined with the relations

$$p = Q - K, \quad |m| = 2K - 1. \tag{5.82}$$

The result is found to be given by the more insightful expressions

$$\hat{K}_+ u_{p,m} = u_{p+1,m}\sqrt{(p+1)(p+|m|+1)},$$
$$\hat{K}_- u_{p,m} = u_{p-1,m}\sqrt{p(p+|m|)}. \tag{5.83}$$

Since the operators $\hat{K}_\pm$ do not change the value of the mode index m, they are pure step operators for the index p. We stress that we have obtained these equations (5.83) by purely algebraic techniques. Similar results have been derived in Refs. [7,8] by an analytical treatment, starting from the explicit expressions for the HG modes. The effect of these step operators on LG modes is also illustrated in Fig. 5.4.

5.8 Conclusions

When the electromagnetic field operators are expanded in modes, the quantum operators for the basic conserved quantities energy, momentum and angular momentum of the radiation field take the form given in Eqs. (5.9) and (5.10). These operators are also generators of the symmetry transformations time evolution, translation and rotation. The quantum nature of these expressions arises from the photon creation and annihilation operators. Moreover, these expressions contain matrix elements of operators acting on mode space.

The paraxial limit of the field operators is justified when they act on states with photons only in paraxial modes. The (annihilation part of the) operators for the electric and the magnetic field can be written in this limit as a mode expansion as in Eq. (5.26) and (5.27). The paraxial modes are given in Eq. (5.23) as the product of a carrier wave $w(z, \omega)$ in the propagation direction, a polarization vector $\mathbf{e}_\nu$ in the transverse plane, and a scalar transverse paraxial mode $\langle \mathbf{R} | u_\mu(z, \omega) \rangle$. We use the Dirac notation only for the latter degree of freedom. The direct connection between quantum operators and mode operators for conserved quantities simplifies in the paraxial limit.

In Section 5.5 the quantum operators for symmetry transformations in the paraxial limit are expressed in terms of mode operators that act on paraxial modes. The quantum operators are given in Eq. (5.48). The mode operators are reminiscent of quantum operators in Schrödinger quantum mechanics. However, these operators are just as classical as the modes themselves, in spite of the fact that the mode operators for orbital and spin angular momentum have non-commuting components and discrete eigenvalues.

Standard Gaussian paraxial modes can be described with algebraic techniques in terms of bosonic ladder operators that act on scalar paraxial modes. The exact correspondence between the states of a plane quantum-mechanical harmonic oscillator and paraxial modes as expressed by Eq. (5.42) is the basis of this description. The ladder operators depend on z, and when they act on an arbitrary solution of the paraxial wave equation, the result is again a solution. In the special case of Gaussian modes the ladder operators provide a convenient way to find operators on paraxial modes that have the mode indices as eigenvalues. This is rather straightforward for the indices n_x and n_y of the HG modes, and more subtle and more interesting for the indices p and m of the LG modes. In particular the radial mode index p has received attention in recent years [7–9,12]. The use of the bosonic ladder operators gives a simple and compact way to derive the form of the operator that has the index p as eigenvalue.

The correspondence between paraxial modes and solutions of the Schrödinger equation of a plane harmonic oscillator is also a powerful tool to obtain explicit expressions for basis sets of paraxial modes which have a common fundamental mode, and which correspond to stationary states of the harmonic oscillator. These basis sets have the HG and the LG modes as limiting cases. Each basis set can be represented as a point on a sphere, in analogy to the Poincaré sphere of polarization. The modes are self-similar, and only scale with the width under propagation.

The algebra of a set of two pairs of bosonic operators can give rise to a representation of the algebra su(2) and also of the algebra su(1, 1) [34]. We apply these Schwinger representations to the LG modes. It is noteworthy that the representation of su(2) produces raising and lowering operators for the mode index m, and the representation of su(1, 1) does the same for the mode index p.

This chapter focusses on the algebraic description of structured paraxial propagation of light. Experiments in quantum communication with photons often use states with only a few photons that form an entangled state. Such states can be created in non-linear crystals. Then a quantum description is needed. When a beam of light with frequency ω_p passes such a crystal, parametric down-conversion can occur, in which absorption of one photon can give rise to the creation of a photon pair, each with frequency $\omega_p/2$ in an entangled state [39]. In a carefully designed setup, the photons travel from the source along different paths to two detectors, where coincident detection takes place. In this way, interference can occur between different histories of the photon pair where different paths from source to the detectors are possible. Information contained in the transverse beam profiles can also plays a crucial role. Paraxial propagation in between the various optical elements is a common ingredient of the process.

References

[1] A. Zeilinger, Experiment and the foundation of quantum physics, Rev. Mod. Phys. 71 (1999) S288–S297.
[2] I. Bialynicki-Birula, Photon wave function, in: E. Wolf (Ed.), Progress in Optics, Vol. 36, Elsevier, Amsterdam, 1996, pp. 245–294.
[3] I. Fernandez-Corbaton, X. Zambrana-Puyalto, G. Molina-Terriza, Helicity and angular momentum: a symmetry-based framework for the study of light–matter interactions, J. Opt. Soc. Am. B 31 (2014) 2136–2141.
[4] A.E. Siegman, Lasers, University Science Books, Mill Valley, CA, 1986.
[5] L. Allen, M.W. Beijersbergen, R.J.C. Spreeuw, J.P. Woerdman, Orbital angular momentum of light and the transformation of Laguerre–Gaussian laser modes, Phys. Rev. A 45 (1992) 8185–8189.
[6] R. Fickler, R. Lapkiewicz, W.N. Plick, M. Krenn, C. Schaeff, S. Ramelov, A. Zeilinger, Quantum entanglement of high angular momenta, Science 338 (2012) 640–643.
[7] E. Karimi, E. Santamato, Radial coherent and intelligent states of paraxial wave equation, Opt. Lett. 37 (2012) 2484–2486.
[8] E. Karimi, R.W. Boyd, P. de la Hoz, H. de Guise, J. Rehácek, Z. Hradil, A. Aiello, G. Leuchs, L.L. Sánchez-Soto, Radial quantum number of Laguerre–Gauss modes, Phys. Rev. A 89 (2014) 063813.
[9] W.N. Plick, M. Krenn, Physical meaning of the radial index of Laguerre–Gauss beams, Phys. Rev. A 92 (2015) 63841.
[10] J. Visser, G. Nienhuis, Interference between entangled photon states in space and time, Eur. Phys. J. D 29 (2004) 301–308.
[11] V.D. Salakhutdinov, E.R. Eliel, W. Löffler, Full-field quantum correlations of spatially entangled photons, Phys. Rev. Lett. 108 (2012) 173604.
[12] D.K. Zhang, X.D. Qiu, W.H. Zhang, L.X. Chen, Violation of a Bell inequality in two-dimensional state spaces for radial quantum number, Phys. Rev. A 98 (2018) 042134.
[13] C. Cohen-Tannoudji, J. Dupont-Roc, G. Grynberg, Photons et Atomes, InterEditions CNRS, Paris, 1987; Photons and Atoms, Wiley, New York, 1992.
[14] L. Mandel, E. Wolf, Optical Coherence and Quantum Optics, Cambridge University Press, 1995.
[15] R. Loudon, The Quantum Theory of Light, 3rd edition, Oxford University Press, 2000.
[16] G. Nienhuis, Conservation laws and symmetry transformations of the electromagnetic field with sources, Phys. Rev. A 93 (2016) 023840.
[17] S.J. van Enk, G. Nienhuis, Commutation rules and eigenvalues of spin and orbital angular momentum of radiation fields, J. Mod. Opt. 41 (1994) 963–977.

[18] S.M. Barnett, Rotation of electromagnetic fields and the nature of optical angular momentum, J. Mod. Opt. 57 (2010) 1339–1343.
[19] S.J. van Enk, G. Nienhuis, Spin and orbital angular momentum of photons, Europhys. Lett. 25 (1994) 497–501.
[20] A. Aiello, J. Visser, G. Nienhuis, J.P. Woerdman, Angular spectrum of quantized light beams, Opt. Lett. 31 (2006) 525–527.
[21] H.A. Haus, Waves and Fields in Optoelectronics, Prentice Hall, Englewood Cliffs, NJ, 1984.
[22] M. Lax, W.H. Louisell, W.B. McKnight, From Maxwell to paraxial wave optics, Phys. Rev. A 11 (1975) 1365–1370.
[23] I.H. Deutsch, J.C. Garrison, Paraxial quantum propagation, Phys. Rev. A 43 (1991) 2498–2513.
[24] D.L. Andrews, K.A. Forbes, Quantum features in the orthogonality of optical modes for structured and plane-wave light, Opt. Lett. 43 (2018) 3249–3252.
[25] H. Kogelnik, T. Li, Laser beams and resonators, Appl. Opt. 5 (1966) 1550–1567.
[26] G. Nienhuis, L. Allen, Paraxial wave optics and harmonic operators, Phys. Rev. A 48 (1993) 656–665.
[27] A. Messiah, Mécanique Quantique, Dunod, Paris, 1964.
[28] G. Nienhuis, J. Visser, Angular momentum and vortices in paraxial beams, J. Opt. A, Pure Appl. Opt. 6 (2004) S248–S250.
[29] M.J. Padgett, J. Courtial, Poincaré-sphere equivalent for light beams containing orbital angular momentum, Opt. Lett. 24 (1999) 430–432.
[30] G. Milione, H.I. Sztul, D.A. Nolan, R.R. Alfano, Higher-order Poincaré sphere, Stokes parameters, and the angular momentum of light, Phys. Rev. Lett. 107 (2011) 053601.
[31] M.A. Bandres, J.C. Gutiérrez-Vega, Ince-Gaussian beams, Opt. Lett. 29 (2004) 144–146.
[32] M.A. Bandres, J.C. Gutiérrez-Vega, J. Opt. Soc. Am. A 21 (2004) 873–880.
[33] W.M. Plick, M. Krenn, R. Fickler, S. Ramelov, A. Zeilinger, Quantum orbital angular momentum of elliptically symmetric light, Phys. Rev. A 87 (2013) 033806.
[34] J. Schwinger, On angular momentum, in: L. Biedenharn, H. van Dam (Eds.), Quantum Theory of Angular Momentum, Academic Press, New York, 1965, p. 229.
[35] B. Yurke, S.L. McCall, J.R. Klauder, SU(2) and SU(1, 1) interferometers, Phys. Rev. A 33 (1986) 4033–4054.
[36] C.G. Gerry, Phase operators for SU(1, 1): application to the squeezed vacuum, Phys. Rev. A 38 (1986) 1734–1738.
[37] A. Vourdas, SU(2) and SU(1, 1) phase states, Phys. Rev. A 41 (1990) 1653–1661.
[38] G.S. Agarwal, J. Bannerji, Reconstruction of SU(1, 1) states, Phys. Rev. A 64 (2001) 023815.
[39] C.K. Hong, L. Mandel, Theory of parametric frequency down-conversion of light, Phys. Rev. A 31 (1985) 2409.

CHAPTER 6

Quantum cryptography with structured photons

Alicia Sit, Felix Hufnagel, and Ebrahim Karimi
University of Ottawa, Department of Physics, Ottawa, ON, Canada

Contents

6.1 Introduction 139
6.2 Generation and detection 143
    6.2.1 Polarization 143
    6.2.2 Holography 145
    6.2.3 Pancharatnam–Berry optical elements 147
6.3 High-dimensional quantum information 149
    6.3.1 Optimal quantum cloning 149
    6.3.2 Protocols 152
    6.3.3 Quantum process tomography 155
6.4 Quantum key distribution implementations 159
    6.4.1 Optical fiber 159
    6.4.2 Free-space 162
    6.4.3 Underwater 167
6.5 Conclusion 173
Acknowledgment 173
References 173

> *Human subtlety will never devise an invention more beautiful, more simple or more direct than does nature because in her inventions nothing is lacking, and nothing is superfluous.*
>   ***Leonardo da Vinci***

6.1 Introduction

Electromagnetic waves are widely used in classical and quantum communications since they propagate with the fastest possible speed and barely interact with the environment. The latter arises from the fact that electromagnetic waves—and their quanta, photons, in the quantum regime—do not possess a rest mass or electric charge. Even a hundred years after the invention of the radio, modulating the frequency or amplitude (number of photons) of electromagnetic waves is a preferred method to broadcast and share information around the globe and between satellites and ground stations. With access to proper antennae, anyone can receive the signal, but not necessarily the information. In addition to simple modulations, the signal is often encrypted based on complex and hard mathematical problems, e.g. finding prime numbers. Decrypting the information—i.e. the ciphertext, cypher-voice or cypher-video—would require a backdoor (such as using proper protocols and a key, e.g. the Rivest–Shamir–Adleman protocol [1], or extremely fast computers, which are computationally expensive tools. The latter might require several years, or even centuries, to be accomplished when the information is not anymore private in any case. However, the supremacy of quantum computers in outperforming classical computers places our current communication and encryption schemes at risk.

A simple example is Morse code, wherein all letters, numbers and prosigns are expressed in terms of dots •, dashes —, and spaces; American and German Morse code are two variations. A predefined key allows two parties to encode and decode (share) a message securely. However, decrypting the code by checking all permutations of dots and dashes within all possible languages is just a matter of time and is feasible even with a personal computer. In 1984, Bennett and Brassard proposed a revolutionary method to share a key to encode/decode information using single photons [2]. The technique is referred to as BB84 nowadays. Using the laws of quantum mechanics, i.e. superposition and contextuality, BB84 is able to generate and share a key between two parties, namely Alice and Bob, in such a way that it is protected from eavesdroppers. Indeed, the presence of an eavesdropper, namely Eve, attempting to obtain the information will result in detectable errors. Thus, Alice and Bob can prevent Eve from obtaining their secret key by monitoring the amount of errors within their shared key and imposing a threshold on said errors. BB84, as well as several other techniques, will be discussed in detail in Section 6.3.2.

Frequency and photon number are not the only parameters that determine a photon's quantum state nor the only ones that can be used to encode information. As a solution to Maxwell's equations, photons possess a vectorial nature: a polarization that describes how the electric and magnetic fields oscillate in space and time upon the propagation. Photons are transverse entities, which means that the electric and magnetic fields are in a plane that is orthogonal to the propagation direction. Therefore, only two independent vectors are required to define the polarization state of a single photon, for example, $|L\rangle_\pi$ (left-handed) and $|R\rangle_\pi$ (right-handed) in the circular polarization basis. A photon's spatial mode, satisfying the wave equation, is as well quantized. Depending on the choice of the coordinates, i.e. symmetry and boundary conditions, the quantization parameters are either continuous or discrete, and can be bounded or even unbounded. For instance, in Cartesian coordinates with the boundaries set at infinity, the photon's mode creation (annihilation) operator is labeled by $\hat{a}^\dagger_{\omega,\pi,k_x,k_y}$ ($\hat{a}_{\omega,\pi,k_x,k_y}$), and consequently the photon state $|\psi\rangle$ is given by

$$|\psi\rangle = \int d\omega \int dk_x \int dk_y \sum_{N,\pi} c^{N,\pi}(\omega, k_x, k_y) |N, \omega, \pi, k_x, k_y\rangle, \qquad (6.1)$$

with $c^{N,\pi}(\omega, k_x, k_y) = C^{N,\pi}(\omega, k_x, k_y) \exp\left(i z \sqrt{\left(\frac{\omega}{c}\right)^2 - k_x^2 - k_y^2}\right)$, where $C^{N,\pi}(\omega, k_x, k_y)$ is the expansion coefficient, N is the number of photons, k_x and k_y are real parameters defining the transverse wavevector. There are different notations used in the quantum optics community; here, we use the following interchangeably: $|N\rangle_{\omega,\pi,k_x,k_y} := |N, \omega, \pi, k_x, k_y\rangle = |N\rangle \otimes |\omega\rangle \otimes |\pi\rangle \otimes |k_x\rangle \otimes |k_y\rangle$, indicating N photons at frequency ω in a polarization state of π, and transverse wavevector of k_x and k_y. The time t and position $\mathbf{r}$ representation of $|\omega\rangle$, $|k_x\rangle$ and $|k_y\rangle$ are, respectively, $\langle t|\omega\rangle = \exp(-i\omega t)$ and $\langle \mathbf{r}|k_x, k_y\rangle = \exp(i(k_x x + k_y y))$. Eq. (6.1) is not the only representation of $|\psi\rangle$; indeed, the photon state can be expressed in terms of any complete or over-complete set of modes. For instance, one may use Hermite–Gauss modes, $\langle \mathbf{r}|m, n\rangle =: \mathrm{HG}_{m,n}(\mathbf{r})$ to express the photon state in the Cartesian coordinates, i.e.,

$$|\psi\rangle = \int d\omega \sum_{N,\pi,m,n} c^{N,\pi}_{m,n}(\omega) |N, \omega, \pi, m, n\rangle, \qquad (6.2)$$

where m, n are positive integers, indicating the photon spatial distribution along the x and y direction, respectively. HG modes are mutually orthogonal solutions to the scalar paraxial wave equation that form a complete basis, i.e. $\langle m, n|m', n'\rangle = \delta_{m,m'}\delta_{n,n'}$ and $\sum_{m,n} |m, n\rangle\langle m, n| = \mathbb{1}$, where $\delta_{i,j}$ and $\mathbb{1}$ are the Kronecker delta and identity operator, respectively.

Due to the cylindrical symmetry of laboratory optics, such as lenses, mirrors, etc., it is often more convenient to use (power) normalized and complete modes that also hold cylindrical symmetry. The Laguerre–Gauss mode set, $\langle \mathbf{r}|p, \ell\rangle = \mathrm{LG}_{p,\ell}(\mathbf{r})$, is an example that is widely used in the quantum optics community. The integer indices, $p > 0$ and ℓ, determine the photon probability and phase distributions along the radial ρ and azimuthal ϕ coordinates, respectively. The photon state is invariant under a 2π rotation, and thus the azimuthal index ℓ is integer, and the position representation of the state is $\exp(i\ell\phi)$. Photons possessing this helical wavefront, i.e. $\exp(i\ell\phi)$, carry $\ell\hbar$ units of angular momentum per photon along the propagation direction; $\hbar$ is the reduced Planck constant. The photon state in cylindrical coordinates (ρ, ϕ, z) is given by

$$|\psi\rangle = \int d\omega \sum_{N,\pi,p,\ell} c^{N,\pi}_{p,\ell}(\omega) |N, \omega, \pi, p, \ell\rangle. \qquad (6.3)$$

142   Chapter 6

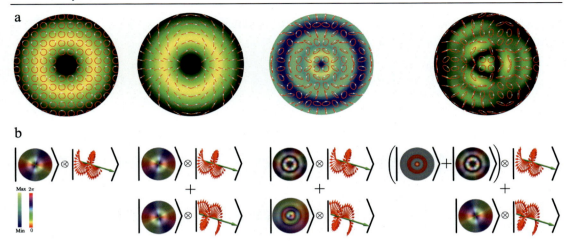

Figure 6.1: *A few examples of structured photons.* (a) Intensity and polarization distributions of photons in superposition of different spatial and polarization modes are shown. Their decomposition in the Laguerre–Gauss modes and the circular polarization basis is shown in (b).

It is noteworthy that the frequency ω, the continuous associate conjugate of time, can be linked to the longitudinal component of the coordinates $(z - ct)$; c is the speed of light in vacuum. Thus, the photon state along the propagation direction can be expressed in terms of the longitudinal modes, cf. [3]. Photons in the above superposition state are known as *structured photons*, and can be categorized into several different classes. Fig. 6.1 shows polarization, probability and phase distributions of a few examples of structured photons in the cylindrical coordinate system. Learning how to generate, manipulate and detect all these photonic degrees of freedom, i.e. photon number, frequency, polarization and spatial modes, provides ways to encode more information onto a single information carrier (photon). In addition to increasing the communication channel capacity, employing all (or some of) these degrees of freedom enhances the communication security. This will be shown and discussed in Section 6.3.2.

In the paraxial regime, polarization lies in a plane and thus can be given as the superposition of two basis vectors, $\{|L\rangle, |R\rangle\}$ or $\{|H\rangle, |V\rangle\}$ or $\{|A\rangle, |D\rangle\}$—L, R, H, V and A, D stand for left-circular, right-circular, horizontal-linear, vertical-linear, antidiagonal-linear and diagonal-linear polarization states. The polarization degree of freedom provides a vector space to generate an arbitrary qubit state, $|\psi\rangle_\pi = \alpha |L\rangle + \beta |R\rangle$, where α and β are arbitrary complex numbers with $|\alpha|^2 + |\beta|^2 = 1$. Polarization qubits can be generated and sorted by means of polarization optics, i.e. waveplates and polarizers. Spatial modes, in contrast, provide an unbounded vector space. The radial p and azimuthal ℓ indices of Laguerre–Gauss, the orthonormal modes in the cylindrical coordinates, span positive integer and integer vector spaces, and thus individually provide the unbounded vector space $|p, \ell\rangle$. Both radial and azimuthal indices of LG modes can be separately used to generate photonic qudits, i.e.,

$$|\psi\rangle_{\text{radial qudit}} = a_0 |p=0\rangle + a_1 |p=1\rangle + a_2 |p=2\rangle + \ldots,$$
$$|\psi\rangle_{\text{azimuthal qudit}} = \ldots + b_{-1} |\ell=-1\rangle + b_0 |\ell=0\rangle + b_1 |\ell=+1\rangle + \ldots, \qquad (6.4)$$

where a_i and b_j are complex parameters. A general spatial mode qudit is then

$$|\psi\rangle_{\text{spatial qudit}} = \sum_{p,\ell} c_{p,\ell} |p, \ell\rangle$$
$$= \ldots + c_{0,-1} |0, -1\rangle + c_{0,0} |0, 0\rangle + c_{1,1} |+1, +1\rangle + \ldots, \qquad (6.5)$$

where c_i are complex numbers normalized to 1, $\sum_{p,\ell} |c_{p,\ell}|^2 = 1$. Combining polarization and spatial modes doubles the dimension of the vector space and the amount of the information that a photon can carry. Generation and detection of photons possessing specific spatial modes will be the subject of Section 6.2. The above formalism is applied to discrete vector spaces, i.e. polarization, spatial modes, as well as photon number. Frequency and wavevector Hilbert spaces, instead, are continuous. Different formalisms are used to describe and use continuous variable photonic degrees of freedom, which will not be covered in this chapter.

In the context of quantum communication, these photonic quantum states are generated at a sender (Alice), transmitted through a channel and detected at a receiver (Bob). One may use different channels for communication and sharing the information, for example, optical fibers, free-space and underwater channels, or any other medium that connects Alice and Bob together. The photonic degrees of freedom (polarization, frequency and spatial modes) that are used for sharing the information might be affected and altered by the communication channel. For instance, fibers alter all of the above degrees of freedom, while only a few might be affected in the free-space propagation. In Section 6.4, the possibility of sharing information using structured photonics via optical fibers, free-space and underwater channels will be discussed. In addition, we explore a few trivial attacks (such as intercept-resend and optimal quantum cloning) to a communication channel in Section 6.3, and will analyze the introduced errors to the communication system.

6.2 Generation and detection

6.2.1 Polarization

There are several different approaches to generating and detecting photonic orbital and spin angular momentum states, which apply to both classical (coherent) and quantum (single and

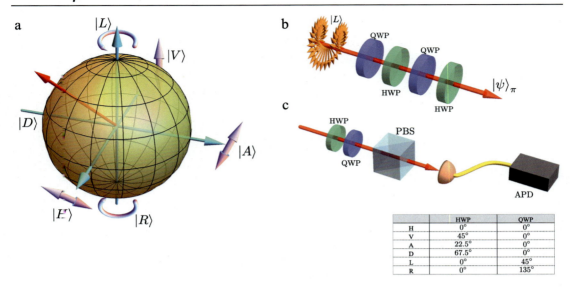

Figure 6.2: *Generation and detection of photonic polarization state.* (a) Polarization vector shown on the Poincaré sphere. (b) Combination of quarter-wave (QWP) and half-wave plates (HWP) which perform an arbitrary SU(2) rotation on the Poincaré sphere. Note that a QWP and HWP are sufficient to move a given state to an arbitrary polarization state. (c) A QWP and a HWP followed by a polarizing beamsplitter (PBS) and detector allow one to project a given polarization state onto an arbitrary polarization state. The table gives the needed angles to perform quantum state tomography.

entangled photons) regimes. We will start with a spin angular momentum (polarization) quantum state i.e. polarization qubit state $|\psi\rangle_\pi = \alpha |L\rangle + \beta |R\rangle$. Let us assume that a single photon is in the horizontal polarization state $|H\rangle$. This can be generated by setting a polarizer at $0°$, e.g. along the horizontal axis in the lab frame; thus, whenever a photon traverses through the polarizer, its polarization state is set to horizontal $|H\rangle$. The combination of a half-wave plate (HWP) and a quarter-wave plate (QWP) will provide the needed unitary operators to move the polarization state $|H\rangle$ to an arbitrary state on the Poincaré sphere; see Fig. 6.2. Here, the axes of the Poincaré sphere are such that H/V lies along the S_1 axis, A/D along the S_2 axis, and L/R along the S_3 axis; any pure polarization state resides on the surface of the sphere. The combination of two quarter-wave and two half-wave plates, in the sequence of $\text{QWP}_{90°}\text{HWP}_{-\gamma/4}\text{QWP}_{0°}\text{HWP}_{90°+\delta/4}$, can also be used to perform arbitrary polarization transformations for a subset of input states; here, the waveplate angles are given in the subscript. This sequence of waveplates rotates the input polarization state around the S_2-axis by an angle γ and then around the S_3-axis by an angle δ. For instance, it transforms $|L\rangle$ to be,

$$|L\rangle \xrightarrow{\text{QHQH}} \exp\left(-i\frac{\delta}{4}\right)\cos\left(\frac{\gamma}{2}\right)|L\rangle + \exp\left(i\frac{\delta}{4}\right)\sin\left(\frac{\gamma}{2}\right)|R\rangle, \qquad (6.6)$$

which is exactly the polarization qubit $|\psi\rangle_\pi$ with,

$$\alpha = \exp\left(-i\frac{\delta}{4}\right)\cos\left(\frac{\gamma}{2}\right), \text{ and } \beta = \exp\left(i\frac{\delta}{4}\right)\sin\left(\frac{\gamma}{2}\right).$$

An inverted process, followed by a quantum (single-photon detector) or classical (power meter) detector, can be used to detect photonic polarization states. For instance, a polarizer set at $0°$, $45°$, $90°$ or $135°$ (followed by the proper detector) can be used to project the polarization onto a linearly horizontal $|H\rangle$, linearly anti-diagonal $|A\rangle$, linearly vertical $|V\rangle$ or linearly diagonal $|D\rangle$ polarization states, respectively. Alternatively, in order to project a polarization quantum state onto an arbitrary polarization state, a set of QWP and HWP followed by a polarizing beam splitter is required; this ensures a constant loss due to reflection from the optical elements. Setting the angles of the HWP and QWP defines the state projection on the Poincaré sphere. Fig. 6.2(c) shows the configuration and settings of the QWP and HWP angles for polarization state tomography [4]. Polarization state tomography, which will be discussed later, is a process used to reconstruct and determine an unknown polarization state by means of a set of measurements.

6.2.2 Holography

Unlike polarization that is related to the vectorial nature of a photon, spatial modes determine its spatial phase and intensity (probability) distributions. One of the easiest methods to generate a light beam with a well-defined spatial mode is with holography. In holography, the desired spatial mode $E_{\text{des}}(\mathbf{r})$ is interfered with a reference beam (in most cases, a tilted Gaussian or plane wave), i.e. $E_{\text{ref}}(\mathbf{r}) \exp(i2\pi x/\Lambda)$; here, Λ defines the tilting angle of the reference beam. The interference fringes, i.e. intensity,

$$I(\mathbf{r}) = (E_{\text{des}}(\mathbf{r}) + E_{\text{ref}}(\mathbf{r})\exp(i2\pi x/\Lambda)) \cdot (E_{\text{des}}(\mathbf{r}) + E_{\text{ref}}(\mathbf{r})\exp(i2\pi x/\Lambda))^*,$$

can be achieved offline (via computer) without actual generation of both the reference and desired beams in the laboratory. The computer generated hologram, sometimes referred to as a kinoform, either will be used directly as an amplitude $I(\mathbf{r})$, phase $\exp(i2\pi I(\mathbf{r}))$ or amplitude–phase mask to generate the desired beam when it is illuminated with the reference beam in the laboratory. The generated hologram, due to the presence of $\exp(i2\pi x/\Lambda)$, is periodic in space along the x-direction. When the hologram is illuminated with the reference beam, the diffraction pattern is generated in the far-field, where the first order of diffraction possesses the desired beam information; in the case of an amplitude hologram, it is the desired field. Between amplitude, phase, and amplitude–phase holograms, phase holograms have the highest efficiency at the first order of diffraction. However, it only carries the desired beam's phase

146 Chapter 6

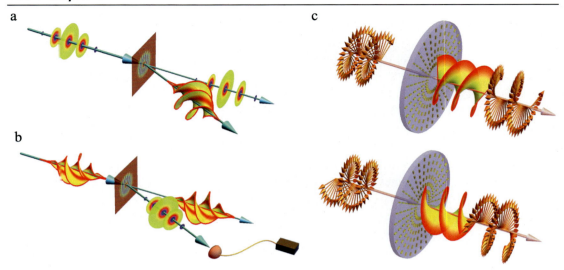

Figure 6.3: Generation and detection of photonic spatial modes. (a) A portion of the Gaussian beam is converted into a LG mode with radial and azimuthal indices of $p = 0$ and $\ell = +3$. The generated LG mode is diffracted at the zero-order of diffraction. (b) A hologram in combination with a single mode optical fiber can be used to detect a specific transverse mode. In this case, the hologram in (a) is used to detect the LG mode with azimuthal index of $\ell = -3$. (c) Action of $q = +1$-plate in the circular polarization basis.

information. Different types of periodic functions result in different efficiencies at the first order of diffraction—see Ref. [5] for more details. Here, we focus on the most efficient and practical way to generate a light beam with a given spatial profile $E_{\text{des}}(\mathbf{r}) = A(\mathbf{r})\exp(i\chi(\mathbf{r}))$, where $A(\mathbf{r})$ and $\exp(i\chi(\mathbf{r}))$ are the beam amplitude and phase, respectively. The phase hologram $\exp(i\,\text{Hologram}(\mathbf{r}))$, where,

$$\text{Hologram}(\mathbf{r}) = \mathcal{M}(\mathbf{r})\,\text{Mod}\left(\chi(\mathbf{r}) - \pi\mathcal{M}(\mathbf{r}) + \frac{2\pi x}{\Lambda}, 2\pi\right), \quad (6.7)$$

generates a photon with $A(\mathbf{r})\exp(i\chi(\mathbf{r}))$ spatial profile at the first order of diffraction when it is illuminated with a uniform light beam. Here, $\mathcal{M}(\mathbf{r}) = \left(1 + \frac{1}{\pi}\text{sinc}^{-1}(A(\mathbf{r}))\right)$, sinc^{-1} is the inverse of the sinc function in the interval of $[0, \pi]$, and $\text{Mod}(a, b)$ is the remainder of a/b [6]. Fig. 6.3(a) shows a hologram that generates a LG mode with radial and azimuthal indices 0 and 3 when it is illuminated with a Gaussian beam.

In the laboratory, these holograms can be created dynamically by means of a phase-only *spatial light modulator* (SLM), a liquid-crystal based device. When a light beam with a given polarization traverses through (or gets reflected), it gains a spatial phase which is determined by the spatial resolution of the SLM and the applied voltage to each pixel. After the SLM, the mode is generated at the first order of diffraction and must be spatially isolated from the other orders, which is usually done at the far-field. This means that the hologram (plus spatial filtering) action is not unitary. Thus, although the mode quality is high, most of the power (photons) is lost in this technique. Similar to the case of polarization, the combination of a hologram that is designed to generate a specific spatial state and a single mode optical fiber can be used to project an unknown spatial profile onto a specific spatial mode state; see Fig. 6.3(b). This technique, known as phase flattening, was first proposed and tested in determining bi-photon OAM quantum states generated in spontaneous parametric down conversion (SPDC) [7]. The phase flattening technique, due to diffraction in free-space, is biased towards a specific quantum state which is determined by the geometry of the experimental apparatus [8], and in some cases (when the radial mode needs to be detected) causes significant crosstalk. Intensity flattening, where the single mode optical fiber probes a solid angle much wider than the hologram surface, can be used to improve the quality of state projection [9]. It greatly reduces the crosstalk between different modes, but at the expense of a huge decrease in the detection efficiency. Whenever it is needed to determine the OAM spectrum (or its Fourier conjugate), an ingenious device, referred to as an OAM sorter, is particularly useful [10]. OAM sorters can reach higher than 80% efficiency in sorting different OAM with a crosstalk below 5% when it is used in combination with a fan-out [11].

SLMs are very slow, typically below a 1 kHz refresh rate, since they are based on nematic liquid-crystals which have a response time of a few milliseconds to external electric or magnetic fields. Therefore, their use in the generation and detection of photonic spatial modes is limited to ≤ 1 kHz. Digital Micromirror Devices (DMDs), which act as amplitude only holograms with arrays of micromirrors that can individually be either *on* or *off*, are suggested for a quicker modulation of spatial modes, one order of magnitude higher than SLMs, i.e. 10 kHz. However, DMDs have a much lower diffraction efficiency ($\simeq 10\%$ at the first order of diffraction) and are incapable of full-control of the spatial modes. The speed of these devices to control spatial modes is much lower than current technologies for controlling polarization states, which can be controlled at GHz speed via suitable electro-optic devices. Thus, there is a need for a faster control of spatial modes.

6.2.3 Pancharatnam–Berry optical elements

Pancharatnam–Berry Optical Elements (PBOEs), including dielectric [12], liquid crystal [13] and plasmonic [14], have been created to couple polarization and spatial modes. Such a coupling allows one to modulate the input polarization state with GHz speed, and, consequently, modulate the spatial mode with GHz speed. Among these PBOEs, q-plates (the liquid-crystal version) are widely used by different research teams in many classical and quantum optics experiments. Its birefringence can be easily adjusted by applying an external electric field, thus making it tunable and functional for different wavelengths. q-plates, and PBOEs in general, are spatially structured birefringent plates, where the axes of birefringence, α, varies across the plate, i.e. $\alpha := \alpha(\mathbf{r}_\perp)$; here, $\mathbf{r}_\perp$ stands for the transverse coordinates. For PBOEs, $\alpha(\mathbf{r}_\perp)$ is an arbitrary function of the transverse coordinates; while for a q-plate, α is a function of the azimuthal angle ϕ in polar coordinates, i.e. $\alpha(\mathbf{r}_\perp) = q\phi + \alpha_0$. Here, the continuity of the birefringent optical axes dictates that q (the topological charge of the plate) is a half integer, and α_0 is the angle of the liquid crystal molecules at $\phi = 0$, i.e. along the x-axis.

When a q-plate is illuminated with a circularly polarized photon/light beam, the outgoing beam becomes

$$\begin{pmatrix} |L\rangle \\ |R\rangle \end{pmatrix} \xrightarrow{q-\text{plate}} \cos\left(\frac{\delta}{2}\right) \begin{pmatrix} |L\rangle \\ |R\rangle \end{pmatrix} + i \sin\left(\frac{\delta}{2}\right) \begin{pmatrix} 0 & e^{i2\alpha(\mathbf{r}_\perp)} \\ e^{-i2\alpha(\mathbf{r}_\perp)} & 0 \end{pmatrix} \cdot \begin{pmatrix} |L\rangle \\ |R\rangle \end{pmatrix}, \quad (6.8)$$

where δ is the plate birefringence. For a PBOE with half-wave retardation $\delta = \pi$, the plate flips the handedness of the input beam polarization and adds a global phase of $|2\alpha(\mathbf{r}_\perp)|$. This can be easily understood using the geometrical picture by means of state evolution on the Poincaré sphere. The action of this plate on circular polarization states can be illustrated as a pole-to-pole path on the Poincaré sphere, e.g. left-handed circular polarization $|L\rangle$ is adiabatically switched to right-handed circular polarization. When the orientation of the plate is rotated, say by $\alpha(\mathbf{r}_\perp)$, the plate action on the input left-handed polarization state is still a pole-to-pole trajectory on the Poincaré sphere, but on a different path. The initial and final states of polarization on the Poincaré sphere for these two cases are identical, but the states evolve on different paths. Thus, the final state gains a geometrical phase [15], which in this case is half of the solid angle probed by these two different trajectories, i.e. $|2\alpha(\mathbf{r}_\perp)|$. An example of a q-plate with topological charge of $q = 1$, and its action in the circular polarization basis, is shown in Fig. 6.3(c). Indeed, the output photon/beam gains a $\pm 2\alpha(\mathbf{r}_\perp)$ phase depending on its input polarization state. For a q-plate, the output beam phase-front is helical and given by $\pm 2q\phi$—the constant $2\alpha_0$ is not considered here. Therefore, the output beam possesses an OAM state of $|\ell = \pm 2q\rangle$, depending on its input polarization state, $|L\rangle \xrightarrow[\delta=\pi]{q-\text{plate}} |R\rangle|+2q\rangle$, and $|R\rangle \xrightarrow[\delta=\pi]{q-\text{plate}} |L\rangle|-2q\rangle$. The fast switching of polarization states results in a fast generation or detection of spatial modes using PBOEs—this can reach to GHz depending on the available technologies for polarization manipulation.

PBOEs are linear devices, and thus preserve coherence. Whenever they are fed with a coherent superposition of left- and right-handed circular polarization, the generated state is in a coherent superposition of $|L\rangle|2\alpha(\mathbf{r}_\perp)\rangle$ and $|R\rangle|-2\alpha(\mathbf{r}_\perp)\rangle$. This coupling between polarization and spatial modes can generate very interesting structured photon states; e.g. for a tuned

($\delta = \pi$) q-plate, they are cylindrical vector beams, i.e.,

$$|\psi\rangle = \alpha |L\rangle + \beta |R\rangle \xrightarrow[\delta=\pi]{q-\text{plate}} |\psi\rangle_{\text{spin-orbit}} = \alpha |R\rangle |+2q\rangle + \beta |L\rangle |-2q\rangle. \quad (6.9)$$

The above state (for a single photon) is known as single-photon spin–orbit state [16]. A $q = 1/2$-plate transfers the linear polarization states $|H\rangle$ and $|V\rangle$ into radially and azimuthally polarized states—we assume $\alpha_0 = 0$. Proper waveplates allow for the manipulation of the output polarization state in Eq. (6.9); consequently, the vector space can be extended to $\{|L\rangle, |R\rangle\} \otimes \{|+2q\rangle, |-2q\rangle\}$. The applications of these states in high-dimensional quantum cryptography are discussed in Section 6.4. When the plate retardation is not set at $\delta = \pi$, the output beam contains a much richer polarization topology, and the vector space is extended to $\{|L\rangle, |R\rangle\} \otimes \{|+2q\rangle, |0\rangle, |-2q\rangle\}$, which is a six-dimensional space.

6.3 High-dimensional quantum information

6.3.1 Optimal quantum cloning

The world of quantum mechanics is very different from the classical counterpart. These differences are not limited to the philosophical aspects, such as its indeterminism or non-local nature, but to physical limitations and puzzles, among others, such as the uncertainty principle and measurement problem. One of these fundamental differences is the no-cloning theorem, which is one of the building blocks of quantum information sciences. In classical physics, it is easy to perfectly copy information—it is just a matter of precision and time since the state of a classical system can be determined perfectly. Determining a quantum state is impossible in the quantum world, and our measurements are limited by the Heisenberg uncertainty principle. This is the basics for an important no-go theorem in quantum mechanics: the *no-cloning theorem*. No-go theorems are theorems stating that specific actions are prohibited by the laws of physics. The no-cloning theorem states that it is impossible to make a perfect copy of an arbitrary quantum state. It was mathematically proven almost 40 years ago, in 1970, by Park [17], and also was independently discussed for a particular case by Giancarlo Ghirardi [18] when Asher Peres and Giancarlo Ghirardi reviewed a manuscript [19] for Foundations of Physics. Later on, in 1982, William Wootters and Wojciech Zurek ingeniously showed that there are no unitary operators capable of cloning an arbitrary (pure) quantum state [20].

Consider a unitary operator $\hat{U}_{\text{clone}}$ capable of cloning an arbitrary pure quantum state $|\psi\rangle$ using an ancillary quantum state $|\text{ancilla}\rangle$, i.e., $\hat{U}_{\text{clone}} \cdot (|\psi\rangle \otimes |\text{ancilla}\rangle) = (|\psi\rangle \otimes |\psi\rangle)$. This operator must be independent of the state $|\psi\rangle$, and should work for any other arbitrary (non-orthogonal) state $|\phi\rangle$, i.e. $\langle \phi | \psi \rangle \neq 0$, which means $\hat{U}_{\text{clone}} \cdot (|\phi\rangle \otimes |\text{ancilla}\rangle) = (|\phi\rangle \otimes |\phi\rangle)$.

Since $\hat{U}_{\text{clone}}$ is unitary, $\hat{U}_{\text{clone}} \cdot \hat{U}_{\text{clone}}^{-1} = \hat{U}_{\text{clone}}^{-1} \cdot \hat{U}_{\text{clone}} = 1$, we have

$$((\langle\text{ancilla}| \otimes \langle\psi|) \hat{U}_{\text{clone}}^{-1} \cdot \hat{U}_{\text{clone}} (|\phi\rangle \otimes |\text{ancilla}\rangle)) = |\langle\psi|\phi\rangle|^2,$$
$$|\langle\psi|\phi\rangle| = |\langle\psi|\phi\rangle|^2. \tag{6.10}$$

The above identity, Eq. (6.10), holds only for two specific cases: (1) when $|\phi\rangle$ is orthogonal to $|\psi\rangle$, i.e. $\langle\psi|\phi\rangle = 0$, and (2) when $|\phi\rangle$ is identical to $|\psi\rangle$, i.e. $\langle\psi|\phi\rangle = 1$. Thus, there is no such unitary operator that makes a perfect copy of an arbitrary quantum state. One may reach the same conclusion (no-cloning) via a different approach. Let us assume that one does have a quantum cloning machine that can make an additional copy of an arbitrary quantum state. The experimenter now can use these copies and perform measurements on two conjugate quantities, say position and momentum. Since the copies are perfect, the outcome of these independent measurements will be given precisely without any actual limitations. This is against the Heisenberg uncertainty principle, which assigns a fundamental limit to the precision of determining conjugate quantities. Now, one may still ask the following question:

> *"The laws of physics (in this case, the Heisenberg uncertainty principle) pose a fundamental limit in determining a quantum state. Thus, making a perfect copy is forbidden by the laws of quantum mechanics. However, how good can a copy of a quantum state be?"*

This limit is known as *optimal quantum cloning*. The copy of the state that is obtained via an optimal quantum cloning machine is not pure anymore and is partially mixed. So, the states of the final copies (labeled 1 and 2) obtained from an optimal clone of state $|\psi\rangle$ are given by the density of states (operators) $\hat{\rho}_1$ and $\hat{\rho}_2$. When the cloning is symmetric, the two copies possess the same amount of information and thus $\hat{\rho}_1 = \hat{\rho}_2$. The fidelity of cloned states for a d-dimensional quantum state (qudit) $|\psi\rangle$, in the case of the symmetric cloning $\mathcal{F}_{\text{optimal cloning}} = \langle\psi|\hat{\rho}_1|\psi\rangle = \langle\psi|\hat{\rho}_2|\psi\rangle$, is given by

$$\mathcal{F}_{\text{optimal cloning}} = \frac{1}{2} + \frac{1}{d+1}. \tag{6.11}$$

Interestingly, the fidelity of the cloned state depends on the state dimension, and it decreases for high-dimensional quantum states. Surprisingly, the experimental realization of a photonic optimal quantum cloning machine is quite simple, and it can be achieved by means of a 50:50 beam splitter (BS) and an ancillary photon in a totally mixed state [21]. Figure 6.4 illustrates the sketch of an optimal quantum cloning machine. In this machine, a photon with the arbitrary quantum state $|\psi\rangle$ is interfered with an ancillary photon in the mixed state $(|\psi\rangle\langle\psi| + \sum_{i=1}^{d-1} |\psi_i\rangle_\perp \langle\psi_i|)/d$, where $|\psi_i\rangle_\perp$ are $(d-1)$ states orthogonal to $|\psi\rangle$. When the two photons possess identical quantum information $|\psi\rangle$ (assuming one is reflected), the two photons interfere and exit from the same exit port 50% of the time. This effect is known as

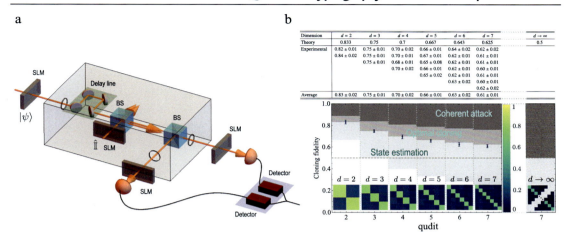

Figure 6.4: *Optimal quantum cloning machine.* (a) The experimental setup to perform high-dimensional optimal quantum cloning with structured photons. (b) Associated experimental results and measured state fidelity for dimensions up to $d = 7$ [22].

two-photon interference, or the Hong-Ou–Mandel (HOM) effect [23]. HOM is at the core of many quantum information schemes and protocols [24]. However, the ancillary photon state is $(d-1)$ times completely orthogonal to the state of $|\psi\rangle$, i.e. both photons (original and ancillary) are completely distinguishable, and thus 25% of the time they exit from the same port with a fidelity of 1/2. Thus, the overall optimal cloning fidelity is $\frac{2}{d+1} \times 1 + \frac{d-1}{d+1} \times \frac{1}{2} = \frac{1}{2} + \frac{1}{d+1}$. For instance, the cloning fidelity for qubits ($d = 2$), qutrits ($d = 3$), and quoctals ($d = 8$) are, respectively, 0.833, 0.750, and 0.611.

As mentioned before, making a good copy of a quantum state for higher dimensions becomes more difficult. Therefore, employing high-dimensional quantum states for quantum communication makes the communication more resistant against any sort of potential cloning attack by an eavesdropper. The optimal quantum cloning machine used by the eavesdropper introduces noise to the channel, turning a pure quantum state into a particular mixed state; the level of the introduced noise depends on the dimension of quantum state and is $\frac{1}{2} - \frac{1}{d+1}$. The amount of errors introduced by the optimal quantum cloning machine into a quantum communication channel with qubit ($d = 2$), qutrit ($d = 3$), and quoctal ($d = 8$) states are, respectively, 0.167, 0.250 and 0.389. By monitoring the noise, one can impose a threshold which guarantees the security against any sort of attack, not only against optimal quantum cloning. The threshold for different attacks is shown in Fig. 6.4(b).

6.3.2 Protocols

Quantum key distribution (QKD) is a promising avenue to continue the secure transmission of our sensitive information, relying on the principles of quantum mechanics such as the no-cloning theorem and quantum contextuality. In accordance with its name, QKD protocols concern themselves with simply generating and distributing a random secret key between two distant parties, colloquially named Alice and Bob, using quantum states. In this way, if an eavesdropper, Eve, were to attempt to gain information, she will introduce detectable errors above a threshold unique to the given protocol, as discussed in Section 6.3.1. As we will see here, QKD protocols employing high-dimensional qudits are advantageous in terms of this error threshold and information capacity. For a comprehensive review of quantum cryptography, we refer the reader to [25,26].

The first QKD protocol, known as BB84, was developed by Bennett and Brassard in 1984 [2]. Its security is based precisely on the notion of conjugate coding: quantum uncertainty prevents the gain of adequate information via measurements between conjugate quantities. We describe here the original 2-dimensional BB84 protocol using polarization qubits; d-dimensional extensions are accomplished using the same protocol but instead with qudits. First, Alice and Bob agree on two so-called *mutually unbiased bases* (MUB) from which they will use to encode bits; for example, let the two MUB be $\mathcal{M}_0 = \{|H\rangle, |V\rangle\}$ and $\mathcal{M}_1 = \{|A\rangle, |D\rangle\}$. In this way, $|H\rangle$ and $|A\rangle$ correspond to the classical bit '0', and $|V\rangle$ and $|D\rangle$ correspond to the classical bit '1'. Alice starts the QKD protocol by randomly choosing a classical bit ('0' or '1'), then randomly chooses a MUB ($\mathcal{M}_0$ or $\mathcal{M}_1$) to accordingly prepare a photon with the correct polarization. She sends her prepared photon over a designated untrusted quantum channel to Bob. Bob can then randomly choose a MUB to measure the photon in and records the result—either '0' or '1' depending on which state the photon was projected onto. This process is repeated for a set number of photons.

Alice and Bob can now communicate over an authenticated classical channel in order to distill their random secret key by sharing with each other *only* which MUB they generated and measured with, respectively. This step is known as sifting as Alice and Bob will discard the results for which they used differing MUB, since Bob's results in these cases will be completely random. For an infinitely long key, 1/2 of the photons are sifted out. From the remaining shared string of classical bits, Alice and Bob must determine if Eve was present during the QKD protocol. They do this by sacrificing a small portion of their key and publicly checking if each bit they share is correct; in this way, they statistically determine how many errors are present in the total key. In $d = 2$, if the quantum bit error rate (QBER) is more than 11%, then an eavesdropper might have been present; Alice and Bob subsequently discard their key and try again. If the QBER is below this error threshold, the generated key can be used in a one-time pad protocol to encrypt a message. Error correction and privacy amplification protocols can also

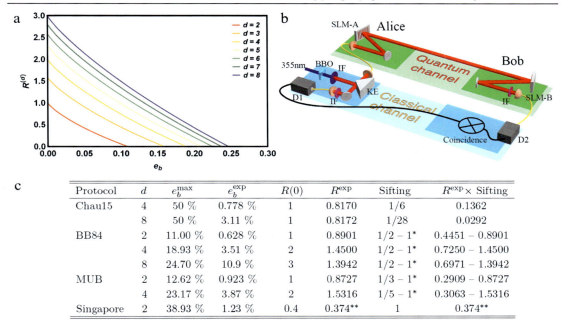

Figure 6.5: High-dimensional quantum key distribution. (a) The secret key rate for BB84 in dimensions $d = \{2, 3, 4, 5, 6, 7, 8\}$ are shown for a range of QBER. The error threshold for a given dimension is defined by where its respective curve crosses the horizontal axis ($R^{(d)}(e_b) = 0$). (b) A general experimental setup for testing different QKD protocols using OAM to encode information. (c) A comparison of experimentally obtained QBER $e_b^{\exp}$ and key rates $R^{\exp}$ for different protocols performed using the setup in (b) (obtained from [27]).

be used to further refine the key to eliminate any remaining errors or mutual information with Eve.

The secret key rate in dimension d as a function of error rate e_b is analytically given as

$$R^{(d)}(e_b) = \log_2(d) - 2h^{(d)}(e_b), \tag{6.12}$$

where,

$$h^{(d)}(x) = -x \log_2\left(\frac{x}{d-1}\right) - (1-x)\log_2(1-x), \tag{6.13}$$

is the d-dimensional Shannon entropy; Fig. 6.5(a) plots Eq. (6.12) for dimensions $d = \{2, 3, 4, 5, 6, 7, 8\}$. If theoretically zero errors were present, the maximum capacity for the channel is $\log_2(d)$; thus, higher dimensions can potentially transmit more information. We see that where the curves cross the horizontal axis defines the error threshold for a given dimension, i.e. the point where zero secret keys will be generated. Therefore, the amount of tolerable errors must be below each threshold to have a positive secret key rate. In particular, this error threshold increases with increasing dimension [28]. The first high-dimensional BB84 scheme using OAM states was performed in [29], achieving 2.05 bits per sifted photon in dimension 7. The BB84 protocol can be further extended by increasing the number of MUB that Alice and Bob choose from—for example, using 3 MUB in dimension 2 is known as the *six-state* protocol. In higher dimensions that are powers of primes, all possible $(d+1)$ MUB can be used to create tomographic protocols [30], named from the fact that quantum state tomography (as discussed in Section 6.3.3) could be performed using the resulting measurements. Note that for non-prime or non-powers of prime numbers, the number of MUB is still unknown; however, three MUB can always be found for any dimension. Tomographic protocols have a secret key rate of

$$R^{(d)}(e_b) = \log_2(d) - h^{(d)}\left(\frac{d+1}{d}e_b\right) - \frac{d+1}{d}e_b \log_2(d+1). \qquad (6.14)$$

However the sifting efficiency for MUB protocols is $1/d$. Symmetric informationally complete positive operator-valued measures (SIC-POVMs), which can perform quantum state tomography (QST) the most efficiently, can be used instead of MUB. This type of encoding scheme is known as the *Singapore* protocol, which also has high-dimensional extensions [31,32].

One protocol that is uniquely suited to high-dimensional quantum states is the *round robin differential phase shift* (RRDPS) protocol [33], which is an extension to *differential phase shift* QKD [34]. Unlike the high-dimensional extensions of the original 2-dimensional BB84, the fundamental idea of RRDPS requires high-dimensional states. Alice prepares a d-dimensional superposition state,

$$|\Psi\rangle = \frac{1}{\sqrt{d}} \sum_{k=1}^{d} e^{i\phi_k} |k\rangle, \qquad (6.15)$$

where $\phi_k \in \{0, \pi\}$ and $|k\rangle$ is the kth state in the given Hilbert space. The original protocol uses a train of coherent optical pulses. Each pulse in the packet is given a random phase shift of 0 or π. The average photon number is brought below 1 such that any particular pulse is not guaranteed to hold a photon upon measurement. At the receiver, Bob performs an interferometric measurement. A delayed interferometer with a randomly chosen delay determines which pulses will overlap to interfere with each other. In this way, Bob is able to determine phase differences between these randomly selected pulses. The attractive feature of the RRDPS protocol is that the necessary amount of privacy amplification is given by $h^{(2)}(1/(d-1))$, which is independent of the error rate in the channel. This bounds the amount of information leaked to Eve, thus removing the need to monitor the errors or signal disturbances in the QBER. The resultant secret key rate for the infinite-key approximation is

$$R = 1 - h^{(2)}(e_b) - h(\frac{1}{d-1}), \tag{6.16}$$

where e_b is the bit error rate. This protocol becomes difficult to scale to high-dimensions as an additional interferometer is required for each dimension that is added in the time-bin approach. A new approach was developed by Bouchard et al. to perform the RRDPS protocol using OAM states [35]. In this way, the measurement technique is greatly simplified, requiring only one phase hologram.

A drawback to RRDPS QKD is the scalability of the protocol in terms of generation and detection as d becomes large. For this, "qubit-like" superpositions formed from a d-dimensional encoding space can be used instead using the *Chau15* protocol [36,37]. In this way, instead of having a train of d optical pulses like in Eq. (6.15), the qubit-like states are $\left|\phi_{ij}^{\pm}\right\rangle = (|i\rangle \pm |j\rangle)/\sqrt{2}$ in a 2^n-dimensional space, $n \geq 2$. Again, the information is encoded in the relative phase. Figure 6.5(b) details a simple setup with which all of the above protocols can be tested using the OAM degree of freedom, as experimentally performed in [27]. This work compared the obtained QBER and key rates, shown in Fig. 6.5(c).

6.3.3 Quantum process tomography

Measuring quantum states has always been one of the fundamental problems in quantum mechanics. It is well known that the measurement process impacts the states being measured, and thus must be dealt with carefully. The primary difficulty in measuring quantum states is that we generally cannot perform consecutive measurements on a single quantum state. The first measurement will change the quantum system, rendering later measurements useless for understanding the original state. In particular, measuring a photon almost always results in the destruction of the photon, meaning any further measurements are impossible. The protocol of quantum state tomography has been developed to overcome these difficulties [38]. In quantum state tomography, an ensemble of identical quantum states is prepared such that different measurements can be made on different copies of the system. The difficulty then becomes selecting a set of optimal measurements that will give all of the desired information about the quantum system. These desired sets of measurements are called informationally complete. When the measurement probabilities are known, then the density matrix can be given in terms of conventional parameters.

The goal with quantum state tomography is to determine the d-dimensional density matrix,

$$\hat{\rho} = \sum_{i=1}^{d} p_i |\psi_i\rangle \langle\psi_i|, \qquad (6.17)$$

which holds all of the necessary information about the quantum system. Here, the $|\psi_i\rangle$ are the pure states with probabilities p_i. The density matrix must have unit trace and be Hermitian.

We can begin by describing the procedure for quantum state tomography in a 2-dimensional Hilbert space. The polarization state of a photon is one such 2-dimensional state. We choose a set of projective measurements in orthogonal bases to find our informationally complete set. In the case of polarization, one such informationally complete set is $\{\hat{\Pi}_H, \hat{\Pi}_V, \hat{\Pi}_A, \hat{\Pi}_D, \hat{\Pi}_L, \hat{\Pi}_R\}$, where $\hat{\Pi}_\psi = |\psi\rangle\langle\psi|$. These projective measurements give the corresponding probabilities, P_ψ, which are used to determine the Stokes parameters:

$$\begin{aligned} S_0 &= P_L + P_R = P_H + P_V = P_A + P_D, \\ S_1 &= P_H - P_V, \\ S_2 &= P_A - P_D, \\ S_3 &= P_L - P_R. \end{aligned} \qquad (6.18)$$

Here, the Stokes parameters (S_1, S_2, S_3) correspond to the coordinates of a polarization vector of length S_0 on the Poincaré sphere, as discussed in Section 6.2.1, and can be used to calculate the density matrix as

$$\hat{\rho} = \frac{1}{2} \sum_{i=0}^{3} \frac{S_i}{S_0} \sigma_i, \qquad (6.19)$$

where σ_i are the Pauli matrices.

An analogous strategy is used in quantum process tomography. However, to understand the action of a quantum process, we must characterize any possible input and output. This requires a tomographically complete set of input states in addition to the complete set of measurements at the output [39]. We can use the Kraus representation with operators E_i to describe a quantum process as a map $\mathcal{E}$ acting on any state $\hat{\rho}$ by

$$\mathcal{E}(\rho) = \sum_i E_i \, \hat{\rho} \, E_i^\dagger. \qquad (6.20)$$

The operators E_i must satisfy $\sum_i E_i^\dagger E_i = \mathbb{1}$. By ensuring that this condition holds, we guarantee that $\mathcal{E}$ is trace preserving on $\hat{\rho}$. The operators E_i act on the system; thus, we want to relate these to a measurable set of operators A_i acting on the state space. This relation is given by $E_i = \sum_m a_{im} A_m$, where a_{im} are complex numbers. We can then reform Eq. (6.20) as

$$\mathcal{E}(\rho) = \sum_{mn} A_m \hat{\rho} A_n^\dagger \chi_{mn}, \tag{6.21}$$

where $\chi_{mn} = \sum_i a_{im} a_{in}^*$ are the elements of the process matrix. To form a tomographically complete set of input states, we should form a basis for the d-dimensional Hilbert space. Take the states ρ_k as linearly independent $d \times d$ matrices. After the action of the quantum process, we can write the output as

$$\mathcal{E}(\rho_k) = \sum_j \lambda_{kj} \rho_j. \tag{6.22}$$

Here, the values λ_{kj} are found experimentally by quantum state tomography on the output state $\mathcal{E}(\rho)$. Next we can write

$$A_m \rho_j A_n^\dagger = \sum_k \beta_{jk}^{mn} \rho_k, \tag{6.23}$$

which allows us to form the relation

$$\sum_{mn} \beta_{jk}^{mn} \chi_{mn} = \lambda_{jk}, \tag{6.24}$$

between the desired χ_{mn} and the experimental values λ_{jk}. The inverse matrix of β_{jk}^{mn} is required to find the process matrix. With τ_{jk}^{mn} as this inverse matrix, the process matrix χ is given by

$$\chi_{mn} = \sum_{jk} \tau_{jk}^{mn} \lambda_{jk}. \tag{6.25}$$

For high-dimensional Hilbert spaces, the Gell-Mann matrices—high-dimensional generalizations of the Stokes parameters—can be used to create the required bases. It is also interesting to consider using MUB, as these can provide a tomographically complete set in those dimensions where d is a power of a prime number. These MUB can be useful as they are commonly used in quantum information. One application of quantum process tomography is to characterize quantum communication channels [40]. The minimum requirement for most quantum communication protocols is the characterization of the quantum bit error rates. Typically, this gives the worst case scenario information to be used for privacy amplification protocols. If more information is known about the channel though, through process tomography, then it is possible to achieve higher key rates from these communication links. One could also use the characterization of the channel to determine which QKD protocol should be used.

158   Chapter 6

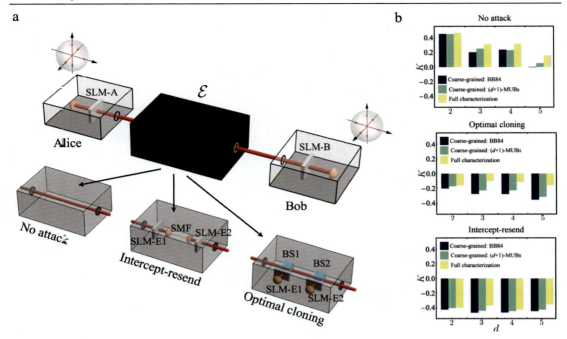

Figure 6.6: Quantum process tomography. (a) Process tomography is performed on a quantum channel between Alice and Bob with when there is no attack, an optimal cloning attack and an intercept-resend attack. (b) The key rates for the quantum channel in dimensions 2 to 5 are shown. The errors introduced by eavesdropping result in negative calculated key rates for the optimal cloning attack and the intercept-resend attack [40].

Different eavesdropping attacks on a quantum channel will also show themselves in the process tomography reconstruction of the channel in [40]. Quantum process tomography was done on a high-dimensional QKD channel (Fig. 6.6(a)) by sending BB84-type states in mutually unbiased bases. The OAM states with $\ell = \{-2, -1, 0, +1, +2\}$ were used to probe the quantum channel for dimensions 2 to 5. The optimal quantum channel with no eavesdropping was measured showing a nearly perfect fidelity between the input and output states. Next, two types of eavesdropping attacks were implemented. The first eavesdropping attack is optimal quantum cloning. Perfect quantum cloning is impossible as was shown in Section 6.3.1. This inability to clone quantum states is necessary for the security of quantum communication. It is thus impossible to take an arbitrary quantum state as an input and give two identical quantum states as the output. However, one can consider what the optimal output in terms of fidelity would be from such a machine. This was shown by Bouchard et al. for high-dimensional quantum states [22]. It is shown that as the dimensionality of the quantum states increases, the optimal cloning fidelity decreases. This is part of the reason that security in quantum channels increases as the dimensionality increases. Process tomography was done to characterize a quantum channel under an optimal cloning attack. Here, the loss in fidelity is not as significant as in the intercept-resend attack, since it is a more discrete attack. The errors are again introduced on the diagonal of the process matrix, preserving the trace.

The next attack was an intercept-resend attack in which Eve receives the photon sent from Alice and makes a random projective measurement using the same approach as would Bob. This is the simplest and most naive strategy for attacking a quantum channel. On average, Eve will make her measurement in the wrong basis 50% of the time for a standard 2-dimensional BB84 protocol. Consequently, the state she sends to Bob in the incorrect basis will introduce errors at a rate of 25%. These errors increase with the state's dimension, as her guesses will be wrong more than 50% of the time. This is precisely what is seen in the process tomography of the channel. The state fidelity with respect to the optimal channel is around 75%, and the errors are introduced on the diagonal of the process matrix, maintaining the trace of 1.

When process tomography is conducted to give full information about the quantum channel, this extra information can be used to achieve a lower bound on the information obtained by Eve than can be done with the simple error rate measurements. This lower bound on Eve's information reduces the necessary amount of privacy amplification, thus increasing the final secret key rate. This is shown in Fig. 6.6(b).

6.4 Quantum key distribution implementations

6.4.1 Optical fiber

Optical fibers have revolutionized our telecommunication infrastructures. At telecommunication wavelengths (1550 nm), they exhibit remarkably low losses on the order of 0.2 dB/km. Conventionally, an optical fiber is designed as a cylindrical dielectric silica waveguide with a step-index profile, where the core has a higher refractive index than the surrounding cladding layer. High transmission capacities, on the order of several tens of Gbps, are routinely achieved using wavelength-division multiplexing (WDM), wherein separate data channels are multiplexed through a single fiber using a set of different wavelengths. Along with erbium-doped fiber amplifiers which amplify over a broad spectrum, our classical optical fiber communication networks are able to transmit information at incredibly high rates over global distances. Nonetheless, there is a growing need to increase the current bandwidth as current fiber limits are being reached—a promising avenue is space-division multiplexing (SDM). Analogous to WDM, SDM is a technique that multiplexes different transverse optical modes over a single fiber. The number of transverse modes supported by a given fiber is determined by the core size, producing the familiar single- and multi-mode fibers. These modes are known as the linearly polarized (LP) modes—the lowest LP$_{01}$ mode possessing a Gaussian-like intensity profile. SDM can thus be achieved in a variety of ways: (1) construct a fiber with a single core which can support multiple LP modes; or (2) construct a fiber with multiple cores, i.e., multi-core fibers, which support several spatially separated transverse modes. Due to limited space, we will concentrate on single-core fibers and special variations thereof; for an in-depth review of SDM fibers, we refer the reader to [41].

In order to effectively use SDM with single-core fibers which support multiple modes—i.e. multi-mode fibers (MMF)—it is important that the crosstalk between modes remains small to mitigate high error rates. Additionally, it can be problematic if too many modes are used for SDM as the complexity of multi-input multi-output detection schemes increases. There has been progress with few-mode fibers which minimizes both of these aspects. However, whereas conventional MMFs support LP modes with minimal problems, the same cannot be said for structured states of light, particularly in the context of quantum communication tasks which require coherence between different modes. The problem arises as follows [42,43]. For a step-index cylindrical waveguide with core refractive index n_c and cladding refractive index n, the LP$_{lm}$ modes are the transverse electric field solutions in the weakly guided regime ($n_c \simeq n$). In the core, these solutions have the form of Bessel functions of the first kind of order $l \geq 0$; the index $m > 1$ denotes the number of zeros of the Bessel function that are supported in the core. The intensity distributions of the first two LP mode groups are shown in Fig. 6.7(a). Each mode group corresponds to a particular effective refractive index n_{eff}, which determines how fast they propagate through the fiber. However, when solved using the full vector wave equation, each scalar LP mode consists of a group of vector vortex mode solutions, revealing a more complex hierarchy of vector modes according to distinct n_{eff}. As shown in Fig. 6.7(a), some vector modes are degenerate in terms of effective refractive index, while others are not. Notably, quadrupole (HE$_{2,1}^{\text{even}}$) and clover (HE$_{2,1}^{\text{odd}}$) modes are strictly degenerate, but radial (TM$_{01}$) and azimuthal (TE$_{01}$) modes are not. However, the three distinct n_{eff} within the LP$_{11}$ group are near-degenerate, causing the modes to be unstable and mix upon propagation through a long fiber. It is thus not practically feasible to use a standard step-index fiber for quantum communication tasks using structured photons as superpositions of different modes will decohere with propagation.

The solution, therefore, is to engineer a new type of optical fiber which sufficiently separates the effective refractive indices of the vector modes such that the modes do not mix. Indeed, this is the approach taken for the development polarization-maintaining fibers, accomplished by breaking the cylindrical symmetry. Here, so-called *ring-core* or *vortex* fibers are able to coherently guide a desired set of structured states in the form of vector vortex modes, named after the high refractive index ring profile of the core; see Fig. 6.7(b, c). This ring lifts the near-degeneracy in n_{eff} between neighboring modes, ensuring low crosstalk. Of note, there are two classes of vector vortex modes that are supported by vortex fibers: spin–orbit-aligned and

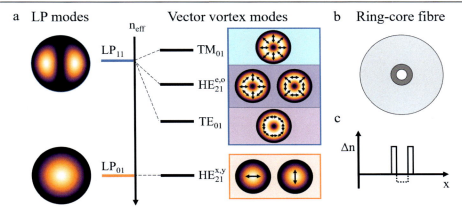

Figure 6.7: Optical fiber modes. (a) The linearly polarized (LP) modes are the weakly guided solutions inside an optical fiber. The intensity distributions of the two lowest orders (LP$_{01}$ and LP$_{11}$) are shown on the left. Each mode group is comprised of several vector vortex modes (shown right, arrows indicate polarization). The effective refractive index n_{eff} of these vector modes are near-degenerate in standard step index fibers; the splitting between the modes is exaggerated for clarity. (b) The transverse cross section of a ring-core fiber. (c) The refractive index profile of a ring-core fiber, where Δn is with respect to the cladding refractive index. For air-core fibers, the central core has a lower refractive index (dashed line).

spin–orbit-anti-aligned states. Aligned (anti-aligned) states refer to the fact that the spin and orbital angular momenta have the same (opposite) sign; for example, $\{|L, +|\ell|\rangle, |R, -|\ell|\rangle\}$ are aligned, whereas $\{|R, +|\ell|\rangle, |L, -|\ell|\rangle\}$ are anti-aligned. Except for the LP$_{11}$ mode group, higher order LP$_{lm}$ mode groups contain pairs of aligned and anti-aligned states that are degenerate within the pair but non-degenerate with respect to the opposite alignment. In this way, degenerate pairs can be used as the encoding states for (quantum) communication schemes.

Air-core vortex fibers have been shown to reach Tbit/s data rates using SDM in the classical regime [44]. Certain vortex fibers have been shown to support up to 9 orders of OAM and 36 states [45]. Quantum key distribution with vortex fibers have also recently been demonstrated. The first was a proof-of-principle scheme of a 2-dimensional BB84 protocol with heralded single photons through a 60-m solid-core vortex fiber that coherently supported $\{|R, +1\rangle, |L, -1\rangle\}$ [46]. A practical 4-dimensional version was subsequently demonstrated with weak coherent pulses through a 1.2-km air-core vortex fiber using the aligned mode set $\{|L, +6\rangle, |R, -6\rangle, |L, +7\rangle, |R, -7\rangle\}$ [47]; see Fig. 6.8 for the experimental setup. Since the modes of different $|\ell|$ in this vortex fiber had different group velocities ($\approx$15 ns between $\ell = |6|$ and $\ell = |7|$), a delay line was introduced in the generation stage to compensate for any time difference at the detection stage. With real-time preparation and a parity-based OAM sorter, a secret rate of 37.85 kbit/s was achieved with a 4D BB84 protocol—a 69% increase

162 Chapter 6

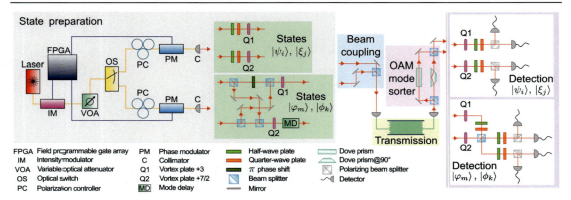

Figure 6.8: *4-dimensional QKD through a vortex fiber.* Experimental setup; see [47] for more details. *Source:* ©2019 APS Physics.

from a 2D protocol run for comparison despite the 4D case having a higher QBER. These results demonstrate the advantage of using higher-dimensional QKD schemes. Such vortex fibers have also been successfully used as a quantum channel to distribute hybrid vector vortex-polarization entangled states [48].

6.4.2 Free-space

An unfortunate drawback of optical fiber networks, particularly in the quantum regime, is that they experience exponential loss with propagation. Despite the best fibers having a low loss, information signals still need to be amplified in classical communication in order to reach global scales. However, a quantum signal cannot be amplified, as prohibited by the no-cloning theorem; therefore, quantum protocols are limited to a distance of around 500 km before background and detector noise dominate the signal. While quantum fiber networks will be beneficial for short range communication, it is necessary to develop the infrastructure for quantum communication through free-space, e.g. satellite-to-ground and *vice versa*, where there is only linear loss. Free-space quantum communication has made tremendous progress with feasible demonstrations of entanglement distribution [49] and secure QKD [50,51] between satellite-to-ground over the last few years. So far, classical communication with orbital angular momentum has been conducted across ground free-space links with direct line of sight up to 143 km [52]. Recent demonstrations have also shown the feasibility of using structured photons for QKD in 2- and 4-dimensions over several hundred meters [53,54]—to be described in more detail shortly.

A major challenge when transmitting optical signals across any free-space channel, regardless of the photonic degree of freedom utilized, is the effect of atmospheric turbulence. Commonly seen as the innocent twinkling of stars, atmospheric turbulence can have a profound effect on practical quantum communication schemes by introducing excess noise and errors. At its core, turbulence manifests from local variations in refractive index, primarily caused by temperature (density) gradients, which is then mechanically mixed by wind; this creates a spatially inhomogeneous and temporally varying refractive index profile. Indeed, the speed of fluctuations can occur on the order of 100 Hz. A plane wave that passes through a turbulent medium will thus become distorted since different parts of the wavefront will take slightly different paths. Thus, turbulence can be a particularly destructive effect on structured light, e.g. modes carrying OAM, as we wish for the transverse spatial profile to remain intact after propagation.

A comprehensive review of turbulence is beyond the scope of this chapter and we refer the reader to other sources [55]; we thus describe here a few necessary parameters for characterization. According to Kolmogorov theory [56], turbulence is locally isotropic and homogeneous on the small scale; fluctuations in a given parameter between a point $\mathbf{x}$ and $\mathbf{x}'$ can be described statistically, depending only on the distance between them. This gives rise to so-called *structure functions*; for example, the refractive index structure function is given by

$$D_n(\mathbf{r}) = \langle [n(\mathbf{x}) - n(\mathbf{x}+\mathbf{r})]^2 \rangle = C_n^2(h) r^{2/3}, \tag{6.26}$$

where $\langle \cdot \rangle$ is an ensemble average, and $r = |\mathbf{r}| = |\mathbf{x}' - \mathbf{x}|$. The refractive index structure constant, $C_n^2(h)$, thus gives a measure of the strength of the refractive index fluctuations at a height h. Over a pathlength of L, we can thus determine the average scale of air cells with homogeneous refractive index, i.e. the Fried parameter r_0 [57],

$$r_0 = \left[0.423 \left(\frac{2\pi}{\lambda} \right)^2 \int_0^L C_n^2(h) \left(\frac{L-h}{L} \right)^{5/3} dh \right]^{-3/5}, \tag{6.27}$$

or simply

$$r_0 = \left[0.423 \left(\frac{2\pi}{\lambda} \right)^2 C_n^2 L \right]^{-3/5}, \tag{6.28}$$

for horizontal propagation near ground level. We observe that small r_0 (large $C_n^2 \approx 10^{-13}$ m$^{-2/3}$) corresponds to strong turbulence, and large r_0 (small $C_n^2 \approx 10^{-17}$ m$^{-2/3}$) corresponds to weak turbulence. There are several ways to experimentally measure r_0 and C_n^2. For example, C_n^2 can be directly measured using a scintillometer—a device which monitors the fluctuations in refractive index by measuring spatial intensity fluctuations of a reference light source. Another approach is to calculate r_0 by measuring a series of short exposure images of a reference optical beam. Due to turbulence, the center of the beam will be slightly deflected by a different angle β in each image. The average deflection angle β_{avg} is sometimes termed the seeing, and it can be related back to the Fried parameter as

$$\beta_{avg} = 0.98 \frac{\lambda}{r_0}. \quad (6.29)$$

For horizontal links, C_n^2 can then be extrapolated using Eq. (6.28).

While determining the turbulence conditions in a free-space channel is important as a diagnostic tool, further knowledge and parameters are required in order to build a practical quantum communication infrastructure. For example, though the Fried parameter gives an idea of beam wander and beam spread, it can be beneficial to know the higher order distortions that turbulence applies to a wavefront. Indeed, a given wavefront (phase profile), $\Phi(r, \phi)$, can be decomposed in terms of the so-called Zernike polynomials, $Z_j(r, \phi)$, which form an orthonormal set of polynomials on the unit disk, such that $\Phi(r, \phi) = \sum_j a_j Z_j(r, \phi)$. Using the Noll indexing $j = 1 + (n(n + 2) + m)/2$ with the radial n and azimuthal m integer degrees ($m \leq n$, $n - |m|$ even), the Zernike polynomials are [58]

$$Z_j(r, \phi) = \begin{cases} \sqrt{2(n+1)} R_n^m(r) \cos(m\phi), & m \neq 0, \ j \text{ even} \\ \sqrt{2(n+1)} R_n^m(r) \sin(m\phi), & m \neq 0, \ j \text{ odd} \\ \sqrt{n+1} R_n^0(r), & m = 0 \end{cases} \quad (6.30)$$

where $R_n^m(r)$ are the radial polynomials defined by

$$R_n^m(r) = \begin{cases} \sum_{b=0}^{(n-m)/2} \frac{(-1)^b (n-b)!}{b!\left(\frac{n+m}{2}-b\right)!\left(\frac{n-m}{2}-b\right)!} r^{n-2b}, & (n-m) \text{ even}, \\ 0, & (n-m) \text{ odd}. \end{cases} \quad (6.31)$$

We thus see that each Zernike polynomial uniquely represents a different type of distortion to a wavefront; for example, Z_2 and Z_3 represent tip and tilt effects, respectively, whereas Z_4 is defocusing, and Z_5 and Z_6 are oblique and vertical astigmatism effects. Figure 6.9 displays the phase profile of the first 16 Zernike polynomials. Again, there are several ways to experimentally measure the distorted wavefront and then determine the coefficients a_j. The easiest method is to use a Shack–Hartmann wavefront sensor (SH-WFS) which is capable of directly measuring the local displacement of an incoming wavefront. A SH-WFS consists of an array of lenslets which locally focuses the beam onto a camera, effectively taking a Fourier transform: if the focused spot is displaced from the center of the lenslet (the references for a plane wave), then the wavefront of that portion of the beam is tilted in accordance to the direction of the displacement. The reconstructed wavefront can then be decomposed in terms of Zernike polynomials. The knowledge of the Zernike polynomial spectrum for a distorted wavefront can be utilized in an active compensation system, such as an adaptive optics system; the inverse of the calculated or measured distorted wavefront can be applied to a deformable mirror,

Quantum cryptography with structured photons 165

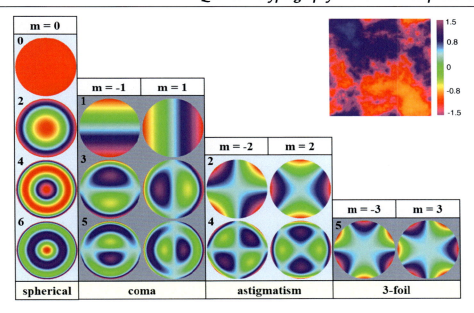

Figure 6.9: Zernike polynomials and turbulence. The Zernike polynomials corresponding to spherical, coma, astigmatism and 3-foil aberrations are plotted on the unit disk. The azimuthal degree, $m = \{0, \pm 1, \pm 2, \pm 3\}$, is indicated above each column; the radial degree, $n = \{0, 1, 2, 3, 4, 5, 6\}$, is indicated for each row. A turbulence phase screen, simulated according to Kolmogorov theory [59,60], is shown as an inset.

thus correcting the distortions. Adaptive optics, widely used in the astronomy community, warrants its own review beyond the scope of this chapter. A few studies have explored the effect of atmospheric turbulence and use of adaptive optics in a quantum communication scheme with polarization and OAM [61–64]. In addition to the lower order effects of beam wander, OAM states have been observed to undergo vortex-splitting for $|\ell| > 1$ caused by the higher order distortions, which can be problematic for adaptive compensation schemes.

With the effects of atmospheric turbulence in mind, let us now describe several free-space quantum key distribution experiments using spatially structured photons. Apart from the challenge of proper beam pointing/tracking and wavefront distortions over long distances or between moving objects, an additional challenge with free-space experiments is the need for a shared reference frame between Alice and Bob. Indeed, quantum communication schemes using polarization, without any form of compensation, are susceptible to errors from a relative rotation in the frames. Though the circular polarizations are rotationally invariant about the beam axis, linear polarizations will add a $\sin^2(\theta)$ error, giving an overall QBER of $\frac{1}{2}\sin^2(\theta)$, where θ is the relative angle between Alice and Bob's frames. A protocol was thus proposed in which particular modes structured in OAM and polarization, possessing zero total angular

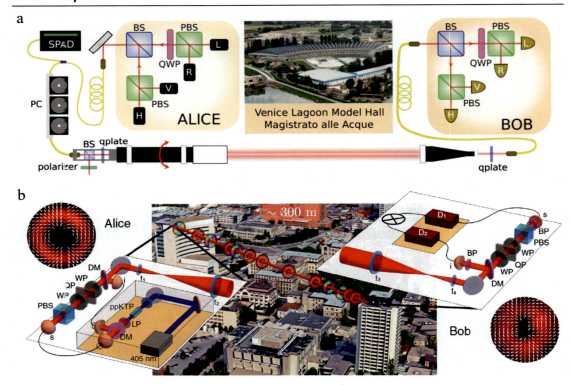

Figure 6.10: Free-space QKD with structured photons. Experimental setups for (a) a 2-dimensional BB84 scheme using rotationally invariant structured photons across a 210-m link (taken from [53]), and (b) a 4-dimensional BB84 scheme using structured photons across a 300-m intra-city link (taken from [54]). *Source: ©2017 Optical Society of America.*

momentum, can be exploited to create alignment-free experiments [65]. Radial and azimuthal modes are the simplest examples which are easily generated using a $q = 1/2$ plate. These rotationally invariant states were implemented in a 2D BB84 scheme with strongly attenuated lasers across a 210-m free-space link in Padua, Italy [53]; see Fig. 6.10(a). A QBER ranging between 3.8% and 6.9% using a decoy protocol was achieved over a 60° rotation of the transmitter without adjustment at the receiver, demonstrating minimal rotational dependence. This was the first demonstration of QKD using 2D structured photons outside, beyond the laboratory scale. However, in terms of turbulence, a relatively large Fried parameter was measured at $r_0 = 17$ cm and $C_n^2 \simeq 4 \times 10^{-15}$ m$^{-2/3}$ at $\lambda = 850$ nm, giving rise to minimal beam wandering and OAM scattering for a transmitted beam radius of 15 mm.

A few years later, the first proof-of-principle high-dimensional QKD scheme using structured photons was demonstrated across a 300-m intra-city free-space link in Ottawa, Canada [54];

see Fig. 6.10(b). Note, the 4-dimensional vector vortex states used for encoding are not rotationally invariant. Similar to [53], a C_n^2 ranging between 2.5×10^{-15} m$^{-2/3}$ and 6.4×10^{-16} m$^{-2/3}$ was measured at $\lambda = 850$ nm for a beam radius of 20 mm at the receiver. However, a heralded single-photon source was implemented ($\lambda_s = 850$ nm, $\lambda_i = 775$ nm) instead of weak coherent states. The term heralded photon source describes the use of single-photon pairs, in such a way that the only information on the second photon (idler) is about the generation timing, and it is used to reduce the noise by measuring the g^2 function, i.e. coincidence measurements. Here, only the signal photon was encoded with information; the idler photon was left "blank" as a Gaussian. The idler photon was thus utilized as a "target" beam with the idea that, under ideal conditions, it would couple optimally to the SMF at the receiver. Any deviation from optimal coupling—observed as a drop in single-photon counts—could thus be attributed to beam wandering induced by the turbulence. Since the signal and idler photons propagated along the same path, and thus experienced the same turbulence, the coincidence events could be post-selected according to when the idler photon counts were near optimal. This effectively discarded the events that were the worst affected by beam wandering. A comparison between a 2D and 4D BB84 protocol was performed using structured photons; raw QBER of $e_b^{2D} = 5\%$ and $e_b^{4D} = 14\%$ were obtained, corresponding to secret key rates of approximately 0.43 and 0.39 bits per sifted photon, respectively. Note, the 4D BB84 threshold is 19%. With the target beam for compensation, the 4D QBER improved to 11%, corresponding to 0.65 bits per sifted photon. This again demonstrates the advantage of high-dimensional encoding.

Separately, 4-dimensional entanglement distribution across a 1.2 km intra-city free-space link in Vienna, Austria was also demonstrated using photons hyperentangled in polarization and time-energy [66]. This first proof-of-concept experiment established a Bell-state fidelity of 94.19% with a lower bound of 1.4671 ebits of entanglement of formation. As a note, while information encoded with the time degree of freedom—either discrete time-bin or continuous energy-time—does not immediately suffer from turbulence upon transmission, it can become problematic to measure the states as it involves interferometry, where matching wavefronts/spatial distributions are necessary at the output. However, it has been shown that this can be overcome by relay-imaging techniques, thus making time encoding feasible for turbulent environments [67,68]. High-dimensional entanglement distribution shares the same advantages of high-dimensional QKD and is thus an important avenue to explore. We refer the reader to [69] for a comprehensive review on high-dimensional entanglement.

6.4.3 Underwater

The absorption spectrum of water has made underwater communication generally very difficult. Over the last 100 years, as advances have been made in all other forms of communication, underwater communication solutions have remained largely unchanged, relying

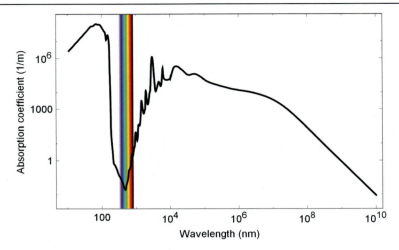

Figure 6.11: Absorption spectrum of water. Here, we see the transmission window in the optical blue-green wavelengths. The wavelength with the lowest absorption is around 480 nm with an absorption coefficient of $a = 0.0186$ m^{-1}. This minimum will shift depending on the type of water. *Source: Data from [71].*

primarily upon acoustic modulation. Though acoustic communication is well suited to the underwater environment due to the great distance which sound can travel through water, it comes with many drawbacks including limited signal rates and security. Water is highly absorptive for the electromagnetic spectrum outside of a transmission window in the blue-green wavelength regime. The wavelengths used for the majority of wireless communication, notably infrared/telecom and radio waves, can propagate no more than a few millimeters to a few meters, respectively [70]. Restricted to this blue-green window, we can consider using collimated laser beams for classical and quantum communication. We must consider absorption, scattering, background noise, optical turbulence, and the photonic degree of freedom that will be used when considering maximum channel lengths which can be achieved for secure quantum communication.

There are two primary considerations that will determine the channel lengths that can be achieved channel attenuation, and errors resulting in lowered state fidelity. The three factors impacting the errors and losses in a channel are absorption, scattering, and turbulence. The absorption will result only in losses in the optical channel. Wavelength-dependent absorption is due to the molecular properties of water and also particles in the water that will absorb light. The absorption spectrum of pure water is shown in Fig. 6.11. It is clear that the blue-green optical wavelengths provide the only viable choice for achieving long distances in the underwater environment. Scattering is also responsible for significant attenuation in the underwater environment. There are many different types of suspended particles present in underwater channels that one does not need to consider in the case of free-space communication links. Mie scattering from these particles in the water must be considered when determining the losses that will be experienced by an optical channel. From a general view of the total magnitude of scattering, water can be characterized by the Jerlov scattering types I–III defining the scattering regimes from most clear to most murky ocean conditions [72]. This scattering has a large impact on the total losses in the channel. In fact in a Jerlov Type III environment, the impact of scattering on channel attenuation will far outweigh losses from the water's absorption at wavelengths in the blue-green window. Simulations of the scattering introduced by various floating particles can be done using the Monte-Carlo method. Using these simulations, we can gain information about how beam propagation will be affected by factors such as the densities of large particles including phytoplankton, which are dependent on the channel depth below the water surface, and the densities of other small particles. We can define the attenuation in a channel generally by the combination of an absorption coefficient $a(\lambda)$ and a scattering coefficient $b(\lambda)$, both of which have a wavelength dependence. This gives the total attenuation coefficient as $c(\lambda) = a(\lambda) + b(\lambda)$, which is used with Beer's law,

$$I(d) = I_0 \exp(-c(\lambda) L), \tag{6.32}$$

to give us the attenuation for a distance L. The most prominent factor influencing scattering and absorption is the chlorophyll concentration which can be separated into specific subsets: fulvic acid, humic acid, and small and large particulate matter. A one-variable model has been developed to determine the concentrations of each of these factors in terms of one chlorophyll concentration value [73]. The chlorophyll concentration in the oceans varies significantly with respect to various factors such as being near the equator or being near a shoreline where nutrients are more plentiful. The total absorption coefficient is given in terms of wavelength-dependent factors, a_i, and concentration C_i for water, fulvic acid, humic acid, and chlorophyll, respectively, by

$$a(\lambda, z) = a_w(\lambda) + a_f(\lambda) C_f(z) + a_h(\lambda) C_h(z) + a_c(\lambda) C_c(z). \tag{6.33}$$

The scattering coefficient is dependent on the water, small particulate matter, and large particulate matter. It is described similarly to the absorption with a wavelength-dependent coefficient, b_i, and the concentration of the particulate matter C_i by

$$b(\lambda, z) = b_w(\lambda) + b_s(\lambda) C_s(z) + b_l(\lambda) C_l(z). \tag{6.34}$$

The concentrations C_i of fulvic acid, humic acid, large particulate matter, and small particulate matter are given by unique literature based factors with respect to the chlorophyll concentration $C_c(z)$ by $C_i(z) = \alpha_i C_c(z) e^{\beta_i C_c(z)}$. The chlorophyll concentration is given by a Gaussian distribution centered on the peak z_{max} where there is the optimal combination of sunlight and nutrients,

$$C_c(z) \propto B_0 + Sz + \frac{h}{\sigma\sqrt{2\pi}} \exp\left(-\frac{(z-z_{max})^2}{2\sigma^2}\right). \tag{6.35}$$

Here, B_0 is the background surface chlorophyll concentration, S is the negative vertical concentration gradient, h is the total chlorophyll above background, and σ is the standard deviation of the chlorophyll gradient. The specific factors and variables are discussed in depth in [73]. The key is to note that the chlorophyll concentration follows a linear decline with depth by Sz and a Gaussian contribution with the peak at z_{max}. In locations which have a high level of chlorophyll at the surface, the maxima of the distribution are around 10–20 meters as sunlight cannot penetrate to deeper depths. Locations with lower surface levels of chlorophyll had maximum concentrations around 50–100 meters below the surface. Combining these models for scattering and absorption from the chlorophyll concentration with the wavelength-dependent absorption of water, one can model the total attenuation to determine the optimal wavelength for optical communication. This study found that 490 nm was the optimal wavelength. Even with the optimal wavelength selected, there is significant variation in the attenuation for different chlorophyll levels, between 0.0428 m^{-1} and 0.428 m^{-1} for the sampled regions.

A number of theoretical investigations have been carried out to see how polarization states specifically will be affected by propagation through an underwater channel. We would imagine that as water is a homogeneous isotropic medium, the polarization states should be largely unaffected by propagation through underwater channels. Monte Carlo simulations can be performed to determine the specific impact scattering will have on polarization states. Scattering does not have a large impact on the error rates observed at least in polarization channels [74]. The polarization of photons can be changed significantly when the scattering angle is large. The majority of the photons which reach the receiving aperture have not been scattered; those photons received which are scattered experience a very small scattering angle and thus the polarization state remains relatively unchanged. This was shown experimentally in the lab for the first time in 2017 in a 3-meter water tank [75]. Here, six sea water samples were taken and the fidelity of the polarization states was found to be above 99% for all samples.

We are left to consider optical turbulence as the primary hindrance to quantum state fidelity in underwater channels. Optical turbulence can be observed in underwater channels in a similar fashion to that seen in free-space communication channels discussed in Section 6.4.2. As mentioned earlier, optical turbulence is created by changes in the refractive index along the beam's path of propagation. A refractive index gradient transverse to the direction of propagation can result in tip-tilt fluctuations and beam wandering, or in the alteration of the profile of the beam when these fluctuations are small with respect to the beam waist. In free-space

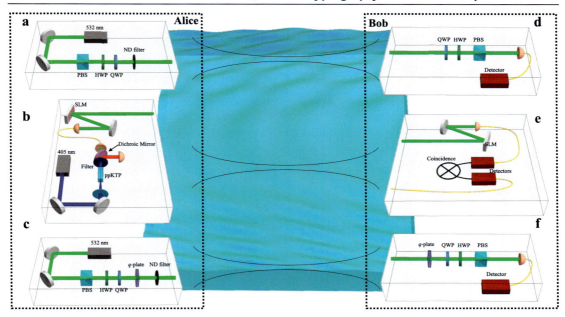

Figure 6.12: Underwater experimental setups. Here we show 3 different experimental approaches to underwater quantum communication. The sender and receiver are shown for: polarization (a, d); orbital angular momentum (b, e); vector vortex modes (c, f). *Source:* [78] [77].

channels, temperature and pressure are the primary factors in producing turbulence. Underwater channels do not suffer from significant localized pressure changes; however, they do experience temperature fluctuations and changes in salinity which can result in a change in the index of refraction. Due to the high viscosity of water in comparison with air, we can expect that beam fluctuations from turbulence will be at a lower frequency in underwater channels. In Tarantino et al., the effect of turbulence on losses in the channel was investigated in detail [76]. They also took into account the effect of background light on the key rates in different scenarios. The simulations determined that quantum communication could be established at distances up to 120 m.

Naturally, with the success of QKD through optical fibers and free-space links, recent efforts have pushed to extend these demonstrations to the underwater environment. Following the entanglement distribution of polarization states through 3 meters of water, several experiments were performed extending the distance and making use of different degrees of freedom. Exiting the laboratory environment, OAM states were used in a 3-meter swimming pool channel [77]. The experimental setup is shown in Fig. 6.12(b) and (e), for Alice and Bob, respectively. Here an SLM is used to generate the photon states. The phase flattening technique is used to detect the states on Bob's side. The OAM states with $\ell = \{-1, 0, +1\}$ were used to characterize a channel for the 3-dimensional BB84 protocol, thus demonstrating the first high-dimensional QKD channel in the underwater environment with a QBER of $e_b^{3D} = 11.73\%$, below the error threshold $e_b^{3D,max} = 15.95\%$. Testing was then expanded to use 4-dimensional states from the OAM space of $\ell = \{-2, -1, +1, +2\}$. Here, the error rate of $e_b^{4D} = 29.77\%$ exceeds the threshold $e_b^{4D,max} = 18.93\%$ to allow for secure communication. To a large extent the errors introduced by turbulence will have some dependence on the size of the modes being sent. A large transverse spatial profile allows for more mode degradation from different Fried cells in the propagation path. We see in this protocol a significant jump in the error rate in the transition from the $\ell = \pm 1$ states to the spatially larger $\ell = \pm 2$ states. A similar experiment was done to show the possibility of using structured vector vortex modes in underwater quantum communication. The experimental setup is shown in Fig. 6.12(c) and (f). The radial and azimuthal states defined by

$$|\Psi_i\rangle \in \left\{ \frac{(|L,-1\rangle + |R,+1\rangle)}{\sqrt{2}}, \frac{(|L,-1\rangle - |R,+1\rangle)}{\sqrt{2}} \right\}, \qquad (6.36)$$

form the first MUB. The second MUB contains the clockwise and counterclockwise sinks given by

$$|\Phi_i\rangle \in \left\{ \frac{(|L,-1\rangle + i|R,+1\rangle)}{\sqrt{2}}, \frac{(|L,-1\rangle - i|R,+1\rangle)}{\sqrt{2}} \right\}. \qquad (6.37)$$

The vector vortex states were used for various channel lengths up to 10 meters, allowing for a study on the effect of channel length on key rates. Polarization states were also studied for channels up to 30 meters in length, with the experimental setup shown in Fig. 6.12(a) and (d). The vector vortex states experience more errors from turbulence in the channel than the polarization states. The error rates for the structured modes were $\approx 1-3\%$ for channel lengths of 1.5, 5.5, and 10.5 m, while the polarization errors were $< 1\%$ for channel lengths up to 30.5 m [78].

In another work, Hu et al. demonstrated the transmission of polarization encoded photons across a 55-meter water channel [79]. An attenuated 532 nm laser is used as the signal. Despite having significant background noises in the detection system, an average state distinguishability of 90.7% was obtained, corresponding to an error rate of 9.3%. Again in a 55-meter underwater channel, Chen et al. observed the propagation of single-photon OAM states [80]. A vortex phase plate placed inside a Sagnac loop is used to generate the OAM states. Depending on the input polarization state, this Sagnac loop can generate pure states $|\ell\rangle$ and $|-\ell\rangle$, as well the superposition states of the form $(|-\ell\rangle + e^{i\phi}|\ell\rangle)/\sqrt{2}$. These states are detected using intensity measurements from a single-photon sensitive ICCD camera. The ICCD measurements show the beam dynamics upon propagation through the turbulent channel.

6.5 Conclusion

Structured states of light provide easy access to high-dimensional Hilbert spaces, and thus increase the available alphabet for quantum information tasks. For example, as shown here, it is more difficult for an eavesdropper to optimally clone high-dimensional systems; thus, high-dimensional encryption schemes can tolerate more noise and channel disturbances. With liquid-crystal based devices, such as SLMs and q-plates, photons can be readily and efficiently structured in the optical (spin and orbit) angular momentum degrees of freedom for applications in quantum communication. Moving forward, the practical implementation of quantum key distribution protocols must be tested beyond the laboratory scale, where uncontrolled factors can introduce excess noise and errors. Though optical fibers can be maintained in a stable environment, specially designed vortex fibers are required in order to coherently transmit structured photons. Furthermore, attenuation in fiber channels limits the possible distances of quantum communication. Thus, free-space infrastructure (such as satellite-to-ground or ground-to-ground hubs) must be constructed in order to globally distribute quantum information. In this free-space environment, high-dimensional protocols operating with structured light are an attractive solution, though methods to correct for atmospheric turbulence still require further exploration. Finally, recent studies have shown that transmitting information optically through underwater channels is a promising avenue for the secure communication between underwater equipment and vehicles. As in the free-space case, further work needs to be done to investigate the compensation of underwater turbulence, particularly in the challenging case of the air-water boundary.

Acknowledgment

This work was supported by Canada Research Chairs (CRC), Canada First Excellence Research Fund (CFREF), and Ontario's Early Researcher Award. We would like to thank our colleagues, and in particular Dr. Frédéric Bouchard with whom we conducted most of this research.

References

[1] Ronald L. Rivest, Adi Shamir, Leonard M. Adleman, A method for obtaining digital signatures and public-key cryptosystems, Commun. ACM 21 (2) (1978) 120–126.
[2] C.H. Bennett, G. Brassard, in: Proceedings of the IEEE International Conference on Computers, Systems, and Signal Processing, Bangalore, India, 1984, 1984.
[3] K.Y. Bliokh, Y.P. Bliokh, S. Savel'Ev, F. Nori, Semiclassical dynamics of electron wave packet states with phase vortices, Phys. Rev. Lett. 99 (2007) 190404.
[4] D.F.V. James, P.G. Kwiat, W.J. Munro, A.G. White, Measurement of qubits, Phys. Rev. A 64 (2001) 052312.
[5] E. Karimi, Generation and manipulation of laser beams carrying orbital angular momentum for classical and quantum information applications, Ph.D. dissertation, Universitá degli studi di Napoli "Federico II", 2009.
[6] E. Bolduc, N. Bent, E. Santamato, E. Karimi, R.W. Boyd, Exact solution to simultaneous intensity and phase encryption with a single phase-only hologram, Opt. Lett. 38 (2013) 3546.

[7] A. Mair, A. Vaziri, G. Weihs, A. Zeilinger, Entanglement of the orbital angular momentum states of photons, Nature 412 (2001) 313–316.
[8] H. Qassim, et al., Limitations to the determination of a Laguerre–Gauss spectrum via projective, phase-flattening measurement, JOSA B 31 (2014) A20–A23.
[9] F. Bouchard, et al., Measuring azimuthal and radial modes of photons, Opt. Express 26 (2018) 31925–31941.
[10] G.C. Berkhout, M.P. Lavery, J. Courtial, M.W. Beijersbergen, M.J. Padgett, Efficient sorting of orbital angular momentum states of light, Phys. Rev. Lett. 105 (2010) 153601.
[11] M. Mirhosseini, M. Malik, Z. Shi, R.W. Boyd, Efficient separation of the orbital angular momentum eigenstates of light, Nat. Commun. 4 (2013) 2781.
[12] Z. Bomzon, G. Biener, V. Kleiner, E. Hasman, Space-variant Pancharatnam–Berry phase optical elements with computer-generated subwavelength gratings, Opt. Lett. 27 (2002) 1141–1143.
[13] L. Marrucci, C. Manzo, D. Paparo, Optical spin-to-orbital angular momentum conversion in inhomogeneous anisotropic media, Phys. Rev. Lett. 96 (2006) 163905.
[14] E. Karimi, et al., Generating optical orbital angular momentum at visible wavelengths using a plasmonic metasurface, Light Sci. Appl. 3 (2014) e167.
[15] E. Cohen, et al., Geometric phase from Aharonov–Bohm to Pancharatnam–Berry and beyond, Nat. Rev. Phys. 1 (2019) 437–449.
[16] E. Karimi, et al., Spin–orbit hybrid entanglement of photons and quantum contextuality, Phys. Rev. A 82 (2010) 022115.
[17] J.L. Park, The concept of transition in quantum mechanics, Found. Phys. 1 (1970) 23–33.
[18] A. Peres, How the no-cloning theorem got its name, Fortschr. Phys. 51 (2003) 458–461.
[19] N. Herbert, Flash—a superluminal communicator based upon a new kind of quantum measurement, Found. Phys. 12 (1982) 1171–1179.
[20] W.K. Wootters, W.H. Zurek, A single quantum cannot be cloned, Nature 299 (1982) 802–803.
[21] E. Nagali, et al., Optimal quantum cloning of orbital angular momentum photon qubits through Hong–Ou–Mandel coalescence, Nat. Photonics 3 (2009) 720–723.
[22] F. Bouchard, R. Fickler, R.W. Boyd, E. Karimi, High-dimensional quantum cloning and applications to quantum hacking, Sci. Adv. 3 (2017) e1601915.
[23] C.-K. Hong, Z.-Y. Ou, L. Mandel, Measurement of subpicosecond time intervals between two photons by interference, Phys. Rev. Lett. 59 (1987) 2044.
[24] F. Bouchard, et al., Two-photon interference: the Hong-Ou-Mandel effect, arXiv preprint, arXiv:2006.09335, 2020.
[25] N. Gisin, G. Ribordy, W. Tittel, H. Zbinden, Quantum cryptography, Rev. Mod. Phys. 74 (2002) 145.
[26] S. Pirandola, et al., Advances in quantum cryptography, arXiv preprint, arXiv:1906.01645, 2019.
[27] F. Bouchard, et al., Experimental investigation of high-dimensional quantum key distribution protocols with twisted photons, Quantum 2 (2018) 111.
[28] N.J. Cerf, M. Bourennane, A. Karlsson, N. Gisin, Security of quantum key distribution using d-level systems, Phys. Rev. Lett. 88 (2002) 127902.
[29] M. Mirhosseini, et al., High-dimensional quantum cryptography with twisted light, New J. Phys. 17 (2015) 033033.
[30] Y.C. Liang, D. Kaszlikowski, B.-G. Englert, L.C. Kwek, C.H. Oh, Tomographic quantum cryptography, Phys. Rev. A 68 (2003) 022324.
[31] B.-G. Englert, et al., Efficient and robust quantum key distribution with minimal state tomography, arXiv preprint, arXiv:quant-ph/0412075, 2008.
[32] N. Bent, et al., Experimental realization of quantum tomography of photonic qubits via symmetric informationally complete positive operator-valued measures, Phys. Rev. X 5 (2015) 041006.
[33] T. Sasaki, Y. Tamamoto, M. Koashi, Practical quantum key distribution protocol without monitoring signal disturbance, Nature 509 (2014) 475–478.
[34] K. Inoue, E. Waks, Y. Yamamoto, Differential phase shift quantum key distribution, Phys. Rev. Lett. 89 (2002) 037902.

[35] F. Bouchard, A. Sit, K. Heshami, R. Fickler, E. Karimi, Round-Robin differential-phase-shift quantum key distribution with twisted photons, Phys. Rev. A 98 (2018) 010301.
[36] H. Chau, Quantum key distribution using qudits that each encode one bit of raw key, Phys. Rev. A 92 (2015) 062324.
[37] H. Chau, Q. Wang, C. Wong, Experimentally feasible quantum-key-distribution scheme using qubit-like qudits and its comparison with existing qubit- and qudit-based protocols, Phys. Rev. A 95 (2017) 022311.
[38] U. Fano, Description of states in quantum mechanics by density matrix and operator techniques, Rev. Mod. Phys. 29 (1957) 74.
[39] I.L. Chuang, M.A. Nielsen, Prescription for experimental determination of the dynamics of a quantum black box, J. Mod. Opt. 44 (1997) 2455–2467.
[40] F. Bouchard, et al., Quantum process tomography of a high-dimensional quantum communication channel, Quantum 3 (2019) 138.
[41] G.B. Xavier, G. Lima, Quantum information processing with space-division multiplexing optical fibres, Commun. Phys. 3 (2020) 1–11.
[42] S. Ramachandran, P. Kristensen, Optical vortices in fiber, Nanophotonics 2 (2013) 455–474.
[43] B. Ndagano, R. Brüuning, M. McLaren, M. Duparré, A. Forbes, Fiber propagation of vector modes, Opt. Express 23 (2015) 17330–17336.
[44] N. Bozinovic, et al., Terabit-scale orbital angular momentum mode division multiplexing in fibers, Science 340 (2013) 1545–1548.
[45] C. Brunet, P. Vaity, Y. Messaddeq, S. LaRochelle, L.A. Rusch, Design fabrication and validation of an oam fiber supporting 36 states, Opt. Express 22 (2014) 26117–26127.
[46] A. Sit, et al., Quantum cryptography with structured photon through a vortex fiber, Opt. Lett. 43 (2018) 4108–4111.
[47] D. Cozzolino, et al., Orbital angular momentum states enabling fiber-based high-dimensional quantum communication, Phys. Rev. Appl. 11 (2019) 064058.
[48] D. Cozzolino, et al., Air-core fiber distribution of hybrid vector vortex-polarization entangled states, Adv. Photonics 1 (2019) 046004.
[49] J. Yin, et al., Satellite-based entanglement distribution over 1200 kilometers, Science 356 (2017) 1140–1144.
[50] G. Vallone, et al., Experimental satellite quantum communications, Phys. Rev. Lett. 115 (2015) 040502.
[51] S.-K. Liao, et al., Satellite-to-ground quantum key distribution, Nature 549 (2017) 43–47.
[52] M. Krenn, et al., Twisted light transmission over 143 km, Proc. Natl. Acad. Sci. 113 (2017) 13648–13653.
[53] G. Vallone, et al., Free-space quantum key distribution by rotation-invariant twisted photons, Phys. Rev. Lett. 113 (2014) 060503.
[54] A. Sit, et al., High-dimensional intracity quantum cryptography with structured photons, Optica 4 (2017) 1006.
[55] L.C. Andrews, R.L. Philips, Laser Beam Propagation Through Random Media, 2nd edition, SPIE Optical Engineering Press, Bellingham, Wash, 2005.
[56] A. Kolmogorov, The local structure of turbulence in incompressible viscous fluid for very large Reynolds' numbers, Dokl. Akad. Nauk SSSR 30 (1941) 301.
[57] F. Dios, J.A. Rubio, A. Rodrìguez, A. Comerón, Scintillation and beam-wander analysis in an optical ground station-satellite uplink, Appl. Opt. 43 (2004) 3866–3873.
[58] R.J. Noll, Zernike polynomials and atmospheric turbulence, J. Opt. Soc. Am. 66 (1976) 207–211.
[59] R.G. Lane, A. Glindemann, J.C. Dainty, Simulations of a Kolmogorov phase screen, Waves Random Media 2 (1992) 209.
[60] I. Toselli, O. Korotkova, X. Xiao, D.G. Voelz, Slm-based laboratory simulations of Kolmogorov and non-Kolmogorov anisotropic, turbulence, Appl. Opt. 54 (2015) 4740.
[61] Y. Ren, et al., Turbulence compensation of an orbital angular momentum and polarization-multiplexed link using a data-carrying beacon on a separate wavelength, Opt. Lett. 40 (2015) 2249–2252.
[62] G. Xie, et al., Phase correction for a distorted orbital angular momentum beam using a Zernike polynomials-based stochastic-parallel-gradient-descent algorithm, Opt. Lett. 40 (2015) 1197–1200.

[63] M.A. Cox, C. Rosales-Guzmán, M.P.J. Lavery, D.J. Versfeld, A. Forbes, On the resilience of scalar and vector vortex modes in turbulence, Opt. Express 24 (2016) 18105–18113.
[64] M.P.J. Lavery, et al., Free-space propagation of high-dimensional structured optical fields in an urban environment, Sci. Adv. 3 (2017) e1700552.
[65] V. D'Ambrosio, et al., Complete experimental toolbox for alignment-free quantum communication, Nat. Commun. 3 (2012) 961.
[66] F. Steinlechner, et al., Distribution of high-dimensional entanglement via an intra-city free-space link, Nat. Commun. 8 (2017) 15971.
[67] G. Vallone, et al., Interference at the single photon level along satellite-ground channels, Phys. Rev. Lett. 116 (2016) 253601.
[68] J. Jin, et al., Demonstration of analyzers for multimode photonic time-bin qubits, Phys. Rev. A 97 (2018) 043847.
[69] M. Erhard, M. Krenn, A. Zeilinger, Advances in high-dimensional quantum entanglement, Nat. Rev. Phys. 2 (2020) 365–381.
[70] L. Lanbo, Z. Shengli, C. Jun-Hong, Prospects and problems of wireless communication for underwater sensor networks, Wirel. Commun. Mob. Comput. 8 (2008) 977–994.
[71] D.J. Segelstein, The complex refractive index of water, Master's thesis, University of Missouri-Kansas City, 1981.
[72] N.G. Jerlov, Optical Oceanography, American Elsevier Publ. Co., Inc., New York, 1968, pp. 1–194, https://www.sciencedirect.com/bookseries/elsevier-oceanography-series/vol/5.
[73] L.J. Johnson, R.J. Green, M.S. Leeson, Underwater optical wireless communications: depth dependent variations in attenuation, Appl. Opt. 52 (2013) 7867–7873.
[74] P. Shi, S.-C. Zhao, Y.-J. Gu, W.-D. Li, Channel analysis for single photon underwater free space quantum key distribution, JOSA A 32 (2015) 349–356.
[75] L. Ji, et al., Towards quantum communications in free-space seawater, Opt. Express 25 (2017) 19795–19806.
[76] S. Tarantino, B. Da Lio, D. Cozzolino, D. Bacco, Feasibility of quantum communications in aquatic scenarios, Optik 216 (2020) 164639.
[77] F. Bouchard, et al., Quantum cryptography with twisted photons through an outdoor underwater channel, Opt. Express 26 (2018) 22563.
[78] F. Hufnagel, et al., Underwater quantum communication over a 30-m flume tank, arXiv preprint, arXiv:2004.04821, 2020.
[79] C.-Q. Hu, et al., Transmission of photonic polarization states through 55-meter water: towards air-to-sea quantum communication, Photon. Res. 7 (2019) A40–A44.
[80] Y. Chen, et al., Underwater transmission of high-dimensional twisted photons over 55 meters, PhotoniX 1 (2020) 1–11.

CHAPTER 7

Spin and orbital angular momentum coupling

Lorenzo Marrucci

Dipartimento di Fisica "Ettore Pancini", Università di Napoli Federico II, Napoli, Italy

Contents

7.1 Introduction 177
7.2 Paraxial spin-orbit coupling: q-plates, meta-surfaces and similar devices 181
7.3 Non-paraxial spin-orbit coupling: spin Hall effect of light and optical fibers 187
7.4 Applications to optical communication 193
7.5 Conclusions 200
References 200

7.1 Introduction

Just as they carry energy and momentum, electromagnetic waves may transport angular momentum. The general expression of the total angular momentum **J** associated with a given electromagnetic field can be written as follows [1,2]:

$$\mathbf{J} = \varepsilon_0 \int \mathbf{r} \times (\mathbf{E} \times \mathbf{B}) \, d^3 r, \tag{7.1}$$

where **E**, **B** are the electric and magnetic fields, **r** is the position vector, and ε_0 is the vacuum permittivity. As for other conserved quantities, the importance of this quantity lies in the fact that, when combined with the angular momentum associated with matter, it defines a conservation law having universal validity. Moreover, as we know from quantum physics or from Noether's theorem, **J** is related to the transformation properties of the system for global rotations. Owing to the vector nature of the electromagnetic fields, the total angular momentum **J** of the electromagnetic field can be further split into two "parts", a spin **S** and an orbital **L**, so that $\mathbf{J} = \mathbf{S} + \mathbf{L}$ [3–8]. Following Refs. [3,4], we can give the following general expressions for these two terms[1]:

$$\mathbf{S} = \varepsilon_0 \int \mathbf{E} \times \mathbf{A}_\perp d^3 r,$$
$$\mathbf{L} = \varepsilon_0 \int \sum_h E_h (\mathbf{r} \times \nabla) A_{\perp,h} d^3 r, \quad (7.2)$$

where $\mathbf{A}_\perp$ is the transverse (zero divergence) part of the vector potential and the index h of the sum refers to the three Cartesian coordinates. As explained, for example, in Refs. [7,9], although not evident from Eqs. (7.2). The spin $\mathbf{S}$ is related to the transformation of the system for a rotation of all field vectors "on the spot", while $\mathbf{L}$ is related to the rotation of the whole field spatial distribution around the origin of the coordinate system, while keeping constant all field vector orientations of the moving points. Owing to the transversality condition of electromagnetic fields, however, these separate rotations, when taken alone, do not generate valid electromagnetic fields, and hence a further adjustment of the fields is needed after the rotation to obtain a physically valid transformation (see, e.g., Refs. [7,9] for further details). For this reason, unlike $\mathbf{J}$, in general $\mathbf{S}$ and $\mathbf{L}$ do not separately behave as "true" angular momenta [3,4]. Yet, they do represent separately conserved quantities for electromagnetic fields in free space and in homogeneous transparent isotropic media. Moreover, when electromagnetic waves interact with matter, the resulting changes of $\mathbf{S}$ and $\mathbf{L}$ can often be associated with distinct properties of matter and distinct phenomena.

In the important regime of paraxial waves, with a main propagation axis z, the z-component of each angular momentum term acquires a clear, unambiguous physical meaning. Indeed, the *spin angular momentum* S_z (SAM) is associated with the polarization state of the wave, and in particular with the polarization ellipticity (or, equivalently, the normalized Stokes parameter s_3). SAM eigenstates correspond to circular polarizations, with $S_z = s\hbar$ per photon for left/right handedness, with $s = \pm 1$ (with the naming convention based on the receiver point-of-view). The *orbital angular momentum* L_z (OAM), on the other hand, is associated with the phase and amplitude distribution of the wave around the z axis. OAM eigenstates, in particular, have a helical phase azimuthal distribution as given by the phase factor $e^{im\varphi}$, with $L_z = m\hbar$ per photon, φ being the azimuthal angle around z (while the amplitude must be φ-independent). Hence, the eigenstates of total angular momentum J_z (TAM) can be identified with circularly polarized helical waves, and have $J_z = j\hbar$ per photon, $j = m + s$ being an integer. Notice, however, that left-handed waves with a given OAM m have the same TAM as

---

[1] It should be noted that several alternative choices are possible for the spin and orbital *densities*, that is, for the integrands appearing in Eqs. (7.2), although all choices lead to the same integrated global quantities. The actual physical meaning of these densities and the possible superiority of one expression over the others is debated; see, for example, Refs. [5,6,8–10].

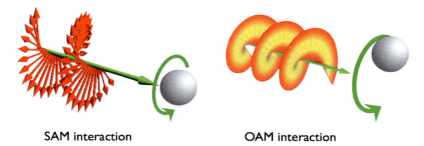

Figure 7.1: Pictorial representation of the two forms of optical angular momentum (for eigenstates): SAM (left panel), corresponding to a helical structure of the electric and magnetic vector fields and OAM (right panel), corresponding to a helical structure of the optical wavefront. When impinging on an absorbing small particle, SAM will induce a rotation of the particle around itself, while OAM will induce a rotation of the particle around the beam axis. Credit: Image by E. Karimi, CC BY-SA 3.0, downloaded from https://commons.wikimedia.org/w/index.php?curid=16630980.

right-handed waves with an OAM $m + 2$ (i.e., these two waves are TAM-degenerate). Arbitrary linear superpositions of these wave pairs are hence also TAM eigenstates, but not SAM and OAM eigenstates. This feature will turn out to be important in certain phenomena, as we shall discuss further below. A pictorial representation of SAM and OAM and of its mechanical effect on a small absorbing particle is shown in Fig. 7.1.

Unlike the general non-paraxial case, in the paraxial limit S_z and L_z are legitimate angular momenta (z-component), because, for rotations around the z axis, field transversality is anyway ensured by the paraxiality of the wave. While these two quantities are separately conserved for propagation in vacuum or in any homogeneous isotropic transparent medium, they may interact and be exchanged with matter in more complex media. In a strict paraxial limit and for transparent materials, SAM is exchanged only with locally anisotropic media, via their birefringence that affects polarization, while OAM may be exchanged only with inhomogeneous media, via the inhomogeneous refractive index that affects the wave phase spatial distribution. Of course, a medium that is both inhomogeneous and anisotropic interacts with the light SAM and OAM at the same time. This in turn opens up the possibility of having also a medium-mediated *spin-orbit interaction* (SOI) (or spin-orbit coupling) between SAM and OAM [11]. This interaction corresponds, for example, to an exchange of electromagnetic OAM that is controlled by the value of SAM and vice versa. One should, however, notice here that *not all* phenomena in which electromagnetic SAM and OAM are simultaneously exchanged with a medium lead to SOI. For example, in ordinary isotropic absorbing materials both SAM and OAM are transferred from the field to the medium, but there is no SOI.

In addition to the important case of inhomogeneous anisotropic media, electromagnetic spin-orbit interaction may arise in other kinds of media owing to non-paraxiality, that is, when S_z and L_z become mixed up with other components of **S** and **L** after a wave deflection [12]. In particular, nanoscale optics is intrinsically non-paraxial, as the local (possibly evanescent) electromagnetic fields develop large wavevector components. As a consequence, systems having structural inhomogeneities at a nanoscale, such as nanoparticles, meta-materials or even the simple abrupt interfaces between adjacent homogeneous media, may strongly affect both SAM and OAM of electromagnetic waves and eventually give rise to spin-orbit coupling, as we shall discuss. Another important non-paraxial phenomenon, which has been fully clarified only recently, is the so-called *transverse spin*, that is, a nonzero **S** density appearing in confined waves which is perpendicular to the propagation direction [12,13]. Transverse spin orientation is always locked to the propagation (or momentum) direction, and this leads to another form of spin-orbit coupling [14].

Besides angular momentum, another fundamental concept to understand spin-orbit interaction of light is that of *geometric phases* [15,16]. These are phases occurring when a vector field, such as the electromagnetic one, is subjected to continuous transformations of its orientation, with a given definition of "parallel transport" for the vector in the control-parameter space. A unified recent treatment of geometric phases in optics is reported in Refs. [17,18]. The best-known limiting cases are the Pancharatnam-Berry (PB) phase, which applies for a sequence of polarization transformations of a paraxial wave, and the Rytov–Vladimirskii–Berry (RVB) phase (also known as "spin-redirection phase"), which can be used when circular polarized waves are subjected to a sequence of deflections in propagation direction (hence, in a non-paraxial case). These geometric phases share the feature that the input value of the electromagnetic spin (or more precisely, of the helicity, that is the spin component along the wavevector **k**, as given by **S** · **k**) determines the sign of the resulting phase acquired in the transformation. The geometric phase in turn contributes to redetermining the OAM, or more generally the spatial structure and subsequent propagation of the wave. Hence, this mechanism corresponds to a spin-orbit interaction or coupling [12].

Given the relation between electromagnetic spin and polarization, it is frequent to find the expression "spin-orbit interaction" used in the recent scientific literature to describe any kind of anisotropic electromagnetic process in which the wave structure and propagation are affected by its polarization. However, this is probably stretching the expression's meaning a bit too far. A proper definition of spin-orbit coupling should be limited to those wave phenomena that can be best explained as being controlled by proper "spin", that is specifically by the circular polarization handedness, as opposed to generic polarization. In most cases, these arise from geometric phases, although one may possibly include in this definition also other phenomena unrelated to geometric phases, such as circular birefringence or the Faraday effect. However, the effects of standard linear birefringence in a homogeneous medium should not be termed a spin-orbit coupling.

An introduction to the main spin-orbit optical phenomena and to their relevance for optical communication will be the main subject of this chapter. In particular, in Section 7.2, the working principle of q-plates and related devices, which are birefringent plates having suitably patterned optic axis, will be illustrated. In Section 7.3, the role of non-paraxial spin-orbit effects, with a specific focus on the spin-Hall effect of light, is discussed mainly in the context of optical fibers. Potential applications of these effects to classical and quantum optical communication will finally be reviewed in Section 7.4.

7.2 Paraxial spin-orbit coupling: q-plates, meta-surfaces and similar devices

Let us consider a slab of birefringent medium having a thickness and material birefringence chosen so as to induce a homogeneous phase retardation of δ, at the working wavelength λ, for light propagation perpendicular to the slab (z axis). The optic axis **n** is assumed to be uniform in the z direction, but inhomogeneous in the (x-y) plane of the slab, according to a prescribed pattern $\mathbf{n}(x, y)$. Such an inhomogeneous birefringent system can be practically realized for example using liquid crystals (as in the case of the first q-plates) [11,19], or liquid-crystalline polymers [20,21], or even by just gluing together differently oriented pieces of solid crystals. An equivalent effective birefringence can also be obtained by patterned sub-wavelength gratings or more general metasurfaces [22–26]. All these cases can be understood in a similar way, as we shall discuss here.[2] Neglecting all diffraction effects occurring in the propagation within the slab, as is valid for sufficiently thin slabs, one can use the Jones formalism to describe the optical propagation through the medium. In the Jones formalism, the optical electric field is generally described by a two-complex-component column vector $|E\rangle = \begin{bmatrix} E_1 \\ E_2 \end{bmatrix}$, here represented also in quantum-like "ket" Dirac notation. In the most common choice of "standard basis", the two components of $|E\rangle$ are identified with linearly polarized fields oscillating as x and y (which we conventionally identify with "horizontal" $|H\rangle$ and "vertical" $|V\rangle$ polarizations), in which case one has $E_1 = E_x$, $E_2 = E_y$ and $|H\rangle = \begin{bmatrix} 1 \\ 0 \end{bmatrix}$, $|V\rangle = \begin{bmatrix} 0 \\ 1 \end{bmatrix}$. In this linear basis, left-right circularly polarizations are then represented by the unit vectors $|L\rangle = \frac{1}{\sqrt{2}} \begin{bmatrix} 1 \\ i \end{bmatrix} = \frac{1}{\sqrt{2}}(|H\rangle + i |V\rangle)$ and $|R\rangle = \frac{1}{\sqrt{2}} \begin{bmatrix} 1 \\ -i \end{bmatrix} = \frac{1}{\sqrt{2}}(|H\rangle - i |V\rangle)$.[3] In the following, we adopt this linear-polarization standard basis.

---

[2] In the case of plasmonic metasurfaces there are also resonance-related dynamic dephasing mechanisms, which are unrelated to geometric phases and will not be discussed here.

[3] On the other hand, other standard-basis choices are also possible, for example circular polarizations themselves can be taken as basis vectors. In the latter case, one has $E_1 = E_L$, $E_2 = E_R$ and $|L\rangle = \begin{bmatrix} 1 \\ 0 \end{bmatrix}$, $|R\rangle = \begin{bmatrix} 0 \\ 1 \end{bmatrix}$, while $|H\rangle = \frac{1}{\sqrt{2}} \begin{bmatrix} 1 \\ 1 \end{bmatrix} = \frac{1}{\sqrt{2}}(|L\rangle + |R\rangle)$ and $|V\rangle = \frac{1}{i\sqrt{2}} \begin{bmatrix} 1 \\ -1 \end{bmatrix} = \frac{1}{i\sqrt{2}}(|L\rangle - |R\rangle)$.

Let us now consider the medium action in the Jones formalism. Let $\alpha(x, y)$ be the angle between $\mathbf{n}(x, y)$ and a fixed reference axis x. The Jones matrix $\hat{M}$ describing the medium action on the field at each transverse position x, y is the same as that of a generic birefringent component and can be written in the following three equivalent ways[4]:

$$\hat{M}(x, y) = \cos\frac{\delta}{2}\begin{bmatrix} 1 & 0 \\ 0 & 1 \end{bmatrix} + i\sin\frac{\delta}{2}\begin{bmatrix} \cos 2\alpha & \sin 2\alpha \\ \sin 2\alpha & -\cos 2\alpha \end{bmatrix} \quad (7.3)$$

$$= \cos\frac{\delta}{2}\hat{I} + i\sin\frac{\delta}{2}\left(\hat{\sigma}_1 \sin 2\alpha + \hat{\sigma}_3 \cos 2\alpha\right) = e^{i\left(\frac{\delta}{2}\right)\mathbf{u}\cdot\hat{\sigma}},$$

where in the second expression we have introduced the identity 2×2 matrix $\hat{I}$ and the Pauli matrices $\hat{\sigma}_1, \hat{\sigma}_2, \hat{\sigma}_3$ with their standard definitions, and in the third expression we have introduced the Pauli 3-vector $\hat{\sigma} = (\hat{\sigma}_1, \hat{\sigma}_2, \hat{\sigma}_3)$ and the unit 3-vector $\mathbf{u} = (\sin 2\alpha, 0, \cos 2\alpha)$.

A generic circularly polarized input wave with complex amplitude $E_0(x, y)$, described by the following ket:

$$|\psi_{in}\rangle = \frac{E_0(x, y)}{\sqrt{2}}(|H\rangle \pm i|V\rangle) = E_0(x, y)|L/R\rangle, \quad (7.4)$$

(where $+$ is for the left-circular case and $-$ for the right-circular one, and $|L/R\rangle$ stands for $|L\rangle$ or $|R\rangle$, accordingly) will be transformed by the action of the medium into the following outgoing field (up to an overall phase):

$$|\psi_{out}\rangle = \hat{M}|\psi_{in}\rangle = E_0(x, y)\cos\frac{\delta}{2}|L/R\rangle + iE_0(x, y)e^{\pm i2\alpha(x,y)}\sin\frac{\delta}{2}|R/L\rangle. \quad (7.5)$$

It is seen from this equation that the output wave is in general the coherent superposition of two circularly polarized waves with opposite circular polarizations. The first term (sometimes named "unconverted" wave) in Eq. (7.5) is identical to the input wave, except for an amplitude reduction by the factor $\cos(\delta/2)$. The second term ("converted" wave) is instead flipped in polarization handedness, rescaled by the amplitude factor $\sin(\delta/2)$, and it has acquired a nonuniform phase retardation given by

$$\Delta\Phi(x, y) = \pm 2\alpha(x, y), \quad (7.6)$$

(plus a uniform $\pi/2$ phase, as given by the i factor). This phase is a PB geometric phase (with an arbitrary additive constant), related only to the geometry of the polarization transformation imparted by the medium.

---

[4] It should also be noticed that we are ignoring here a possible overall phase, and hence the $\hat{M}$ matrix is chosen to be a SU(2) (unit-determinant unitary) matrix.

Spin and orbital angular momentum coupling 183

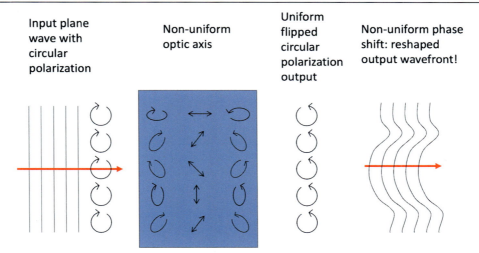

Figure 7.2: Schematic representation of the concept of a PB phase optical element. A circularly polarized plane wave passes through the element and the polarization at each point in the transverse plane undergoes a different trajectory in the Poincaré sphere, owing to the different optic axis orientations. At the exit, the polarization is again uniform and circular, but it is flipped in handedness. The PB geometric phase associated with the polarization manipulation gives rise to a reshaped optical wavefront. *Credit: Original image by the author.*

The second wavefront-reshaped term in Eq. (7.5) is often the desired output, although in certain applications (for example when generating polarization singularities [27]) the whole superposition may be needed. When necessary, the second term can be separated from the first by circular-polarization filtering. Alternatively, by setting $\delta = \pi$ (corresponding to half-wave birefringent retardation) one may entirely cancel the first term and maximize the amplitude of the second (this operation is usually called "tuning", and $\delta = \pi$ then corresponds to "optimal tuning" [28]).

Notice that the described result is very flexible: any wavefront reshaping as specified by the transverse phase retardation $\Delta\Phi(x, y)$ can be generated by a suitable pattern of the birefringent medium. The needed optic-axis geometry is fixed by Eq. (7.6), where the sign is determined by the circular polarization handedness that will be employed. The concept is schematically illustrated in Fig. 7.2. The same device will also generate the conjugate wavefront-reshaping $-\Delta\Phi(x, y)$ if the input polarization handedness is inverted. Even more complex field structures could be in principle created by patterning at the same time both α and δ, although this is currently challenging for most technologies (but see, for example, [26]). In the scientific literature, wavefront-reshaping devices based on this principle have been variedly named "polarization holograms", "Pancharatnam-Berry phase optical elements", "q-plates",

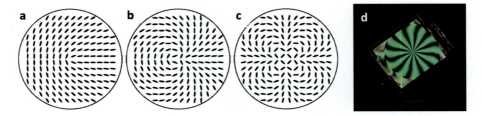

Figure 7.3: (a-c) Examples of optic axis pattern of q-plates for the following values of the charge q: (a) $q = 1/2$, (b) $q = 3/2$, (c) $q = 3$. The local orientation of the axis is shown as small rods. (c) Photograph of a q-plate (with $q = 3$) seen between crossed polarizers, so as to visualize the birefringence pattern. Electrical contacts are used for electric tuning. *Credit: Images by the author and his coworkers. Right-panel photo by Sergei Slussarenko.*

"diffractive waveplates", "metasurfaces", etc., depending on the details of the specific realization and on the degree of generality of the proposed applications [11,19–22,24,29–31]. The polarization-controlled wavefront reshaping can be considered as a manifestation of SOI, as the wavefront alteration can in turn be related to variations in the OAM components of the wave.

Let us now consider the specific optic-axis pattern given by the following law:

$$\alpha(r, \varphi) = \alpha_0 + q\varphi, \tag{7.7}$$

where α_0 and q are constants and r, φ are polar coordinates in the (x-y) plane (see Fig. 7.3 for some examples). A medium with such optic axis pattern is what has been originally named as "q-plate" in the literature, particularly in the case of liquid crystal devices [28,32].[5] The constant q must be an integer or a half-integer to avoid discontinuity surfaces in the medium. Along the z-axis, in $r = 0$, there is a topological singularity line of "charge" q of the optic axis pattern. Inserting Eq. (7.7) in Eq. (7.5), we obtain the field generated by a q-plate for the case of a circularly polarized input wave. Apart from an irrelevant overall phase and assuming that the input field has a uniform phase and that the q-plate has optimal tuning $\delta = \pi$, the outgoing "converted" wave field is given by the following expression:

$$|\psi_{out}\rangle = i E_0 e^{\pm i 2 q \varphi \pm i 2 \alpha_0} |R/L\rangle. \tag{7.8}$$

We see that this output field corresponds to a helical mode with OAM $m = \pm 2q$. In other words, the output OAM sign is controlled by the input polarization handedness, a clear example of SOI (see Fig. 7.4). The OAM magnitude $|m|$ is fixed by the medium geometry, via the parameter q.

---

[5] Similar devices, named as "polarization converters", were originally introduced to pattern the polarization orientation, without explicit reference to OAM or wavefront manipulation. See, for example, Ref. [33].

Spin and orbital angular momentum coupling 185

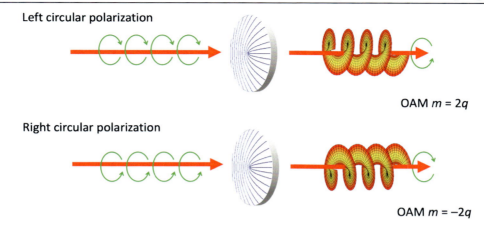

Figure 7.4: Schematic representation of the effect of a q-plate on an input plane wave beam. If the input wave is left-circular-polarized (upper panel) the output wave will be helical, with OAM eigenvalue $m = 2q$. If the input wave is right-circular-polarized (lower panel) the output OAM eigenvalue will be $m = -2q$. In both cases, the output is circularly polarized with opposite handedness. The picture actually corresponds to the case $q = 1/2$, with the corresponding optic axis pattern visible in the q-plate and the outgoing wavefront showing a single helix, because $m = \pm 1$. Credit: Images by the author. Reprinted with permission from [34].

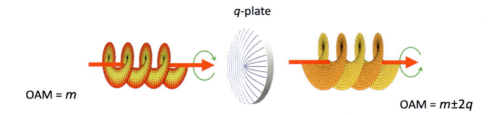

Figure 7.5: Schematic representation of the OAM-shifting behavior of a q-plate when the input has already nonzero OAM. The picture actually corresponds to the case $q = 1/2$ and an input with $m = \pm 1$, for which the shift is by ± 1 and the output has $m = \pm 2$. Credit: Original image by the author.

If the input OAM (encoded in the phase structure of the input wave) is nonzero, then the q-plate transformation leads to $m_{out} = m_{in} \pm 2q$, that is, to an OAM *shift* by the integer $2q$, with the input polarization handedness controlling the shift direction (see Fig. 7.5). Introducing a ket notation also for the OAM state, with $|m\rangle$ representing a phase factor $e^{im\varphi}$, one can write the following q-plate general transformation laws (this is in the case of optimal tuning, but it can be easily generalized):

$$|L, m\rangle \xrightarrow{\text{q-plate}} e^{i(\pi/2 + 2\alpha_0)} |R, m + 2q\rangle,$$
$$|R, m\rangle \xrightarrow{\text{q-plate}} e^{i(\pi/2 - 2\alpha_0)} |L, m - 2q\rangle. \tag{7.9}$$

In these laws, the radial profile is not mentioned explicitly. The laws are actually valid strictly only for ideally thin q-plates and refer to the near field at input and output, which have the same radial profile. The diffraction from the OAM azimuthal phase will rapidly affect the radial profile of the converted wave (for example by developing the vortex amplitude "hole", if it is not present in the input), so that the radial structure of the output wave is actually not the same as that of the input one. The propagation of these q-plate-generated OAM eigenstates, when the input is a Gaussian beam can be well described by the so-called *circular beams* (or by the subfamily of hypergeometric-gaussian modes, when the q-plate is placed at the focal plane of the input Gaussian beam) [35,36].

It is interesting to consider the angular momentum balance for a single circularly polarized photon passing through a (optimal tuned) q-plate [11,37]. The input photon SAM is $\pm\hbar$, while we take the OAM to be $m_{in}\hbar$. The TAM of the photon is hence $(m_{in} \pm 1)\hbar$. At the output of the plate, the polarization handedness is inverted and the OAM shifted, so that the SAM is now $\mp\hbar$ and the OAM is $m_{out}\hbar = (m_{in} \pm 2q)\hbar$. The output TAM is then $[m_{in} \pm (2q - 1)]\hbar$, with a photon TAM overall variation $\Delta J_z = \pm 2(q - 1)\hbar$. Since angular momentum must be conserved, this variation must be exchanged with the medium, that is, with the q-plate, in the form of an optical torque. The case $q = 1$ leads to a special result: the photon crossing the q-plate does not change its TAM, and therefore no torque is generated on the medium. However, it must be emphasized that SAM and OAM of the photon are both varied in the q-plate: the SAM switches sign, for example passing from $+\hbar$ to $-\hbar$, while the OAM passes from zero to $2\hbar$. The two variations exactly cancel each other! This phenomenon in which angular momentum of light changes its nature, exploiting the interaction with the medium but remaining entirely within the optical field, was called "*spin-to-orbital angular momentum conversion*". The reason why the case $q = 1$ is special is related with the rotational symmetry of the medium. It is known, indeed that transparent media that are globally rotationally symmetric around an axis, even if locally anisotropic, cannot exchange angular momentum with light (as long as the coordinate origin is on the symmetry axis).

Before closing this section, let us also consider the behavior of a q-plate when the input wave is not circularly polarized. For a generic polarization, the input wave can be written as follows:

$$|\psi_{in}\rangle = E_0 (\alpha |L\rangle + \beta |R\rangle), \tag{7.10}$$

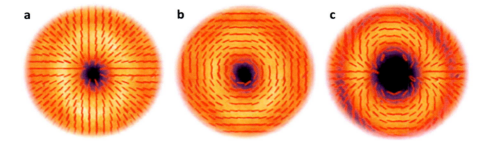

Figure 7.6: Examples of vector-vortex beams generated by q-plates for input Gaussian beams having uniform linear polarization. The pictures are obtained by superimposing a photograph of the light intensity pattern with the experimentally reconstructed local polarizations (small ellipses) as determined by full tomography [38]. (a) Radial beam generated by a q-plate with $q = 1/2$ for an input linear H polarization (the HV polarization basis is chosen here so as to set $\alpha_0 = 0$). (b) Azimuthal beam generated by a q-plate with $q = 1/2$ for an input linear V polarization. (c) Higher-order vector-vortex beam generated by a q-plate with $q = 1$ for an input linear H polarization. *Credit: Image by the author collaborators.*

where α and β are generic complex coefficients such that $|\alpha^2| + |\beta^2| = 1$. After passing through a q-plate, assuming optimal tuning, the wave is converted into the following field:

$$|\psi_{out}\rangle = i E_0 \left(\alpha e^{i2q\varphi + i2\alpha_0} |R\rangle + \beta e^{-i2q\varphi - i2\alpha_0} |L\rangle \right). \tag{7.11}$$

Depending on α and β, this field may exhibit various azimuthal patterns of polarization. In particular, when $|\alpha| = |\beta|$, the input polarization is linear and the output is then also linear everywhere, but with an orientation that varies linearly with φ. This particular field structure is called a *vector-vortex beam*, which exhibits a *correlation* between SAM and OAM (in a quantum language this correlation is actually a form of *entanglement*) [38]. Examples of these vector-vortex beams are given in Fig. 7.6.

7.3 Non-paraxial spin-orbit coupling: spin Hall effect of light and optical fibers

In this section we consider SOI effects arising as a consequence of large wave deflections, divergence, focusing or confinement, which are hence intrinsically non-paraxial. We further split this vast category of SOI effects in three fairly distinct classes. A first class of such effects, today usually known as "spin-Hall effect of light" (SHEL),[6] may for example arise in

---

[6] SHEL is not to be confused with the more specific *optical spin Hall effect*, which refers to a spin-dependent scattering effect occurring in exciton-polariton systems [39].

inhomogeneous media inducing large wave deflections by propagation, such as gradient-index media (in what was originally called the "optical Magnus effect" [40]), for reflection/refraction on dielectric or metallic interfaces, or in scattering from structural inhomogeneities or small particles at a micro or nanoscale [41–43]. The same effect lies at the root of certain SOI non-paraxial phenomena occurring for strong focusing, such as the appearance of OAM in the longitudinal field of a strongly focused circularly polarized beam. A second class of SOI effects may occur in homogeneous birefringent media, when the electromagnetic wave is non-paraxial to start with, for example because it has a non-negligible beam divergence [44–46], or when the divergence is induced by refraction at the entrance surface of the crystal, as for example in the so-called conical diffraction through biaxial crystals (see for example [47] and the references therein). Finally, a third class of non-paraxial SOI phenomena is linked to the so-called *transverse spin*, that is, a nonzero **S** density appearing in confined traveling waves, oriented perpendicular to the propagation direction and locked to the propagation (or momentum) direction [12–14,48,49]. This occurs in free space, close to the focal region, but also in evanescent fields near guided waves or surface waves. As seen, there is a large variety of non-paraxial SOI phenomena and for a full overview we refer the reader to some recent review articles, such as Refs. [12,13,50] and to the other papers cited above. Here, we will instead focus only on the first class, the SHEL, as this is the most relevant one for propagation through optical fibers and hence for optical communication.

The SHEL phenomena can be understood by analyzing the RVB geometric phase associated with wave deflection, whose sign is controlled by the input helicity $\mathbf{S} \cdot \mathbf{k}$. When adopting this theoretical framework, the orbital degree of freedom is usually described in terms of plane waves – or momentum – as opposed to OAM. Plane waves and OAM eigenstates are, however, just two alternative choices of the wavefunction representation basis, so the two descriptions can obviously be linked to each other. In particular, angular momentum conservation law and OAM come naturally into play whenever the system has a global rotational symmetry. For example, a plane surface between two isotropic transparent media is obviously rotationally symmetric around the interface normal. Setting the z-axis parallel to the interface normal, the photon total angular momentum J_z must be conserved in the reflection/refraction, irrespective of any S_z variation occurring in the process. This in turn implies that an OAM variation must necessarily compensate for any SAM variations. For example, in total-internal reflection of circularly polarized light, with incidence plane xz (z is here taken to point into the medium the light is coming from) and incidence angle ϑ, the photon S_z changes from $-s\hbar\cos\vartheta$ to $+s'\hbar\cos\vartheta$, where s and s' denote the input and output helicities (that is, the sign of $\mathbf{S} \cdot \mathbf{k}$), respectively. Hence $\Delta S_z = (s' + s)\hbar\cos\vartheta$. For a collimated Gaussian-like beam, which is not an OAM eigenstate, such a variation of OAM cannot be obtained by generating a helical wave. The (average) **L** variation instead manifests itself as a tiny beam center-of-energy transverse displacement $\Delta \mathbf{r}_\perp$ in the y direction. In combination with the average momentum $\mathbf{P} = \hbar\mathbf{k}$ carried by the beam photons, this displacement results in a mean

Spin and orbital angular momentum coupling **189**

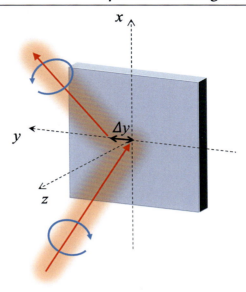

Figure 7.7: Pictorial illustration of the SHEL effect in light reflection. The input and output light beams (shown as dark-red arrows with a halo) are assumed here to be both left-handed circularly polarized (circular blue arrows). The z-component SAM variation (in this case positive) must then be compensated by a lateral shift Δy of the reflected beam along the y axis (in this case in the positive direction), relative to the impinging beam (Imbert-Fedorov shift), so that the reflected beam acquires a nonzero negative mean OAM (although the reflected beam is not an OAM eigenstate) balancing the SAM variation. For opposite polarization handedness, also the shift is in the opposite direction. The shift shown in the figure is greatly exaggerated for the sake of clarity. *Credit: Original image by the author.*

variation $\Delta L_z = (\Delta \mathbf{r}_\perp \times \mathbf{P})_z = -\Delta y \hbar k \sin \vartheta$ per photon that compensates the variation ΔS_z resulting from the beam reflection. Setting $\Delta L_z = -\Delta S_z$ one obtains the following expression for the transverse shift:

$$\Delta y = \frac{(s' + s)}{k} \cot \vartheta. \tag{7.12}$$

See Fig. 7.7 for an illustration of this effect. A similar effect occurs for ordinary (non-total) reflection and for refraction. This spin-dependent transverse shift occurring on reflection/refraction is known as Imbert-Fedorov shift [51]. Equivalently, this beam shift can also be computed from the RVB transverse phase-gradient in **k** space resulting from the reflection/refraction **k** deflection, as for example explained in Ref. [12].

It is well known that optical fibers confine light by a repeated (actually continuous) total internal reflection process. Hence it is natural to expect that SHEL may play an important role also in fiber optics. A helicity-dependent continuous transverse displacement of small light beams spiraling in large confining glass cylinders has been indeed observed [52]. However, we now wish to look at the same phenomenon from the point of view of fiber modes, as this is more relevant for discussing optical communication.

We consider ideal optical fibers having perfect cylindrical symmetry around their main axis z. The translation invariance along z guarantees that propagating wave electric field can be written in full generality as $\mathbf{E}(r,\varphi,z,t) = \mathbf{E}(r,\varphi)e^{i(\beta z - \omega t)}$ (an equivalent expression can be written for the magnetic field), where β is the mode propagation constant. Each fiber mode $\mathbf{E}(r,\varphi)$ and the corresponding propagation constant β are fully specified by a set of mode indices (analogous to quantum numbers), plus the time frequency ω. In the common weak-guidance approximation (WGA), for which SOI and other vector effects are entirely neglected, the modes of the fiber can be approximated as the outer product of a (transverse) polarization vector and a scalar wave. The scalar wave is then determined by the Helmholtz equation written for the given fiber index profile, with β playing the role of an eigenvalue. Modes with different polarizations are β-degenerate, which means that we can choose an arbitrary polarization basis to represent them. A convenient choice for us is to use the circular polarizations, labeled by the SAM (or helicity) eigenvalues $s = \pm 1$. The fiber eigenvalue equation is also subject to rotation invariance around the fiber axis z, so in this WGA approximation the modes can be taken to be OAM eigenfunctions and labeled by the OAM integer eigenvalue m. Another positive integer p is finally used to label different radial profiles of modes having the same m and s. In sum, we can use the three indices (m, p, s) to label any possible fiber mode, in the WGA.[7,8] Owing to the fiber mirror symmetry, it is easy to prove that modes with m and $-m$ must be degenerate, so that β may depend only on the pair $p, |m|$ and for each given β there is always at least a four-fold degenerate mode group ($m = \pm|m|$, $s = \pm 1$), with the exception of $m = 0$, which is doubly degenerate. In the following we will symbolically denote by the ket $|m, p, s\rangle_W$ the WGA fiber mode corresponding to indices (m, p, s) and with $\beta_{W, |m|, p}$ the corresponding WGA propagation constant. In this WGA, no SOI effects are obviously possible.

Let us now add in the role of vector nature of the electromagnetic field. The exact vector modes of a fiber can only be computed numerically (see, e.g., [53]). However, they can be classified and understood qualitatively and even semi-quantitatively by a simple perturbative analytical approach, which starts from the "unperturbed" WGA modes and adds in all vector effects as a small perturbation, that can be represented by an operator $\hat{V}$ to be included in the

---

[7] The commonly used LP basis of fiber modes in WGA uses the same two indices m and p plus a linear polarization basis.

[8] We must warn the readers that many different choices are used in the scientific literature both for the actual mode indices and for their labeling. Hence, one should not generally assume, for example, that the symbol "m" always stands for the OAM eigenvalue, rather than TAM or the radial number.

mode eigenvalue equation [54,55]. In this approach, we can still use indices (m, p, s) to label the exact vector modes $|m, p, s\rangle$.[9] The $\hat{V}$ matrix element between WGA modes can be written as follows (adapted from Eq. 2 of Ref. [55]):

$$_W\langle m', p', s'| \hat{V} |m, p, s\rangle_W = \delta_{m'+s',m+s} \left(A_{|m'|,p',|m|,p} - m's' B_{|m'|,p',|m|,p}\right), \quad (7.13)$$

where $\delta_{j',j}$ is the Kronecker delta and A, B are suitable reduced matrix elements, independent of spin and of OAM sign, which can be written explicitly as radial-mode integrals (see Eq. 2 of Ref. [55] for further details). The $\delta_{m'+s',m+s}$ factor in Eq. (7.13) ensures that terms having different values of the TAM $j = m + s$ remain uncoupled by the vector perturbation. This is the result of the rotational symmetry of the fiber around the z axis, implying that exact vector fiber modes can still be taken as J_z eigenstates, with TAM eigenvalue j. However, as we shall discuss below, they will not be exact OAM and SAM eigenstates anymore.

The most significant effect of the vector perturbation is to lift all degeneracies, unless they are symmetry-enforced. In each mode group ($m = \pm|m|$, $s = \pm 1$) the four degenerate modes have different TAM and hence are *not* cross-coupled by the perturbation, with the notable exception of $(1, -1)$ and $(-1, 1)$, both corresponding to $j = 0$, which must be treated separately. Hence, the first-order perturbation of β is given only by the diagonal matrix element

$$\Delta\beta^2 = \beta^2 - \beta_W^2 = {}_W\langle m, p, s| \hat{V} |m, p, s\rangle_W = A_{|m|,p,|m|,p} - ms B_{|m|,p,|m|,p}, \quad (7.14)$$

leading to a splitting for the SAM-OAM *aligned* and *anti-aligned* cases, which have opposite signs of the ms factor appearing in the second term in Eq. (7.14). This is the most important effect of SOI in an optical fiber and can be ascribed to the SHEL phases arising from the continuous total-internal reflections in the fiber (see Fig. 7.8). The mode pairs $|m, p, s\rangle$ and $|-m, p, -s\rangle$ remain degenerate to each other. This is actually the result of fiber mirror symmetry, which implies that j and $-j$ states must stay degenerate. The case $j = 0$ is special because two degenerate modes are cross-coupled to each other by the perturbation, leading to the need to diagonalize the 2×2 perturbation matrix. It is fairly easy to verify that this leads to defining two new nondegenerate vector modes given by the linear combinations $\frac{1}{\sqrt{2}}\left(|1, p, -1\rangle_W \pm |-1, p, 1\rangle_W\right)$, commonly denoted as $TM_{0,p}$ (transverse magnetic, where the first 0 index corresponds to j) and $TE_{0,p}$ (transverse electric), respectively.

Another important effect of the vector perturbation for $j \neq 0$ is to partly hybridize nondegenerate WGA modes having the same TAM j. The exact vector modes having $j \neq 0$ (often labeled HE_{jp} or EH_{jp} in the fiber literature, where HE refer to $j\langle s\rangle > 0$, or spin-orbit aligned,

---

[9] In numerical approaches to computing exact vector modes it is however more common to use the TAM index j instead of m. Of course, it is immediate to switch from one representation to the other.

192   *Chapter 7*

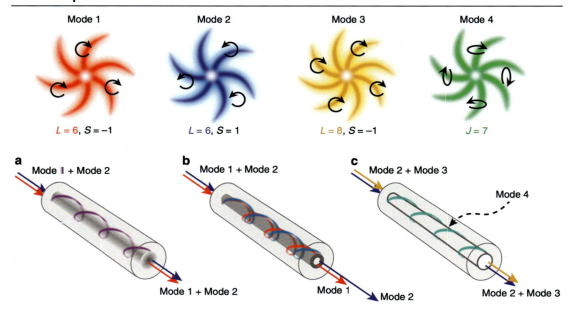

Figure 7.8: Effect of SOI and other vectorial effects on the optical modes of a fiber. OAM is here depicted as a spiral (with handedness giving sign, and the number of arms the value $L = |m|$) and SAM (polarization) as circular black arrows. Colors are used to distinguish different modes and do not indicate the frequency of the light beam. In panel (a) modes are in WGA, so two modes with the same L are β-degenerate and propagate at the same speed, regardless of polarization. In panel (b) SOI splits the degenerations between aligned and anti-aligned SAM-OAM, so the same two modes travel at different speed. In panel (c), the vectorial effects are so strong that two modes with the same j are hybridized and travel as a single mode in the fiber. These modes can be described as the product of a given OAM phase distribution and a space-variant elliptical polarization state (as shown in mode 4), but they are neither OAM nor SAM eigenstates. *Credit: Image reprinted with permission from [55].*

and EH to $j \langle s \rangle < 0$, or spin-orbit anti-aligned, $\langle s \rangle$ being the average spin)[10] can therefore be written as the following linear combination of WGA modes[11]:

$$|m, p, s\rangle = N_{m,p,s,p} |m, p, s\rangle_W + C_{m,p,s,p} |m + 2s, p, -s\rangle_W , \qquad (7.15)$$

---

[10] These are actually quasi-circularly-polarized EH and HE modes. EH and HE modes can equivalently be also defined as the sum or difference of the j and $-j$ degenerate modes, leading to quasi-linearly-polarized EH and HE modes, which are similar to vector-vortex modes. Notice that the HE modes include also the case $m = 0$, for which $j \langle s \rangle \approx 1$, which is often the fundamental mode pair for ordinary fibers. The case $j = 0$ requires a separate treatment, which generalizes the first-order perturbation case we briefly discussed above by including the coupling to higher-order p modes, thus leading to exact TM and TE modes.

[11] Here, for simplicity, we have neglected the vector coupling between different radial modes, that is, different indices p. To include also the latter, each term in (7.15) should be replaced with a sum over p', with coefficients $N_{m,p,s,p'}$ and $C_{m,p,s,p'}$.

where $N_{m,p,s,p}$ and $C_{m,p,s,p}$ are coefficients to be determined numerically. In particular, the 1st-order perturbative expression for the coefficient $C_{m,p,s,p}$ is the following (corresponding to Eq. 3 of Ref. [55]):

$$C_{m,p,s,p} = \frac{A_{|m+2s|,p,|m|,p} + (ms+2) B_{|m+2s|,p,|m|,p}}{\beta^2_{W,|m|,p} - \beta^2_{W,|m+2s|,p}}, \qquad (7.16)$$

while $N_{m,p,s,p} \approx 1$. Owing to this hybridization effect, exact vector modes of the fiber are not eigenstates of OAM and SAM, although they remain eigenstates of TAM. Hence exact vector modes typically have a non-uniform elliptical polarization (see Fig. 7.8). This effect is normally very small, but it can be amplified by a suitable fiber design [55]. It should be noted that this hybridization effect is the result of all vector effects acting in the fiber, and not only of SOI.

7.4 Applications to optical communication

In this final section of the chapter, we will survey various examples of potential applications of the discussed SOI effects in the field of optical communication, both classical and quantum.

In the framework of classical communication, a number of applications of SOI is related to the use of q-plates, metasurfaces and similar PB phase devices for controlling the transverse mode of an optical beam, for example by converting an ordinary input Gaussian beam into a specific transverse mode at output, or vice versa. This capability, in turn, can be used to implement *mode-division multiplexing* (MDM) of the optical signal, that is to utilize different simultaneous channels in the same optical beam, corresponding to different transverse modes, such as, for example, different OAM eigenmodes [56]. The information-carrying signal can be then encoded in the overall amplitude or phase (or frequency, which is equivalent) modulation, for each given transverse mode. MDM can be applied both to free-space communication and fiber-based communication, as long as the fiber is not single-mode, and it can be easily combined with standard wavelength-division multiplexing (WDM) in order to achieve record-high channel transmission capacities [57,58]. MDM can be considered as a subclass of the more general space-division multiplexing approach (SDM), which exploits in various ways the transverse space dimensions for multiplexing, including for example the more straightforward parallel-beams approach (with multicore fibers in the case of fiber-based communication) [59,60].

To operate MDM effectively, one needs in general to be able to generate different transverse modes, including possibly vector modes with a given nontrivial polarization pattern. PB-phase devices are not the only way to do this, obviously; there are alternative technologies such as

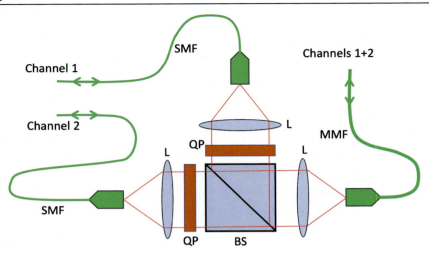

Figure 7.9: Figure representing a simple possible optical layout for mode multiplexing/demultiplexing using q-plates. This setup allows one to send/receive two optical channels traveling in distinct single-mode fibers (SMF) into two distinct spatial-mode channels of a multi-mode fiber (MMF). All the channels could be actually doubled in numbers if MIMO is used for handling degenerate modes. The two modes are generated by two q-plates (QP) having different charge q (for obtaining distinct modes) and then superimposed by a beam-splitter (BS). Lenses (L) are used to couple/decouple to the fibers. *Credit: Original image by the author.*

spatial light modulators (SLMs) or OAM sorters. However, PB-phase devices such as q-plates have certain unique features that make them particularly convenient in prospect, for MDM applications (see Fig. 7.9 for an example of a simple multiplexing/demultiplexing scheme using q-plates). They are very compact and relatively cheap devices (compared for example with SLMs). They can be entirely passive elements or be electrically controlled for active switching of the multiplexed/demultiplexed channel [28]. This switching can be operated at extremely fast rates, if the polarization control is exploited, as polarization can be switched very quickly with electro-optical devices such as Pockels cells [19]. In the case of liquid crystals or dielectric metasurfaces, the transmission and conversion efficiencies can be very high, thus reducing the energy dissipation of the channel.

These advantages are even stronger if vector modes are adopted for multiplexing, as few alternative methods can be used in this case for their generation and usually require interferometric setups. As we have seen, q-plates are particularly well suited for generating vector-vortex modes [38]. Proof-of-principle multiplexing with vector-vector modes for high-rate free-space communication has been demonstrated using liquid-crystal q-plates for multiplexing/demultiplexing [61]. Using vector-vortex beams to define the multiplexing channels can be beneficial as these modes are more robust against turbulence [62]. In the case of fiber-based communication, q-plates are extremely useful for exciting specific higher-order fiber modes, such as for example OAM modes [63,64]. Actually, by properly exploiting the q-plate SOI effect in combination with other optics, it is even possible to generate modes having the same azimuthal structure as exact vector-vortex modes of the fiber, as given in Eq. (7.15), that is, including a balanced superposition of two distinct OAM modes having opposite SAM and the same TAM [55]. This ensures the possibility to achieve minimal cross-coupling in the multiplexing/demultiplexing process.

Besides the possible use of q-plates or similar SOI-based devices as multiplexer/demultiplexer elements outside the fibers, we have seen in the previous section that SOI occurring *within* the fibers themselves can be important and its main effect is to lift the degeneracy between SAM-OAM aligned and anti-aligned modes. This degeneracy lifting is potentially useful for the possible use of MDM in fiber-based communication, for the following reasons. Ideally, one would like to use distinct fiber modes as distinct channels, but degenerate or quasi-degenerate modes always have very strong and random couplings to each other, due for example to small imperfections in the fiber, fiber bending and thermal gradients, leading to strong cross-talk between such modes and mix-up of the relative signals. Hence, a group of degenerate modes must be used as a single communication channel. Alternatively, degenerate modes can be used as distinct channels by adopting a radio-inspired MIMO (multiple-input multiple-output) technology, which is essentially a numerical decoupling of the modes after signal acquisition, based on a previous characterization of the coupling matrix. However, MIMO is computation-intensive and requires extra resources (including time and energy), so it is convenient to minimize its use, whenever possible. Doing MIMO within a lower-dimensional space is less resource-expensive, so it is convenient to design fibers so as to have only low-dimensional degenerate mode groups. OAM-supporting fibers are in particular designed to have an annular profile of the core refractive index, specifically engineered so as to suppress higher-order radial modes while allowing a certain number of nonzero OAM modes [58]. Suppression of higher-order radial modes is typically needed to avoid multiple random quasi-degeneracies. However, in WGA, fiber modes still form four-dimensional degenerate mode groups (except for the fundamental mode group, which is two-dimensional). This is where having a strong SOI in the fiber comes to be useful. By lifting the degeneracy between SAM-OAM aligned and anti-aligned modes, thus reducing the degenerate mode-group dimension from four to two, SOI increases the number of effective distinct channels, when mode groups are used as channels, or it reduces the MIMO required resources, when degenerate modes are used as separate channels [55,65,66].

Let us now discuss the applications of SOI to quantum communication. Once again, the most important application is the use of q-plates and similar devices for converting a photon state corresponding to a given optical mode into another. A particularly important feature here,

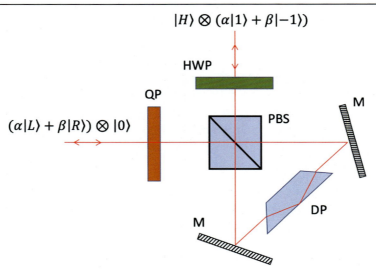

Figure 7.10: Optical layout that can ideally function as a unitary (or deterministic) qubit interface for transferring an arbitrary photonic qubit quantum state $|\varphi\rangle = \alpha|0\rangle + \beta|1\rangle$ initially encoded in the photon polarization (using an LR logical basis) into the same qubit $|\varphi\rangle$ encoded in the OAM two-dimensional subspace with $m = \pm 1$ [68,69]. Notice that the full quantum state has also a zero OAM at input and a fixed linear polarization (H, for "horizontal") at the output. This apparatus can operate both ways, it is fully reversible. Legend: DP – Dove prism rotated by an angle $\theta = \pi/8$ with respect to the horizontal-vertical reference frame; HWP – half-wave plate; M – mirror; PBS – polarizing beam splitter; QP – q-plate with $q = 1/2$. Credit: Original image by the author.

however, is the polarization control of the resulting output mode, as for example, the SAM control on the output OAM sign occurring in a q-plate. This in turn allows for converting a *qubit* encoded in the polarization state of an input photon into the *same* qubit encoded in the OAM state of the output photon, or vice versa. In other words, the quantum information can be shifted from a "polarization register" into an "OAM register" or vice versa, using the q-plate as a quantum interface [67–69] (see Fig. 7.10 for an example of the layout of a unitary quantum interface). This in turn can be used also to facilitate the encoding of *multiple qubits* in the same photon, for example one in polarization and one in OAM, or even more, if addressing different values of OAM, while still doing all read/write operations with polarization, which is much easier to address. This is also equivalent to generating a photonic *qudit*, that is, a high-dimensional quantum state of the photon that can be used to encode a greater amount of information in the same particle [70]. Quantum communication by means of photonic *qudits* can be convenient for various reasons, including a greater density of information, greater robustness to decoherence, greater security in the quantum cryptographic schemes, etc.

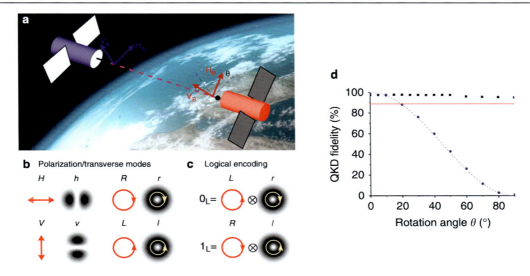

Figure 7.11: Concept of "alignment-free" quantum communication based on rotation-invariant photonic states. (a) Schematics quantum communication between two rotating satellites, each with its own reference frame for the polarization, as defined for example by the orthogonal "horizontal" and "vertical" linear polarizations H and V, or by the relative phases of left (L) and right (R) circular polarizations. (b) Introducing also a corresponding space of OAM transverse modes $m = \pm 1$ (here denoted as l/r) and their linear superpositions h and v (which are modes having the same azimuthal structure as Hermite-Gauss ones), one may combine them into (c) TAM zero modes $|0_L\rangle = |L\rangle \otimes |r\rangle$ and $|1_L\rangle = |R\rangle \otimes |l\rangle$, which can then be used as rotation-invariant logical basis states for quantum communication. Panel (d) shows the fidelity in a proof-of-principle quantum key-distribution experiment as a function of rotation angle θ, for rotation-invariant qubits (black squares) and polarization-encoded qubits (blue dots). *Credit: Image adapted from [71].*

A more specific application in quantum communication exploits the described capabilities of q-plates for converting an ordinary polarization-encoded qubit into a qubit encoded in OAM-SAM combined states having a total TAM of zero. This corresponds to having for example a state with left circular polarization and OAM $m = -1$, or right circular polarization and OAM $m = +1$. These two states can be used as logical 0 and 1 of the qubit (see Fig. 7.11) and can be obtained from an input photon having zero OAM (as in a Gaussian beam) by simply passing through a q-plate with $q = 1/2$. Since q-plates preserve coherence, any qubit formed by arbitrary superpositions of the two polarizations are then converted into the same qubit encoded in these 0-TAM photonic states. The great advantage of these 0-TAM states is that they are rotationally invariant around the beam axis (because the overall phase of a quantum state for a rotation by angle ϑ around the beam axis changes by $e^{ij\vartheta}$, where $j = s + m$ is the TAM eigenvalue, which in this case vanishes). Hence, a quantum communication that is based on these states for encoding will be totally insensitive to any relative rotation of the transmitter and receiver [71,72], as for example illustrated in Fig. 7.11. This feature can be very useful, particularly when considering a possible quantum transmission between distant flying objects, such as for example satellites, or for a future quantum communication between small handheld portable devices. The qubit robustness can be also extended to other rotation axes and to other forms of perturbations (e.g., due to turbulence or partial obstruction of the communication channel) if a suitable filtering operation is included in the channel [71,73].

Just as imposing a TAM $j = 0$ to the photonic states makes them rotation-invariant, imparting a large value of the TAM makes the photon states highly rotation-variant. For a TAM eigenstate the rotation dependence is expressed by the phase factor $e^{ij\vartheta}$, which can be rapidly varying if j is large. A state with a large TAM eigenvalue can be easily obtained by using a single q-plate with a large value of the topological charge q. This large rotation-sensitivity can in turn be exploited to arrange a photonic setup that was dubbed "photonic gear", as it "converts" an input mechanical rotation by angle ϑ into a much larger rotation of the output optical polarization vector [74]. More precisely, the output polarization rotates by the factor $j\vartheta$. When analyzed with a polarizer, this rotating (linear) polarization gives rise to an "accelerated" Malus' law, showing rapidly oscillating intensity fringes as a function of ϑ (see Fig. 7.12). The underlying concept is as follows. In the transmitting unit, a q-plate with large q value is used to convert an input Gaussian beam with linear polarization, which is a superposition of $s = 1$ and $s = -1$ with $m = 0$, into a corresponding superposition of j and $-j$ TAM eigenstates with $j = s + m = -1 + 2q$. A half-wave plate can further increase this to $j = s + m = 1 + 2q$, by flipping again the spin.[12] This superposition state, which is actually a very high-order vector-vortex beam with a rapidly space-variant linear polarization around the beam axis, is then transmitted to the receiving stage. Here, a second q-plate integral with the receiving unit converts the photons back into a linearly polarized state with OAM $m = 0$, whose polarization is hence again uniform. However, the $e^{\pm ij\vartheta}$ phase factors acquired by the j and $-j$ TAM eigenstates owing to the relative rotation of the transmitting and receiving units are retained in the last q-plate conversion (where they are attributed to $s = \pm 1$ states) and give rise to the resulting enhanced rotation of the output linear polarization. This behavior, illustrated in Fig. 7.12, is analogous to the N00N state quantum enhancement of phase metrology, with the angular momentum j playing the role of the number of photons N of the N00N state. A photonic gear can be used for example to optically transfer the information about the relative orientation of a transmitter and a receiver with enhanced sensitivity, even at great distance. This is a very specific form of optical communication, of course, but it can be interesting for specialized applications. The SOI-based preparation and analysis stages and the spin-orbit-correlated nature of the transmitted light state are essential to its working

---

[12] The same result can be obtained without half-wave plate, by using a q-plate with a negative value of q.

Spin and orbital angular momentum coupling 199

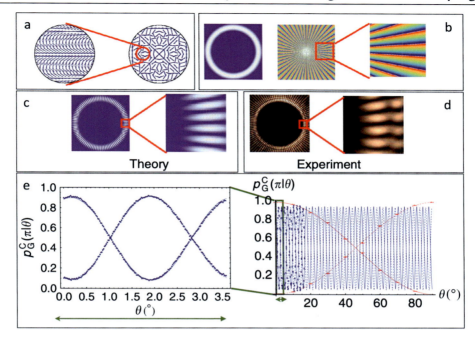

Figure 7.12: Optical behavior of "photonic gears" for the communication of relative orientation information. This example is relative to the case of photonic states with TAM $j = 51$ ($m = 50$) obtained with q-plates having $q = 25$. (a) Optic axis pattern of the q-plate with a zoomed-in area. (b) Intensity and phase distribution of the resulting TAM eigenstates. (c) Theoretical intensity distribution of the photonic state resulting from the superposition of $j = 51$ and $j = -51$. (d) Corresponding experimental intensity distribution. (e) Experimental data on the "super-resolved" Malus' law resulting for a rotation by angle θ of the transmitting and receiving units, with a zoomed-in portion to better show the rapid oscillations. *Credit: Image adapted from [74].*

principle. The described photonic gears are not fundamentally quantum in their working principle, they can be operated also classically, although in the case of quantum states there may be additional interesting opportunities (e.g., distributing entangled quantum information about relative orientations or a further quantum enhancement of the rotational sensitivity).

In conclusion of this section, I would like to briefly mention one last concept that perhaps will have a relevance in the future for the field of optical communication and more generally information technology. This is the possibility of using the geometric phases arising from SOI to laterally confine and hence waveguide light, without ordinary (dynamical) phases arising from modulations of the optical path length and refractive index. This possibility has been demonstrated recently in a series of theoretical papers, accompanied by a proof-of-principle experiment in which the waveguide was built with a sequence of discrete SOI-lensing units

[75–77]. It is, however, quite challenging to move on from this proof-of-principle demonstration to a more potentially useful continuous geometric-phase waveguide based on the same concept, and this has not been experimentally demonstrated yet. For this and other reasons, this concept is currently very far from real-world applications. Yet, it is intriguing and inspiring to find that one can put to use SOI in such unexpected and novel ways.

7.5 Conclusions

In this chapter we have presented a class of phenomena that fall under the name of "spin-orbit interactions" of electromagnetic waves or, equivalently, of photons. After a general introduction, we have focused on two main examples, which we selected owing to their relevance to the topic of optical communication: (i) the paraxial optical phenomena occurring for light propagation across patterned birefringent media, such as q-plates or metasurfaces; (ii) the spin-Hall effect of light taking place in total internal reflection (as well as for other light deflection phenomena) and hence affecting the propagation of light in higher-order optical modes within optical fibers. For these two cases, we described the key physical principles in some detail. Finally, we reviewed a number of examples in which these phenomena are directly relevant for applications in optical communication, both classical and quantum.

References

[1] J.D. Jackson, Classical Electrodynamics, 3rd edition, John Wiley & Sons, 1998.
[2] A. Zangwill, Modern Electrodynamics, Cambridge University Press, 2013.
[3] S.J. van Enk, G. Nienhuis, Spin and orbital angular momentum of photons, Europhys. Lett. 25 (1994) 497–501.
[4] S.J. van Enk, G. Nienhuis, Commutation rules and eigenvalues of spin and orbital angular momentum of radiation fields, J. Mod. Opt. 41 (1994) 963–977.
[5] K.Y. Bliokh, J. Dressel, F. Nori, Conservation of the spin and orbital angular momenta in electromagnetism, New J. Phys. 16 (2014) 093037.
[6] A. Aiello, M.V. Berry, Note on the helicity decomposition of spin and orbital optical currents, J. Opt. 17 (2015) 062001.
[7] S.M. Barnett, L. Allen, R.P. Cameron, C.R. Gilson, M.J. Padgett, F.C. Speirits, A.M. Yao, On the natures of the spin and orbital parts of optical angular momentum, J. Opt. 18 (2016) 064004.
[8] E. Leader, The photon angular momentum controversy: resolution of a conflict between laser optics and particle physics, Phys. Lett. B 756 (2016) 303–308.
[9] S.M. Barnett, Rotation of electromagnetic fields and the nature of optical angular momentum, J. Mod. Opt. 57 (2010) 1339–1343.
[10] R.P. Cameron, F.C. Speirits, C.R. Gilson, L. Allen, S.M. Barnett, The azimuthal component of Poynting's vector and the angular momentum of light, J. Opt. 17 (2015) 125610.
[11] L. Marrucci, C. Manzo, D. Paparo, Optical spin-to-orbital angular momentum conversion in inhomogeneous anisotropic media, Phys. Rev. Lett. 96 (2006) 163905.
[12] K.Y. Bliokh, F.J. Rodríguez-Fortuño, F. Nori, A.V. Zayats, Spin-orbit interactions of light, Nat. Photonics 9 (2015) 796–808.

[13] A. Aiello, P. Banzer, M. Neugebauer, G. Leuchs, From transverse angular momentum to photonic wheels, Nat. Photonics 9 (2015) 789–795.
[14] L. Marrucci, Spin gives direction, Nat. Phys. 11 (2015) 9–10.
[15] M.V. Berry, The adiabatic phase and Pancharatnam's phase for polarized light, J. Mod. Opt. 34 (1987) 1401–1407.
[16] R. Bhandari, Polarization of light and topological phases, Phys. Rep. 281 (1997) 1–64.
[17] K.Y. Bliokh, M.A. Alonso, M.R. Dennis, Geometric phases in 2D and 3D polarized fields: geometrical, dynamical, and topological aspects, Rep. Prog. Phys. 82 (2019) 122401.
[18] M.V. Berry, P. Shukla, Geometry of 3D monochromatic light: local wavevectors, phases, curl forces, and superoscillations, J. Opt. 21 (2019) 064002.
[19] L. Marrucci, C. Manzo, D. Paparo, Pancharatnam-Berry phase optical elements for wavefront shaping in the visible domain: switchable helical modes generation, Appl. Phys. Lett. 88 (2006) 221102.
[20] N.V. Tabiryan, S.R. Nersisyan, D.M. Steeves, B.R. Kimball, The promise of diffractive waveplates, Opt. Photonics News (March 2010) 40–45.
[21] M.J. Escuti, J. Kim, M.W. Kudenov, Geometric-phase holograms, Opt. Photonics News (February 2010) 22–29.
[22] E. Hasman, V. Kleiner, G. Biener, A. Niv, Formation of Pancharatnam–Berry phase optical elements with space-variant subwavelength gratings, Opt. Photonics News (December 2002) 45.
[23] D. Lin, P. Fan, E. Hasman, M.L. Brongersma, Dielectric gradient metasurface optical elements, Science 345 (2014) 298–302.
[24] N. Yu, F. Capasso, Flat optics with designer metasurfaces, Nat. Mater. 13 (2014) 139–150.
[25] E. Karimi, S.A. Schulz, I. De Leon, H. Qassim, J. Upham, R.W. Boyd, Generating optical orbital angular momentum at visible wavelengths using a plasmonic metasurface, Light Sci. Appl. 3 (2014) e167.
[26] R.C. Devlin, A. Ambrosio, N.A. Rubin, J.P.B. Mueller, F. Capasso, Arbitrary spin-to-orbital angular momentum conversion of light, Science 358 (2017) 896–901.
[27] F. Cardano, E. Karimi, L. Marrucci, C. de Lisio, E. Santamato, Generation and dynamics of optical beams with polarization singularities, Opt. Express 21 (2013) 8815–8820.
[28] B. Piccirillo, V. D'Ambrosio, S. Slussarenko, L. Marrucci, E. Santamato, Photon spin-to-orbital angular momentum conversion via an electrically tunable q-plate, Appl. Phys. Lett. 97 (2010) 241104.
[29] S.G. Cloutier, Polarization holography: orthogonal plane-polarized beam configuration with circular vectorial photoinduced anisotropy, J. Phys. D, Appl. Phys. 38 (2005) 3371–3375.
[30] M. Honma, T. Nose, Liquid-crystal Fresnel zone plate fabricated by microrubbing, Jpn. J. Appl. Phys. 44 (2005) 287–290.
[31] G. Biener, A. Niv, V. Kleiner, E. Hasman, Formation of helical beams by use of Pancharatnam–Berry phase optical elements, Opt. Lett. 27 (2002) 1875–1877.
[32] S. Slussarenko, A. Murauski, T. Du, V. Chigrinov, L. Marrucci, E. Santamato, Tunable liquid crystal q-plates with arbitrary topological charge, Opt. Express 19 (2011) 4085–4090.
[33] M. Stalder, M. Schadt, Linear polarized light with axial symmetry generated by liquid-crystal polarization converters, Opt. Lett. 21 (1996) 23–25.
[34] L. Marrucci, The q-plate and its future, J. Nanophotonics 7 (2013) 078598.
[35] E. Karimi, G. Zito, B. Piccirillo, L. Marrucci, E. Santamato, Hypergeometric-Gaussian modes, Opt. Lett. 32 (2007) 3053–3055.
[36] G. Vallone, On the properties of circular beams: normalization, Laguerre–Gauss expansion, and free-space divergence, Opt. Lett. 40 (2015) 1717–1720.
[37] L. Marrucci, Generation of helical modes of light by spin-to-orbital angular momentum conversion in inhomogeneous liquid crystals, Mol. Cryst. Liq. Cryst. 488 (2008) 148–162.
[38] F. Cardano, E. Karimi, S. Slussarenko, L. Marrucci, C. de Lisio, E. Santamato, Polarization pattern of vector vortex beams generated by q-plates with different topological charges, Appl. Opt. 51 (2012) C1–C6.
[39] A. Kavokin, G. Malpuech, M. Glazov, Optical spin Hall effect, Phys. Rev. Lett. 95 (2005) 136601.

[40] V.S. Liberman, B.Y. Zel'dovich, Spin-orbit interaction of a photon in an inhomogeneous medium, Phys. Rev. A 46 (1992) 5199–5207.
[41] M. Onoda, S. Murakami, N. Nagaosa, Hall effect of light, Phys. Rev. Lett. 93 (2004) 083901.
[42] O. Hosten, P. Kwiat, Observation of the spin Hall effect of light via weak measurements, Science 319 (2008) 787–790.
[43] O.G. Rodríguez-Herrera, D. Lara, K.Y. Bliokh, E.A. Ostrovskaya, C. Dainty, Optical nanoprobing via spin-orbit interaction of light, Phys. Rev. Lett. 104 (2010) 253601.
[44] A. Ciattoni, G. Cincotti, C. Palma, Circularly polarized beams and vortex generation in uniaxial media, J. Opt. Soc. Am. A 20 (2003) 163–171.
[45] A. Ciattoni, G. Cincotti, C. Palma, Angular momentum dynamics of a paraxial beam in a uniaxial crystal, Phys. Rev. E 67 (2003) 036618.
[46] E. Brasselet, Y. Izdebskaya, V. Shvedov, A.S. Desyatnikov, W. Krolikowski, Y.S. Kivshar, Dynamics of optical spin-orbit coupling in uniaxial crystals, Opt. Lett. 34 (2009) 1021–1023.
[47] M.V. Berry, Conical diffraction asymptotics: fine structure of Poggendorff rings and axial spike, J. Opt. A, Pure Appl. Opt. 6 (2004) 289–300.
[48] F.J. Rodríguez-Fortuño, G. Marino, P. Ginzburg, D. O'Connor, A. Martínez, G.A. Wurtz, Anatoly V. Zayats, Near-field interference for the unidirectional excitation of electromagnetic guided modes, Science 340 (2013) 328–330.
[49] N. Shitrit, I. Yulevich, E. Maguid, D. Ozeri, D. Veksler, V. Kleiner, E. Hasman, Spin-optical metamaterial route to spin-controlled photonics, Science 340 (2013) 724–726.
[50] F. Cardano, L. Marrucci, Spin-orbit photonics, Nat. Photonics 9 (2015) 776–778.
[51] K.Y. Bliokh, A. Aiello, Goos–Hänchen and Imbert–Fedorov beam shifts: an overview, J. Opt. 15 (2013) 014001.
[52] K.Y. Bliokh, A. Niv, V. Kleiner, E. Hasman, Geometrodynamics of spinning light, Nat. Photonics 2 (2008) 748–753.
[53] M.F. Picardi, K.Y. Bliokh, F.J. Rodríguez-Fortuño, F. Alpeggiani, F. Nori, Angular momenta, helicity, and other properties of dielectric-fiber and metallic-wire modes, Optica 5 (2018) 1016–1026.
[54] S.E. Golowich, S. Ramachandran, Opt. Express 13 (2005) 6870–6877.
[55] P. Gregg, P. Kristensen, A. Rubano, S. Golowich, L. Marrucci, S. Ramachandran, Enhanced spin orbit interaction of light in highly confining optical fibers for mode division multiplexing, Nat. Commun. 10 (2019) 4707.
[56] G. Gibson, J. Courtial, M.J. Padgett, M. Vasnetsov, V. Pas'ko, S.M. Barnett, S. Franke-Arnold, Free-space information transfer using light beams carrying orbital angular momentum, Opt. Express 12 (2004) 5448–5456.
[57] J. Wang, J.-Y. Yang, I.M. Fazal, N. Ahmed, Y. Yan, H. Huang, Y. Ren, Y. Yue, S. Dolinar, M. Tur, A.E. Willner, Terabit free-space data transmission employing orbital angular momentum multiplexing, Nat. Photonics 6 (2012) 488–496.
[58] N. Bozinovic, Y. Yue, Y. Ren, M. Tur, P. Kristensen, H. Huang, A.E. Willner, S. Ramachandran, Terabit-scale orbital angular momentum mode division multiplexing in fibers, Science 340 (2013) 1545–1548.
[59] D.J. Richardson, J.M. Fini, L.E. Nelson, Space-division multiplexing in optical fibres, Nat. Photonics 7 (2013) 354–362.
[60] N. Zhao, X. Li, G. Li, J.M. Kahn, Capacity limits of spatially multiplexed free-space communication, Nat. Photonics 9 (2015) 822–826.
[61] G. Milione, M.P.J. Lavery, H. Huang, Y. Ren, G. Xie, T.A. Nguyen, E. Karimi, L. Marrucci, D.A. Nolan, R.R. Alfano, A.E. Willner, 4×20 Gbit/s mode division multiplexing over free space using vector modes and a q-plate mode (de)multiplexer, Opt. Lett. 40 (2015) 1980–1983.
[62] W. Cheng, J.W. Haus, Q. Zhan, Propagation of vector vortex beams through a turbulent atmosphere, Opt. Express 17 (2009) 17829–17836.
[63] P. Gregg, M. Mirhosseini, A. Rubano, L. Marrucci, E. Karimi, R.W. Boyd, S. Ramachandran, Q-plates as higher order polarization controllers for orbital angular momentum modes of fiber, Opt. Lett. 40 (2015) 1729–1732.

[64] K. Ingerslev, P. Gregg, M. Galili, F. Da Ros, H. Hu, F. Bao, M.A. Usuga Castaneda, P. Kristensen, A. Rubano, L. Marrucci, K. Rottwitt, T. Morioka, S. Ramachandran, L.K. Oxenløwe, 12 mode, WDM, MIMO-free orbital angular momentum transmission, Opt. Express 26 (2018) 20225–20232.

[65] L. Zhu, G. Zhu, A. Wang, L. Wang, J. Ai, S. Chen, C. Du, J. Liu, S. Yu, J. Wang, 18 km low-crosstalk OAM + WDM transmission with 224 individual channels enabled by a ring-core fiber with large high-order mode group separation, Opt. Lett. 43 (2018) 1890–1893.

[66] J. Zhang, J. Liu, L. Shen, L. Zhang, J. Luo, J. Liu, S. Yu, Mode division multiplexed transmission of WDM signals over 100-km single-span OAM fiber, Photon. Res. 8 (7) (2020) 1236–1242.

[67] E. Nagali, F. Sciarrino, F. De Martini, L. Marrucci, B. Piccirillo, E. Karimi, E. Santamato, Quantum information transfer from spin to orbital angular momentum of photons, Phys. Rev. Lett. 103 (2009) 013601.

[68] E. Nagali, F. Sciarrino, F. De Martini, B. Piccirillo, E. Karimi, L. Marrucci, E. Santamato, Polarization control of single photon quantum orbital angular momentum states, Opt. Express 17 (2009) 18745–18759.

[69] V. D'Ambrosio, E. Nagali, C.H. Monken, S. Slussarenko, L. Marrucci, F. Sciarrino, Deterministic qubit transfer between orbital and spin angular momentum of single photons, Opt. Lett. 37 (2012) 172–174.

[70] E. Nagali, L. Sansoni, L. Marrucci, E. Santamato, F. Sciarrino, Experimental generation and characterization of single-photon hybrid ququarts based on polarization and orbital angular momentum encoding, Phys. Rev. A 81 (2010) 052317.

[71] V. D'Ambrosio, E. Nagali, S.P. Walborn, L. Aolita, S. Slussarenko, L. Marrucci, F. Sciarrino, Complete experimental toolbox for alignment-free quantum communication, Nat. Commun. 3 (2012) 961.

[72] G. Vallone, V. D'Ambrosio, A. Sponselli, S. Slussarenko, L. Marrucci, F. Sciarrino, P. Villoresi, Free-space quantum key distribution by rotation-invariant twisted photons, Phys. Rev. Lett. 113 (2014) 060503.

[73] O.J. Farias, V. D'Ambrosio, C. Taballione, F. Bisesto, S. Slussarenko, L. Aolita, L. Marrucci, S.P. Walborn, F. Sciarrino, Resilience of hybrid optical angular momentum qubits to turbulence, Sci. Rep. 5 (2015) 8424.

[74] V. D'Ambrosio, N. Spagnolo, L. Del Re, S. Slussarenko, Y. Li, L.C. Kwek, L. Marrucci, S.P. Walborn, L. Aolita, F. Sciarrino, Photonic polarization gears for ultra-sensitive angular measurements, Nat. Commun. 4 (2013) 2432.

[75] S. Slussarenko, A. Alberucci, C.P. Jisha, B. Piccirillo, E. Santamato, G. Assanto, L. Marrucci, Guiding light via geometric phases, Nat. Photonics 10 (2016) 571–575.

[76] A. Alberucci, C.P. Jisha, L. Marrucci, G. Assanto, Electromagnetic confinement via spin-orbit interaction in anisotropic dielectrics, ACS Photonics 3 (2016) 2249–2254.

[77] C.P. Jisha, A. Alberucci, L. Marrucci, G. Assanto, Interplay between diffraction and the Pancharatnam-Berry phase in inhomogeneously twisted anisotropic media, Phys. Rev. A 95 (2017) 023823.

CHAPTER 8

Quantum communication with structured photons

Robert Fickler and Shashi Prabhakar

Tampere University, Physics Unit, Tampere, Finland

Contents

8.1 Introduction 206
8.2 Quantum protocols 208
    8.2.1 Information capacity, dense coding and noise resistance 208
    8.2.2 Quantum key distribution 209
    8.2.3 Quantum coin tossing 210
    8.2.4 Quantum secret sharing 211
    8.2.5 Layered quantum key distribution 212
8.3 Experimental toolbox 215
    8.3.1 Generation and detection methods 215
    8.3.2 Modulation methods 217
8.4 Quantum network 219
    8.4.1 Entanglement sources 219
        8.4.1.1 *Tuning entanglement via phase matching of SPDC* 221
        8.4.1.2 *Tuning entanglement via pump beam shaping* 221
        8.4.1.3 *Tuning entanglement via path identity* 222
        8.4.1.4 *Multi-partite entanglement sources* 223
    8.4.2 Quantum channels 223
        8.4.2.1 *Fiber quantum channels* 224
        8.4.2.2 *Free-space quantum channels* 226
        8.4.2.3 *Underwater quantum channels* 227
        8.4.2.4 *Characterization of quantum channels* 227
    8.4.3 Quantum repeater 228
        8.4.3.1 *High-dimensional teleportation and entanglement swapping* 228
        8.4.3.2 *Quantum memories* 230
    8.4.4 Quantum interfaces 231
    8.4.5 Quantum router 231
8.5 Conclusion 233
Acknowledgments 233
References 233

Structured Light for Optical Communication
https://doi.org/10.1016/B978-0-12-821510-4.00014-5
Copyright © 2021 Elsevier Inc. All rights reserved.

8.1 Introduction

Quantum communication, the sub-field of quantum information science that studies the exchange and distribution of quantum information, requires quantum systems with an appropriate degree of freedom to encode the quantum state. While in other sub-fields various physical systems have been recommended as information carriers, only photons provide a realistic solution for communication to transmit quantum states from one place to another, separated by large distances. In most of the implementations, polarization has been the degree of freedom (DOF) of choice, due to its evolved and sophisticated technologies to imprint, manipulate and detect the encoded qubits. However, these advantages come at the cost of being restricted to two-dimensional quantum states. In addition to polarization, photons have two continuous DOFs, namely the spatial and temporal domains. Both DOFs offer the possibility to discretize the state space to a theoretically arbitrary large number of discrete quantum levels, which are often called higher dimensions relating to the mathematical dimensionality of the underlying Hilbert space. The quantum information encoded in a d-dimensional state is called *qudit*, in analogy to the more conventional notation of a two-dimensional qubit encoding. While these two aforementioned paradigmatic ways of encoding qudits have both advantages and disadvantages, it is safe to say that both fields are progressing fast, and offer the promise to enhance current quantum communication protocols, and will have a great impact on future quantum networks. This chapter focuses on the spatial domain for which high-dimensional quantum communication is one of the future key quantum technologies. In combination with polarization, these spatially encoded photons are often termed as *structured photons*.

One popular way to discretize the spatial domain is with transverse spatial modes. Amongst the various families of spatial modes (corresponding to different coordinate bases), the most popular one is the Laguerre–Gaussian (LG) mode family, which is the solution to the paraxial wave equation in cylindrical coordinate system. They are popular, not only because their cylindrical symmetry matches the symmetry of most optical components, but they also show a powerful additional physical property, namely an orbital angular momentum (OAM) [1]. The OAM is directly connected to the azimuthal phase variation, which causes the wavefront to twist around the optical beam axis, as shown in Fig. 8.1(a) on the left. Due to the presence of a phase singularity along the beam axis, the probability of detecting a photon there is zero. This feature is revealed if many identical photons are recorded by a single-photon sensitive camera, as shown in Fig. 8.1(a) on the right. Since there is a large number of orthogonal OAM modes, they can be used to encode high-dimensional quantum states. A few of examples of single-photon recordings are shown in Fig. 8.1(b).

The experimental verification of OAM-entanglement by Mair et al. [2] can be seen as the starting point for quantum optics and quantum information experiments using structured photons. Since this seminal work in 2001, quantum information encoded in the transverse

Quantum communication with structured photons 207

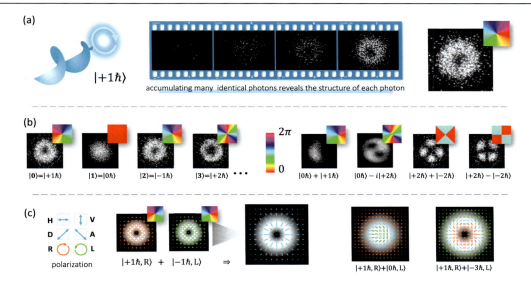

Figure 8.1: Structured photons forming a qudit quantum systems. (a) Graphical depiction of a twisted phase front leading to a photon carrying $1\hbar$ of OAM. Accumulating many of such photons reveals the transverse structure, which resembles a donut. Recordings are taken using single photons. Inset depicts the azimuthal phase structure shown in hue-color. (b) Recordings of photons with different OAM values, which serve as laboratory implementation of qudits. Superpositions of such modes lead to more complex detection patterns. (c) Superposing two modes with orthogonal polarizations lead to modes with complex polarization patterns as can be seen in the theoretical examples shown here.

structure of photons has progressed from using different mode families to realize qudit states to a myriad of experiments performing high-dimensional quantum information tasks. In addition, manifold benefits have been realized over the last decade when the polarization DOF was also included leading to complex transverse polarization structures, examples are shown in Fig. 8.1(c). Note that in the following sections, we use the term "structured photons" as the more general notion for spatially structured photons irrespective of whether the polarization is involved or not. The term "spatial modes" refers to mode structures with a uniform polarization.

Nowadays, the field of quantum information using structured photons can be seen as an established research branch in quantum optics to which the current chapter gives an experimentalist's introduction. However, we do not claim the completeness of all concepts, theoretical details and experimental achievements. As such, it is a subjective compilation of some of the most interesting and promising sub-parts of the field. This chapter focuses on communication schemes and protocols (the ones that can be considered supplemental to quantum cryptography[1]) and outlines how quantum communication might look on a large-scale quantum network. Although no fully functioning network has been realized for structured photons yet, many important key-components are rapidly progressing, such that a large-scale network seems within reach.

In Section 8.2, several protocols crucial to the field of quantum communication are described. Section 8.3 introduces the experimental tools with a strong focus on the quantum-specific requirements. Section 8.4 outlines the key components of future quantum network such as entanglement sources, transmission channels, quantum interfaces, and quantum repeaters. This extended introduction serves the purpose of giving an overview of this wide research field and encourages the reader to inquire further according to one's interests by looking into the provided references.

8.2 Quantum protocols

Quantum communication is based on protocols, which describe how a secure information exchange occurs between two or more parties. The protocols establish a set of rules that describe the required quantum states and the way they have to be transmitted and measured. If the protocol is implemented correctly, the established communication scheme promises to be superior with respect to data capacity and security in comparison to classical information transmissions. The benefits are based on fundamental features of quantum physics, such as superpositions, entanglement, randomness, and the impossibility to perfectly copy a quantum state.

Here, an overview is given on some of the protocols implemented or envisioned for structured photons. These protocols include, but are not limited to, quantum dense coding, quantum coin tossing, quantum key distribution (QKD), quantum secret sharing (QSS), and layered quantum key distribution.

8.2.1 Information capacity, dense coding and noise resistance

Before discussing more complex quantum protocols, it is instructive to set the stage and layout the utilized nomenclature by introducing the most obvious advantages of high-dimensional quantum states for their application in quantum communication. Quantum communication usually relies on qubits as the information carrier. Qubits are mathematically described by two complex orthonormal basis vectors denoted as $|0\rangle$ and $|1\rangle$, in analogy to the classical bits of 0 and 1. When discretizing the spatial domain, e.g. using transverse spatial

---

[1] Note that the quantum cryptography is covered in separate chapters in great detail, thus, we only mention it briefly.

modes of light, not only two values are possible but theoretically infinitely many due to the availability of an unlimited number of orthogonal modes that can serves as basis vectors. The prime example here is Laguerre-Gaussian modes LG_p^ℓ, which are described by the azimuthal quantum number ℓ and the radial quantum number p. Both quantum numbers allow for the encoding of quantum information, such that they offer experimental means to realize qudits. As the two quantum numbers ℓ and p are unbounded, they span a set of quantum states that can be labeled by any positive integer to infinity ($|0\rangle, |1\rangle, |2\rangle, \ldots, |\infty\rangle$) leading to a vastly increased information capacity per single photon. In fact, the information encoded in a single qudit photon is given by $\log_2(d)$ bits per photon. While this increase is theoretically unlimited, in experimental implementations, a natural limit is posed due to the finite size of optical systems.

In addition to the enhanced single-photon channel capacity, quantum physics also allows dense coding of information using entangled bi-partite states. For qubits, the four Bell states, for example, permit the transmission of two bits by sending only one qubit in a so-called dense-coding scheme. By using high-dimensionally entangled states, this enhancement can be further increased. However, as such the scheme requires very challenging generation and modulation techniques an experimental demonstration for structured photons is still awaited [3].

Structured photons acting as a qudit system also provide enhanced robustness to noise. This feature can be understood intuitively when looking at the optimal cloning scheme.[2] Although the no-cloning theorem of quantum mechanics forbids the perfect copying of a quantum state, it is possible to optimally clone a state. In such schemes, the quality of the clone, as well as the original state, is reduced by a minimum amount. This reduction can be expressed in terms of the cloning fidelity, which scales with the dimensionality d of the involved states as $F_{clone}^d = \frac{1}{2} + \frac{1}{1+d}$. Hence, the larger the dimension of the quantum state, the lower the fidelity of the optimal copy allowed by quantum physics with $F_{clone}^{d \to \infty} = 0.5$. In other words, a possible eavesdropper naturally introduces significant errors when cloning a higher-dimensional qudit and thus can be detected more easily, which also means that the tolerable noise levels are increased.

8.2.2 Quantum key distribution

QKD is one of the most popular quantum communication schemes. In general, a QKD protocol aims at establishing an unconditionally secure key between two parties by exchanging quantum states. The distributed key is then used to encrypt messages securely. The advantage of QKD over classical cryptography schemes is that a possible eavesdropper trying to hack

---

[2] This topic is covered in more detail in Chapters 1 and 6.

the quantum channel will always be detected and the generated key is discarded. In addition to the benefits of larger channel capacity and increased noise-resistance of high-dimensional quantum states in the standard QKD protocol of Bennet and Brassard [4], there are various other QKD protocols, which benefit from an increased complexity of the states. Amongst them, for example, is the increased number of mutually unbiased bases available for qudits or the possibility to "hide" lower-dimensional states in larger state spaces, both leading to an increase in noise tolerance.

As a lot of progress has been made in QKD using structured photons, Chapters 1 and 6 are focusing exclusively on the various aspects, implementations and challenges of high-dimensional QKD, such that we abstain from discussing it in more detail here. Instead, we introduce three other quantum communication protocols, which supplement and extend these efforts.

8.2.3 Quantum coin tossing

Analog to quantum key distribution schemes, the quantum coin tossing protocol is used between two parties. However, instead of trying to securely communicate with each other, the two parties, commonly named Alice and Bob, want to decide randomly on a certain task by throwing a coin. If Alice and Bob are spatially separated and do not trust each other, throwing a coin at one location and communicating the outcome via a classical channel, e.g. a phone, does not make either Alice or Bob able to tell whether either has cheated. However, when the coin toss is implemented via the exchange of quantum states, it is possible to detect the dishonesty of one of the involved parties. Additionally, using high-dimensional quantum states, such as qutrits (states having three dimensions), any possible foul play is easier to detect, and the security of the protocol is improved.

In this three-dimensional protocol, Alice state acts as the coin in our example. If Alice bets on heads, she encodes one of the two orthogonal superposition states using a qubit subspace, such as $(|0\rangle + |1\rangle)/\sqrt{2}$ or $(|0\rangle - |1\rangle)/\sqrt{2}$. If she chooses tails, she encodes one of the states $(|0\rangle + |2\rangle)/\sqrt{2}$, $(|0\rangle - |2\rangle)/\sqrt{2}$, which are again mutually orthogonal but have a non-vanishing overlap with the first set of states. She then transmits her state to Bob, who subsequently makes his bet, after which Alice can let him know if he has won or not. To verify that Alice is not cheating, Bob measures the state using the same definition of the bases as Alice for heads and tails. However, he extends both bases by also including a third orthogonal state, i.e. $|2\rangle$ for heads and $|1\rangle$ for tails. After measuring the state, Bob declares the protocol as a success if he identifies the state that Alice predicted, otherwise he regards the virtual throw of the coin as a failure. A deeper analysis reveals that Alice can only successfully cheat without being noticed with a 25% probability, which limits the success probability for detecting a cheater to

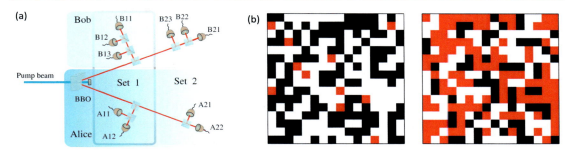

Figure 8.2: Quantum coin tossing. (a) Sketch of the setup of a photonic realization. Using beam splitters, Alice probabilistically projects her photon onto one of the four possible states labeling heads or tails as described in the main text. After that, Bob projects his photon onto one of the six possible states, also labeling heads or tails. (b) Pictorial representation of the performed experiment. Every pixel corresponds to a throw of the coin, i.e. a joint detection of the two structured photons. The color label if heads (black) or tail (white) has been communicated by Alice. The red color depicts the cases in which Bob after measuring concludes a failure of the procedure. In the left image, the result is shown if both parties are being honest. Due to the experimental imperfections there are still failures of around 6%. The right image shows the result if Alice tries to cheat, which causes an increased rate of failures (46%), and as such, Bob will be able to notice it. Reprint from [5].

75%. Assuming that Alice is cheating, for example by sending an incoherent mixture of heads and tails and always pretending that Bob has lost, it is easy to see that her success probability will be only 62.5%. Thus, Bob will be able to detect her cheating. In other words, if the rate of failure events is too high, Bob will know that Alice has cheated.

The described protocol has been demonstrated in an experiment using photons that are high-dimensionally entangled in the OAM degree of freedom, as shown in Fig. 8.2 [5]. The required qutrit states were realized using the OAM numbers $\ell = \{-1, 0, +1\}$. Instead of exchanging a photon between the two parties, an entangled state was shared between Alice and Bob. A projection on Alice's side was used to prepare the state at Bob's side through the virtue of entanglement. Additionally, it was shown that despite some experimental errors, the level of security obtained using qutrits surpassed the maximally possible level of security for qubit states. Thus, using structured photons enabled the implementation of a high-dimensional quantum coin tossing protocol.

8.2.4 Quantum secret sharing

We now turn to protocols that involve more than two parties and the first one is QSS. QSS is a method for sharing a secret message among a group of parties that can retrieve the message if all parties agree to cooperate. While the first proposal relied on multi-partite Greenberger–Horne–Zeilinger (GHZ) states [6], which are difficult to generate, the scheme can be reformulated to work with a single photon, which is sent from a distributor via all participants back to the distributor. Instead of measuring the received quantum system as in the GHZ-based QSS protocol, all participants perform local unitary operations on the photon after receiving it and before passing it on to the next party. While the single-photon scheme is significantly simpler to realize, it was shown that only an implementation using high-dimensional quantum states could be considered secure [7].

The QSS protocol involves d out of $d + 1$ mutually unbiased bases (MUB) and, thus, is limited to the power of prime dimensions where the total number of MUBs are known. At first, a distributor, e.g. Alice, generates a state out of one of the utilized MUBs. She then randomly chooses two different high-dimensional unitary operations, X and Y, and performs the operations on the state. The X-unitary cyclically shifts the state within the same MUB. Hence, there are d different choices of X from which Alice can randomly choose. The Y-unitary is constructed such that it maps the state between the different MUBs. Again, Alice can then randomly select one out of d different possibilities. Alice subsequently sends the photon to the second party, who also chooses randomly one X- and one Y-unitary and performs them on the state. Afterwards, the state is passed on to the next participant and so on before the last participant sends the photon back to Alice. After the arrival of the photon at Alice's location, all participants randomly announce their choices of Y-unitaries from which Alice can deduce in which basis she has to perform the measurement. Using the measurement result, the starting state and her original choice of the X-unitary operation, she obtains the secret information, which is a d-dimensional bit value, through simple calculus. The same value can now be obtained by all participants using their own choices of X-unitaries, but only if they collaborate without cheating.

The main challenge in implementing such a protocol experimentally is the requirement of performing a large set of high-dimensional unitary operations. Using a weak laser and the spatial structure of perfect vortex beams, i.e. a special type of vortex beams whose diameter is independent of the OAM value, an experiment involving 11-dimensional states and ten parties has been demonstrated [8]. The clever idea behind this implementation is that the required unitary operations are translated to simple spatially varying phase modulations, such that using a single phase mask per participant allows the performance of the unitary operations as shown in Fig. 8.3.

8.2.5 Layered quantum key distribution

The final protocol that is discussed has been developed with inspiration from first realizations of multi-partite high-dimensionally entangled quantum states of structured photons [9].

Quantum communication with structured photons 213

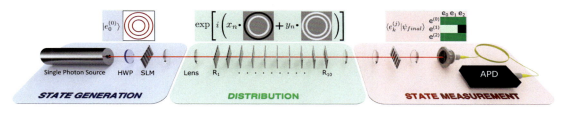

Figure 8.3: Schematic representation of the state generation, distribution and the measurement for a successful implementation of the quantum secret sharing protocol. The insets show examples for an implementation using qutrits, i.e. the generation of the initial state as a superposition of three perfect vortex modes (three red rings), its unitary manipulation during distribution (gray-color phase masks) and the required final projection (squares depicting all nine possible states, with black being the one obtained if measuring in the correct MUB) [8]. Image courtesy of Andrew Forbes.

Compared to coin tossing, standard QKD and QSS, layered QKD involves another level of complexity as it involves asymmetrically entangled states that lead to different layers of secure information shared between the involved parties. Because the generation, as well as the measurements in such a protocol, are extremely challenging, it has only been implemented in a laboratory recently [10]. However, as it shows the great benefit of using multi-particle high-dimensional entangled states, the general idea is outlined in this section. We start with a simple example using a tri-partite state of the form

$$|\psi_{442}\rangle = \frac{1}{2}\left(|0\rangle_A|0\rangle_B|0\rangle_C + |1\rangle_A|1\rangle_B|1\rangle_C + |2\rangle_A|2\rangle_B|0\rangle_C + |3\rangle_A|3\rangle_B|1\rangle_C\right), \tag{8.1}$$

where the indexed numbers 442 labels the local dimensionality of the state and the indexed letters depict the three parties, i.e. Alice, Bob, and Charlie. If such a state is shared between the three parties, a joint measurement of the state leads to four different equally likely outcomes: $|000\rangle$, $|111\rangle$, $|220\rangle$, $|331\rangle$. Alice and Bob can obtain outcomes $|0\rangle$, $|1\rangle$, $|2\rangle$ and $|3\rangle$, i.e. their local states are four-dimensional, while Charlie will only measure a $|0\rangle$ or $|1\rangle$, i.e. his local state is two-dimensional. Alice and Bob can now define two uniform random strings of bits. One string that is perfectly correlated with Charlie's outcome, namely k_{ABC} being 0 for the outcomes $|0\rangle$ and $|2\rangle$ and 1 in all other cases. The other bit string is defined as k_{AB} being 0 for the outcomes $|0\rangle$ and $|1\rangle$, and otherwise 1. Hence, the bit string k_{AB} is uncorrelated with Charlie's outcomes. If the additional assumption is made that copies of pure states have been distributed, both obtained bit strings are also independent of any other data and, thus, the obtained bits for encoding information are not only uniformly distributed but also secure from any eavesdropping attacks. In other words, Alice and Bob can now share secure information with Charlie using one layer, i.e. the bit string k_{ABC}, and they can securely exchange information that is hidden from Charlie using the second layer, i.e. k_{AB} as shown in Fig. 8.4(a) and (b) for a graphical representation.

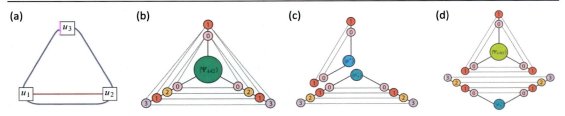

Figure 8.4: Graphical representations of the layered QKD example using the state $|\psi_{442}\rangle$ of Eq. (8.1). (a) Three parties u_1, u_2 and u_3 share the state and thereby establishing a joint layer between all (blue line) and an additional layer between u_1 and u_2 (red line). (b) Graphical representation of the state showing that only party u_1 and u_2 have access to the layer defined by state $|2\rangle$ and $|3\rangle$. (c) Graphical depiction how a two- and four-dimensional bi-partite EPR-like state can achieve a similar structure. (d) Graphical depiction how a GHZ state together with a four-dimensional bi-partite EPR-like state can be used to obtain a similar results. The realization in (c) and (d), however, would have a lower key rate, which shows one of the benefits of the state in (b). Figure adapted from [11].

To turn this simple example into a secure protocol for an arbitrary amount of users and layers, Pivoluska et al. [11] devised a scheme in which the layered state is constructed from qubits, which are then translated to high-dimensionally entangled states. The dimensionality of the state of each party scales with l as $d = 2^l$, where l stands for the number of layers. In each round, every party performs a randomly chosen measurement with 2^l outcomes (depending on the local dimensionality of the state) and sorts the results into the earlier defined "virtual" qubits. For each layer, a key will be established only if all parties have measured in the local computational basis of the qubits, in all other cases, the outcomes can be regarded as test rounds. These test rounds can be used for each of the layers separately to evaluate the required error correction and privacy amplification. In such a protocol, different unconventional situations might appear, such as in some rounds a key will be established only for certain layers, the key rates can vary depending on the layers, or that the implementation and security of every layer is independent of the parties that are not part of this layer. It is interesting to note that the same results could be obtained by using multiple high-dimensionally entangled bi-partite systems or a combination of such states and two-dimensionally entangled GHZ states, as seen from the graphical representation in Fig. 8.4(c) and (d). However, the layered QKD scheme using multi-partite high-dimensionally entangled states seems to be more efficient in terms of achievable key rates [11].

In general, the use of structured photons as high-dimensional quantum states in quantum communications schemes is offering various benefits due to the increased data capacity, possible asymmetric structures as well as the increased complexity of the state. However, these advantages come at the cost of requiring advanced techniques to manipulate the states as well as to distribute them over large distances, both of which we discuss in more detail in the subsequent sections.

8.3 Experimental toolbox

Because a huge progress has been made in the development of a large number of experimental tools and techniques for structured light over the last decades, it is impossible to discuss the currently available full toolbox in every detail. Instead, this section focuses on the most common methods that have been devised and are currently being pushed forward for quantum communication applications.

8.3.1 Generation and detection methods

A key property for structured photons is that they need to be transversely coherent. This is usually ensured by spatial filtering through a single-mode optical fiber, which outputs photons with a fully coherent transverse Gaussian beam profile. Furthermore, it is crucial to have the ability to flexibly encode and detect, not only states of a certain computational basis, but ideally any arbitrary superposition or any other basis state from all available MUBs.

One of the most common generation strategies to encode quantum information is the use of computer generated holograms, as shown in Fig. 8.5(a). These holograms are complex-shaped gratings, where photons in the first diffraction order acquire the desired phase and amplitude modulations. Using holographic techniques not only enables filtering out the unmodulated photons, but it also allows structuring of the amplitude through spatially varying adjustments of the grating efficiency. Both of these features lead to a near-perfect state generation at the cost of increased loss [12]. The holograms are generally implemented using commercially available spatial light modulators, such as liquid crystal displays or digital micromirror devices. While the quality of the generated structures using either device is only limited by its spatial resolution, liquid crystal displays are usually more efficient reaching efficiencies of up to 90%. On the other hand, digital micromirror devices are faster reaching switching rates on the order of tens of kHz.

Another very popular method of generating structured photons are so-called q-plates, which link the polarization degree of freedom to the transverse spatial structure as depicted in Fig. 8.5(a). The general working principle can be described by having a waveplate where the optical axis varies throughout the transverse spatial extent.[3] On one hand, q-plates belong to the most convenient devices for imprinting quantum states encoded in complex polarization patterns. Moreover, they allow polarization to be modulated extremely fast, such that

---

[3] This topic is covered in more detail in Chapter 7.

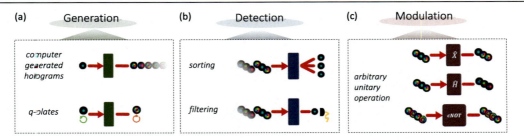

Figure 8.5: Key components of the experimental toolbox for structured photons. (a) The generation of structured photons commonly relies on the modulation by computer generated holograms or q-plates. (b) Detecting the encoded quantum states of structured photons often relies on sorting or filtering, depending on the photon's mode structure. (c) Advanced modulation schemes are capable of performing arbitrary unitary modulations, such as various quantum gates.

switching rates on the order of GHz are possible. On the other hand, linking the modulation to polarization complicates the generation of high-dimensional quantum states due to the two-dimensional nature of polarization. Nevertheless, they constitute one of the leading experimental tools to encode quantum states and have already been realized using liquid crystals [13], plasmonic nanostructures [14] and metasurfaces [15], where the latter allow for a more flexible modulation.

In addition to encoding quantum states, all quantum communication schemes also require a read-out, i.e. an unambiguous detection of the structure of the photon. In general, the schemes can be divided into filtering or single-outcome measurement schemes and sorting or multi-outcome measurements, as shown in Fig. 8.5(b).

In filtering schemes, the measurement device is able to detect the photon only if it is found in a certain state, comparable to a polarizer for polarization. The most common filtering scheme for spatial modes relies on the fact that only photons with a flat phase front couple efficiently into single-mode fibers (SMF). Using appropriate holograms, only the phase front of a specific type of mode is phase flattened and, as such, only this mode couples into the SMF. Hence, the hologram together with the SMF acts as a mode filter. By careful adjustment of the hologram and imaging system [16] or through multiple consecutive phase modulations [17] (see Section 8.3.2 for more information on the technique), it is possible to project a single photon on any possible high-dimensional quantum state in a near-perfect manner. The flexibility and simplicity of most of the filtering schemes make them a very powerful and a popular method to measure the quantum state of single photons and verify the entanglement of structured photon pairs. In fact, the original idea goes back to the paper of Mair et al. in 2001 [2]. Note that it is also possible to realize a mode filter by using the second aforementioned generation technique in reverse, i.e. using q-plates in combination with a polarizer.

A different approach to measuring the spatial structure of single photons is by sorting them into different locations or paths and subsequently detecting them using multiple detectors. Such mode sorters can be considered as the modal analogue of polarizing beam splitters. A popular example is the Cartesian to log-polar coordinate transformation [18], implemented through two simple phase elements. If OAM encoded quantum states are fed into this sorter, the transformation converts their helical phase to a transverse phase gradient, which then leads to an OAM dependent transverse position in the far-field, e.g. obtained by implementing a lens. The main advantage of sorting techniques is the increased efficiency, as all possible outcomes can be obtained without the need to adjust the detection system. Following the same idea of sorting, a multi-plane light conversion (MPLC) technique can be used to sort the full set of Laguerre-Gaussian modes [19]. Since the only requirement for this scheme is that the input modes should be orthogonal, near-perfect mode sorters for all MUBs and mode families can be achieved. The quality of sorting is only limited by the number of phase modulation planes involved (see Section 8.3.2 for more information on MPLC). If losses are not crucial, it is also possible to implement mode sorting schemes using only two phase modulations [20] or one phase modulation in combination with a complex scattering medium [21]. Finally, if the mode structure should remain intact during sorting, schemes have been devised to sort structured light into the two output ports of a specially adjusted interferometer [22–24]. However, if more than two modes are intended to sort, this scheme requires cascading of multiple interferometers, which limits the scalability.

Finally, it is worth noting that a lot of progress in camera technologies has been made over the last decades, such that using a single-photon sensitive camera to observe the structure seems to be the most obvious detection method for structured photons [25]. However, as single-photon detection at a certain pixel does not lead to a conclusion about the whole modal structure, using cameras would either require one to change the basis in which quantum states are encoded from spatial modes to transverse locations, i.e. pixels, or to implement sorting schemes in front of the camera.

With these advanced detection methods, it is possible to reconstruct the complete state of the quantum system through protocols such as the quantum state tomography (QST). In this process, the full density matrix of the received photons is obtained, which can be further used to better quantify the performance when applying different protocols or the effects of quantum channels on the state.

8.3.2 Modulation methods

In addition to the generation and detection of modes, it is also required in quantum communication to modulate and transform structured photons. A mode sorter can be thought of as involving the modulation of transverse spatial modes to different paths. However, the efficient processing of quantum information will require modulation within the set of transverse spatial modes. In general, a device that is able to perform an arbitrary unitary operation on a given set of modal structures, a so-called multiport, is required. By using the symmetry properties of azimuthally structured photons, and clever interferometric arrangements, a general method to build a so-called $\hat{X}$-gate, a quantum gate, which cyclically transforms between the modes, can be devised for any dimension [26]. Together with a mode-dependent phase-shift, implemented by a dove prism, it is possible to realize any arbitrary unitary operation. While experiments demonstrating $\hat{X}$-gates for four-dimensional systems have been realized [27], a scaling to large dimensions becomes very challenging due to the requirement of many optical components and interferometers.

Another method of realizing mode transformations is conceptually much simpler, namely transformations using multiple consecutive phase modulations [28]. Here, any unitary transformation is obtained by sending the structured photons through multiple modulation planes through which the input modes slowly converge to the desired output set of modes. Although only the phase of the photons is modulated at different planes, the free-space propagation between the planes also enables a lossless amplitude modulation. In order to obtain the required phase modulations, the technique of wavefront matching (WFM) can be used [19]. In the WFM scheme, the phase modulations are obtained in an iterative process in which the set of input modes is compared to the respective output modes. The process is repeated consecutively for every plane, and the overlap between the correct mode combinations is maximized. If enough modulation planes are allowed, the input modes are adiabatically transformed into the desired output ones. The main advantage of this scheme is that usually a few modulation planes are enough to obtain very high mode conversion fidelities. Moreover, it does not rely on any symmetry property of the utilized modes, thus it can be used for any mode set and structure. Although more phase modulation planes improve the quality of the unitary transformation, already three modulation planes are enough to perform a broad range of unitary operations, such as high-dimensional $\hat{X}$-gates and Hadamard $\hat{H}$-gates [29]. The latter is especially interesting as it allows one to switch between different MUBs and as such is useful for QSS and quantum cryptographic tasks, where encoding and detecting quantum states in different bases is essential.

Additionally, as there are two independent quantum numbers for each spatial mode, owing to the two-dimensional nature of the spatial domain, it is possible to perform high-dimensional controlled quantum operations on a single carrier. For Laguerre–Gaussian modes for example, the OAM value of a photon can be cyclically shifted to other OAM-values depending on the radial structure, i.e. the p-value, which corresponds to a high-dimensional controlled $\hat{X}$-gate (the qubit version is the well-known controlled NOT-gate).

Finally, having devices at hand that perform any unitary operation on structured photons, will enable the realization of multi-partite interferences required in large-scale quantum networks, as we discuss in the next section.

8.4 Quantum network

While high-dimensional quantum states already allow a larger transmission distance due to higher noise thresholds, a global and fully functioning network will be required in the future for harnessing the full potential of optical quantum technologies. This section discusses how such a network can be implemented and which key components have been or are currently being developed.

Similar to the idea of standard quantum networks utilizing qubit-encoding to enable a global distribution of quantum states, a high-dimensional quantum network implemented with structured photons requires a few key-components. One being a so-called quantum repeater (QR), which mitigates the distance limitations of entanglement distribution. A common realization of a QR consists of a multi-partite interference in an entanglement swapping scheme to map entanglement to particles, which are too far apart such that direct distribution is not possible. In addition, a QR requires a quantum memory, which ensures the appropriate timing for the multi-partite interference to work.

In addition to QRs, a larger network might additionally require a quantum interface for structured photons, which can convert the photon wavelength either to adapt them to the requirements of the memory, to the photons to be interfered with, or to minimize the loss in the quantum channel. A sketch of the network is shown in Fig. 8.6. For all these devices, a lot of progress has been made over the last years such that it is of interest to outline the ideas, developments and future challenges.

8.4.1 Entanglement sources

The most fundamental starting point in the realization of a quantum network is an appropriate entanglement source for structured photons. Interestingly, the first realization of an OAM entangled photon source by Mair et al. [2] using the nonlinear process of spontaneous parametric down-conversion (SPDC) can also be considered the starting point of structured light in quantum optics. Until today, the SPDC process is still the primary source for obtaining two or even more photons that are entangled in their transverse structure. For applications in quantum information, the sources need to fulfill specific requirements, amongst which the quality and dimensionality of the entanglement, the involved subspace of modes, as well as

220 Chapter 8

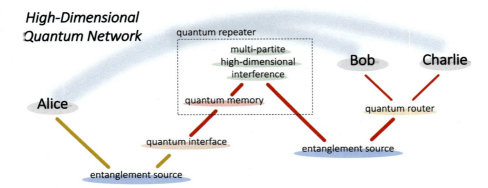

Figure 8.6: Sketch of a possible high dimensional quantum network. Two pairs of high-dimensionally entangled photons are generated at two different locations. Depending on the wavelength of the generated photons, they can then be coherently transformed into a more suitable frequency by a quantum interface and possibly stored in a high-dimensional quantum memory before they are interfered with a photon from the other source. In this process, the entanglement will be swapped to be established between Alice and Bob, or when using a quantum router between Alice, Bob and Charlie (depicted by blue clouds). By repeating such processes with additional sources, entanglement can be established over arbitrarily large distances.

the brightness are of key importance. Especially two specifications, i.e. the obtained quantum state and its physical realizations, would ideally be tunable to the specific setting, e.g. the necessities of the quantum channel or repeater.

The d-dimensionally entangled state generated in SPDC processes is commonly described by the pure state[4]

$$|\psi\rangle = \sum_{n=0}^{d-1} c_{n,n} |n\rangle_A |n\rangle_B , \qquad (8.2)$$

where the indices A, B label the photons shared between Alice and Bob, and n represents the different states. These states are then realized in the laboratories by the respective transverse spatial modes. (When taking the polarization DOF into account, each n labels a combination of spatial mode and polarization.) Additionally, $c_{n,n}$ describes the probability amplitude with which both photons can be found in the respective modes. Hence, if all probability amplitudes are the same, i.e. $c_{n,n} = \frac{1}{\sqrt{d}}$, the state is maximally entangled, which is often the required state

---

[4] Note that in experiments the state generated by an SPDC process is better described as a density matrix. Due to space constraints, we follow the commonly used assumption and use a pure state description when discussing the bi-partite quantum state. However, especially when measures to quantify entanglement are used, deviation from a pure state has to be taken into account.

in quantum communications. Therefore, the general aim is to generate a maximally entangled state with the largest dimensionality possible. Depending on the actual SPDC configurations and the respective laboratory implementations, such a state is challenging to realize, such that various strategies to obtain it have been devised.

8.4.1.1 Tuning entanglement via phase matching of SPDC

In SPDC, the entanglement of structured photons carrying OAM can be intuitively explained by angular momentum conservation arguments together with a common angular birth-zone of the photon pairs. The OAM spectrum obtained through the SPDC process is often characterized by the so-called spiral bandwidth, which describes the probability amplitudes of higher-order OAM-carrying photons that are generated in this process. If the spatial domain is discretized using Laguerre-Gaussian modes, the radial DOF can also be utilized to increase the dimensionality of the generated entangled bi-partite state. Additionally, the bi-partite entangled state generated via SPDC can also be expressed using different transverse spatial mode families [2,30–32].

Irrespective of the actual targeted mode structure, the obtainable modal spectrum, dimensionality, and quality of the quantum states crucially depend on the phase-matching conditions of the SPDC process. Intuitively, this can be explained as follows: On one hand, the wider the phase-matching condition is, the larger the number of high order modes, thus, the higher the dimensionality of the generated state. As the phase matching is usually broader for shorter crystals, it is possible to increase the dimensionality of the state by simply using shorter nonlinear crystals at the cost of the brightness of the source. Moreover, the more precise the linear momentum of the pump photon is defined, the better the photon pairs are correlated in their momenta and, thus, in their transverse spatial modes. Hence, by only loosely focusing the pump beam, a higher quality of the entangled state can be obtained. Both strategies are sketched in Fig. 8.7(a).

Furthermore, one can also influence the modal spectrum by collecting down-converted photon pairs of different beam waists [33] such that the ratio between the beam waists of pump and down-converted photons is the significant value to adjust to the respective needs (see the inset in Fig. 8.7(a)). Finally, the spatial tuning of the phase matching itself, e.g. through tuning of the poling period, is known to affect the modal spectrum generated from the SPDC process [36].

8.4.1.2 Tuning entanglement via pump beam shaping

Another way to achieve tunability is to structure the pump photon to obtain the required quantum state [34,37]. The general idea behind this strategy is to adjust the phase matching such

Entanglement Sources

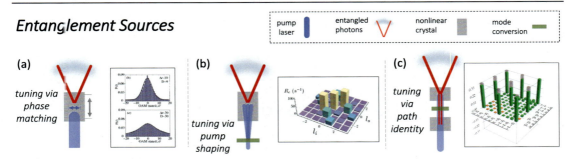

Figure 8.7: SPDC process to generate high-dimensionally entangled structured photons. To obtain the most suitable high-dimensionally entangled quantum states, different mechanism are usually applied, ranging from tuning via phase matching (a), to pump shaping (b) and path identity (c). Insets show experimentally obtained spiral bandwidths and states taken from [33], [34], and [35], respectively.

that when pumped with a single-mode pump beam, e.g. a Gaussian mode, the modal spectrum of SPDC is ideally a single mode. This configuration depends on several parameters, but is usually obtained when the pump beam is focused to a beam waist such that the beam's Rayleigh range is on the order of half of the crystal length. If the pump photon is then structured to be in a superposition of well-chosen modes, the down-converted pair is found to be in multiple modes as well, thereby realizing a high-dimensionally entangled state, as shown in Fig. 8.7(b). While this method is quite simple to implement and allows a lot of control for tuning the generated state in terms of the number of modes as well as their phase relations, it is limited to low dimensionalities as the efficiency of SPDC usually decreases significantly for higher-order modes of the pump photons.

8.4.1.3 Tuning entanglement via path identity

A third approach is based on the general idea of entanglement by path identity [35]. In that scheme multiple consecutive crystals are pumped in a coherent superposition to generate entangled photons. Similar to before, the SPDC process is performed in such a way that, if a photon pair is generated, they are found in a single spatial mode, usually in a Gaussian mode. In the case that the down-conversion process happens, the photons are modulated to the desired spatial structure and are fed with the pump beam into a second nonlinear crystal operated with the same configurations. In case no photon pair was generated in the first crystal, it is possible that a pair of photons is now generated in the lowest order Gaussian mode in the second crystal. If performed correctly, i.e. if it is impossible to determine in which of the two crystals the photons were created, the two possible states of the photon pairs are described by a coherent superposition. Thus, the bi-partite SPDC state is a maximally entangled state, where both photons can be found in either the lowest order mode (if generated in the second crystal) or the imprinted spatial mode structure (if generated in the first crystal), as shown in Fig. 8.7(c). The resulting state is a qubit-entangled state. The dimensionality can then be increased by increasing the number of consecutive crystals. Generating the bi-partite high-dimensional entangled state with this technique allows one to custom-tailor the involved spatial structures without relying on limiting conservation laws in the SPDC process. In addition, it enables controlling the dimensionality of the entanglement by choosing the right amount of consecutively pumped crystals. Further, the scheme enables a control of the phases of the state and can be extended to multi-partite entangled states by pumping multiple crystals in parallel and post-selecting on more than two photons. As such, it can be seen as a promising source for high-dimensionally entangled structured photons and might be one of the building blocks of a future quantum network.

8.4.1.4 Multi-partite entanglement sources

As discussed above, certain quantum communication protocols rely on multi-partite, high-dimensionally entangled states, i.e. photonic quantum states of more than two structured photons. Besides using entanglement by path identity, alternative approaches have been demonstrated that involves two crystals to generate two entangled photon pairs. The four structured photons are then entangled in a complex linear optical setup using multi-photon interferences, mode-specific post-selection, and triggering mechanisms [9]. Another alternative might be using single photons, either from heralded or deterministic single-photon sources, and sending them into a complex linear optical network [38] realized by high-dimensional unitary operations that manipulate the spatial structures.

8.4.2 Quantum channels

To establish a large-scale network, quantum channels used to distribute the photons are essential. The main requirement is that the sent photon is neither lost nor disturbed. The former inevitably lowers the transmission rates, however, brighter photon sources can often compensate the loss up to a certain extent. When brightening the source no longer becomes efficient, quantum repeaters become necessary (see Section 8.4.3). While losses are often independent of the DOF in which the quantum states are encoded, signal disturbances can vary depending on the DOF used in the quantum communication link. In this respect, structured photons often face larger disturbances than polarization encoded photons. A simple model for such disturbances can either be the coupling to other spatial modes (usually higher-order modes), which are not considered in the utilized subspace where the quantum information is encoded, or a mode-dependent coupling to other DOFs, e.g. to the temporal domain via modal dispersion. In either case, the resulting quantum state will either have changed to another OAM state or appear to be degraded. Both effects are discussed in detail in the following subsection in addition to compensation and correction methods.

224   Chapter 8

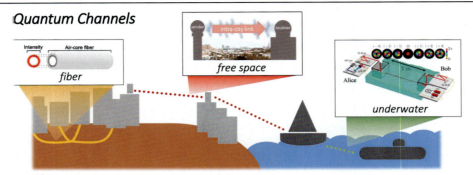

Figure 8.3: Sketch of various transmission channels. Channels for which quantum communication using structured photons has been demonstrated include fibers, free space and water. Reprints of the fiber and underwater channel are taken from [39] and [43], respectively.

8.4.2.1   Fiber quantum channels

Similar to current classical communication networks, the major backbone of a possible quantum network might rely on the transmission of signals through fibers, especially in urban areas. The main advantages of fiber transmission is the guaranteed near-perfect isolation from background light and weather conditions as well as that no line-of-sight is required. These advantages come at the cost of requiring an expensive fiber infrastructure. For structured photons, the current infrastructure cannot be used since special fibers that allow the transmission of higher-order modes are necessary. However, there is a high demand for investments in a more advanced fiber network, since these special fibers are not only needed as quantum channels for secure communication, but also to provide larger data capacities for classical communication systems using structured light.

In general, one can distinguish between two approaches for fiber-based quantum channels. One approach is targeting custom-tailored fibers [39–41], whose eigen-modes fit structured light such that a certain number of higher-order modes can be transmitted without a significant coupling between them. If the core of the fiber, for example resembles the intensity structure of OAM beams, i.e. it is ring-shaped, and has the suitable specifications in terms of dimensions and refractive index, it is able to carry two higher-order modes of the same topological charge, and thus the same OAM value, with opposite sign. If used in combination with the appropriate polarization, such fibers have shown to also work in the quantum domain [42]. Using step-index fibers [41] or air-core fiber [39] as depicted in Fig. 8.8, also do not show any significant inter-modal coupling after 1 km of fiber length, when more than two modes are propagating through it. They have been used in high-dimensional entanglement distribution and cryptography schemes; however, a possible modal dispersion has to be treated carefully.

A modal dispersion would lead to longitudinal walk-off, i.e. a coupling between the spatial mode and the arrival time of the photons. The resulting state thus becomes a non-separable state of spatial modes and arrival times

$$|\phi\rangle_i = \frac{1}{\sqrt{d}} \sum_{n=0}^{d-1} |n\rangle \xrightarrow{\hat{D}} \frac{1}{\sqrt{d}} \sum_{n=0}^{d-1} |n\rangle |t_n\rangle, \qquad (8.3)$$

where $\hat{D}$ depicts a simplified model of a dispersion operator as it couples spatial modes n to arrival times t. The differences in arrival times are often too small to be measurable such that the superposition states appear to have diminished coherence. However, the seemingly introduced decoherence is done through the measurements that neglect the temporal DOF and thereby induce a mixing between the modes (mathematically this corresponds to tracing over the temporal DOF, which requires a density operator description). This coupling can be accounted for through various strategies. Either the modal dispersion, i.e. the temporal shift, is pre- or post-compensated [41], or an appropriate subset of structured modes, which are degenerate in terms of the modal dispersion, are used [42]. Both correction strategies, however, usually limit the number of modes that can be utilized and, therefore, the dimensionality of the quantum state.

Another approach does not rely on special fibers to circumvent intermodal coupling, but rather compensates for it. Here, a standard multimode fiber is used as the quantum channel, which naturally causes a coupling between the spatial modes that are sent into the fiber. After transmission, the state, i.e. the amplitude and phases of all modes, will be randomly scrambled. A simplified model is that each mode n can couple to all other modes n' according to a complex coupling term $c_{n,n'}$, such that

$$|\phi\rangle_i = \frac{1}{\sqrt{d}} \sum_{n=0}^{d-1} |n\rangle \xrightarrow{\hat{T}} \frac{1}{\sqrt{d}} \sum_{n=0}^{d-1} \sum_{n'=0}^{d-1} c_{n,n'} |n'\rangle, \qquad (8.4)$$

where $\hat{T}$ labels a transmission operator that couples all modes of the fiber to each other, thereby scrambling the input state. As before, this process is usually a coherent transformation such that it is possible to compensate for it, e.g. through proper pre- or post-compensation. Moreover, if one photon of an entangled pair is sent through the fiber, entanglement itself can be used to obtain enough information about the transmission operator, such that even high-dimensional quantum correlations can be retrieved [44]. While this approach does not require custom-tailored fibers and, in principle, allows a larger number of modes, it comes at the cost of actively monitoring the modal coupling, which continuously changes due to temperature drifts or fiber movements.

8.4.2.2 Free-space quantum channels

Sending photons via long-distance free-space channels is another valuable possibility for realizing a quantum channel in a larger network. For a strong laser beam of structured light, the record distance has been obtained in an inter-island link over 143 km on the Canary islands [45]. Although this is an impressive distance, it also shows the main challenges that must be overcome or might fundamentally limit the use of structured photons. These challenges are the larger divergence of higher-order modes and the reduced quality of the transmitted state caused by atmospheric turbulence. While the former can be seen as a fundamental limit due to the size constraints of the sending and receiving telescopes, the latter can be compensated (at least in principle) with appropriate adaptive optics.[5] Using similar arguments as in the discussion on multi-mode fiber transmission, the effect of atmospheric turbulences does not necessarily mean an irreversible decoherence but rather a coupling of the transmitted modes to a vast amount of higher-order modes available. The latter might be described in a similar manner to the description of the transmission through multimode fibers, as in Eq. (8.4) for example. Although it should be theoretically possible to compensate for this disturbance entirely, the main challenges are a fast and precise enough optimization in order to cope with fast-changing atmospheric turbulence, which vary on the millisecond time scale. Adaptive optical systems are required to measure the disturbance, e.g. through an auxiliary beam, and include a fast feedback loop in order to correct for wavefront distortions with kHz modulation speed. In addition, if the atmospheric turbulence is too strong, vortex splitting for higher-order OAM modes can be observed [46], such that a correction using a single modulation plane might not be enough.

When a pair of entangled structured photons propagate through atmospheric turbulence, the disturbances cause a decay of the entanglement, which is even stronger for high-dimensionally entangled states [47]. Interestingly, two-photon interferences are hardly affected if one or even both photons are disturbed by the turbulence under weak scintillation conditions. The fundamental reason for this phenomenon is that weak turbulence conditions are not able to convert symmetric states into anti-symmetric states [48].

Additionally, certain mode families might be advantageous depending on the actual conditions. Higher-order Ince–Gaussian beams, for example, do not possess higher-order topological charges but split, singly charged vortices, which seem to make them less vulnerable to turbulence [49]. Entangled Bessel–Gaussian photons are known to be self-healing such that obstructing parts of the beam in transmission may be automatically corrected [30].

Without any active adaptive optical systems in place, the OAM entanglement has been distributed over a distance of 3 km using a detection technique that is relatively insensitive to

---
[5] This topic is covered in greater detail in Chapters 9 and 10.

turbulent distortions, where specific two-dimensional superposition states were measured using slit masks [50]. Further, single photons encoded in spatial polarization structures have been shown to enable a high-dimensional quantum cryptography link over 300 m [51]. As both implementations have been realized in cities, where the turbulence is assumed to be stronger, free-space channels, especially when equipped with adaptive optics, will become a valuable option for quantum channels. Given that high-dimensional quantum communication and entanglement distribution schemes are known to be very noise tolerant [52], one might also expect that such links function during daylight conditions.

8.4.2.3 Underwater quantum channels

Another possibility of a quantum channel is a free transmission under water. Similar to the free-space transmission through air, turbulence-induced distortions can be seen as one of the main hurdles in such a channel [43]. While air-turbulences lead to local refractive index changes due to the pressure and temperature variations, in water, which is incompressible, mainly temperature fluctuations cause the mode of the beam to get distorted. This distortion was found to be already strong after a few meters, such that adaptive optics similar to the free-space correction systems are required. However, the speed at which the beam structure changes is significantly slower than in air, of the order 0.1 s, such that slower spatial light modulators could be used. Equally important, the efficiency of the transmission varies significantly for different wavelengths with the least absorption at around 400-500 nm, such that quantum channels will be required to work in this wavelength range. Nevertheless, recent experiments have shown promising results over distances up to 30 m [53], which might be extendable up to a few hundreds of meters making the underwater channels a practicable option for quantum communication schemes utilizing structured photons.

8.4.2.4 Characterization of quantum channels

The distortions and deformations of the transmitted quantum signal appearing after long-distance channels, some of which were briefly sketched above, require a detailed characterization. On the one hand, it is helpful when correcting or accounting for channel induced alterations of the transmitted light. On the other hand, it might be used to detect a possible eavesdropper and thereby ensure a secure transmission.

One of the most detailed methods of characterizing the channel is a process called quantum process tomography, in which a full reconstruction of the effect on the transmitted quantum state is obtained.[6] To obtain the process matrix from which all effects of the quantum channel can be deduced, the full set of states of a given Hilbert space, together with the states of all

---

[6] As the process tomography is discussed in more detail in Chapter 6 of this book, it is only briefly sketched here for completeness.

MUBs, are sent through the channel. The obtained output state is then characterized by a full tomographic measurement again using all states of all MUBs. After obtaining the process matrix, key figures of merit such as the process fidelity and process purity, describing the quality of the channel and level of induced mixture, can be deduced. Experimentally, it was possible to use structured photons up to five dimensions to perform quantum process tomography [54]. Moreover, it was shown that the data can be used to get more information on possible eavesdropping strategies or induced errors, which leads to the reduction of the quantum bit error rate if the channel is used in a cryptography scheme.

The downside of this scheme is often the time-consuming characterization, especially if the channel is not highly efficient. Here, the use of a structured laser beam in a non-separable state between polarization and spatial modes can be beneficially applied. Instead of sending one photon of the spatially entangled pair through the quantum channel and checking the obtained correlation, the strong characterizing laser beam can be sent through the channel and characterized afterwards. The procedure now allows for the detection of the undisturbed polarization DOF (mimicking the unperturbed partner photon) and measure it against the perturbed spatial mode. The degree of correlation between these two degrees of freedoms can be used to characterize the effect of channel on an entangled state [55]. When involving other high-dimensional degrees of freedom instead of polarization, the scheme can be extended to a high-dimensional channel characterization for quantum information tasks [56].

8.4.3 Quantum repeater

A quantum repeater is one of the key elements in large-scale quantum networks, as it enables, in principle, the distribution of quantum states over arbitrarily large distances. While there are various possible architectures, we focus on the original idea, the so-called first category quantum repeater, which uses quantum memories and entanglement swapping [57]. In such a scheme, the pairs of entangled photons are distributed over the largest possible distance that losses permit. If this distance is not enough, a second entangled photon pair is needed. By taking one photon of each pair and performing an entanglement swapping operation on them, the two untouched photons are consequently entangled. By a proficient positioning of the photon sources, the distance over which entanglement is transmitted is increased. The scheme can be repeated with more entangled photon pairs, such that in principle unlimited distances can be covered. In addition, entanglement swapping relies on multi-partite interferences, such that quantum memories, which ensure the proper arrival times of the photons, are required.

8.4.3.1 High-dimensional teleportation and entanglement swapping

Entanglement swapping fundamentally relies on quantum teleportation, hence, it is instructive to briefly discuss its functioning. In the teleportation scheme, if Alice wants to teleport a state

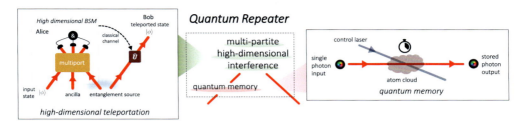

Figure 8.9: Sketch depicting a high-dimensional quantum repeater (first generation), i.e. two of its main components. Left: Scheme for quantum teleportation of a high-dimensional state. Right: Setup to realize a quantum memory for structured photons.

encoded on a photon to the another party, Bob, they both need to share an entangled state first. Alice then has to perform a Bell-state measurement (BSM) between the photon with the state to be teleported and the photon of the entangled pair. To do so, Alice projects the two photons onto one of the four Bell states. After communicating to Bob about which Bell state she obtained, he can perform a set of simple unitary operations on his photon to obtain the state Alice wanted to teleport. As this protocol is independent of the state that Alice teleports to Bob, it also works for one photon that is a part of an entangled pair and, thus, using the same procedure, the entanglement can be swapped to exist between the two photons that never interacted. Hence, entanglement swapping is often called teleportation of entanglement.

As the teleportation requires projection on Bell states through BSM, the situation becomes more challenging for high-dimensional teleportation schemes. Already in three dimensions, nine Bell states exist with three symmetric states and six neither symmetric nor antisymmetric states. Hence, a simple extension of the well-known two-dimensional teleportation scheme, using the symmetry properties of the Bell states, does not exist. In fact, it was proven that a deterministic BSM for high-dimensional quantum states does not exist, when using only linear optics and two photons as in the qubit-BSM [58]. However, it was shown to be possible when using d-2 ancillary photons and a d-dimensional multiport, where d labels the dimensionality of the BSM [59,60]. For the high-dimensional BSM to perform, the ancillary photon, the photon to be teleported, and the photon from the high-dimensionally entangled pair from Alice's side, need to be brought to complex interference in the multiport (shown in Fig. 8.9 on the left). With an appropriate post-selection procedure the teleportation is performed with a success probability of 1/81 (using feed-forward techniques, this probability can be increased to 1/9). Unfortunately, this procedure is extremely challenging and, thus, it has only been demonstrated for the path DOF. In addition, only the teleportation process was realized and not a full entanglement swapping, as an additional photon (the ancillary photon) would be required and the experimental challenges increase tremendously if more than four photons are necessary. However, as novel techniques for multiport realizations are being developed (see

Section 8.3.2) along with improved and tunable entanglement sources (see Section 8.4.1), it might become possible in the near future to extend this scheme to structured photons.

While a high-dimensional entanglement swapping scheme for structured light is still missing, a complex two-dimensional version has been already realized [61]. As many subspaces of a high-dimensionally encoded photon pair are anti-symmetric under particle exchange, it is possible to perform a simultaneous entanglement swapping of a large number of two-dimensional subspaces. If the BSM is performed using two-photon interference at a beam-splitter with high-dimensionally entangled structured photons, the photons anti-bunch only for a specific combination of modes. In consequence, the interference acts as a filter for structured photon pairs. Post-selecting on these cases, where the two photons anti-bunch, will lead to a complex state for the two remaining photons. The resulting state is found to be a statistical mixture of anti-symmetric states including all possible combinations of two-dimensionally entangled photon pairs. While this scheme might not be directly applicable as a high-dimensional quantum repeater, the complex bi-partite entangled state, which is generated in this process, could be beneficial in other multi-partite quantum communication schemes.

8.4.3.2 Quantum memories

As described above, a working high-dimensional quantum repeater requires the two photons and the ancillary photons to arrive at the same time at the device that performs the high-dimensional quantum interference. As most of the sources are probabilistic, it will be necessary in a working quantum network to also have a quantum memory at one's disposal. Ideally, such a memory will store the photon for a given time and releases it once the second (and ancillary photons) are ready to be interfered.

So far, one of the most common implementations of a high-dimensional quantum memory for structured photons utilized a cloud of cold atoms trapped in a magneto-optical trap [62]. Here, the state of the structured photon is mapped into the ensemble of cold atoms employing electromagnetically induced transparency. In this protocol, the transparency of the atomic cloud, with respect to the photon to be stored, is controlled by a strong light field driving another atomic transition. When the photon that needs to be stored is sent into the transparent cloud of atoms, the strong control field will be switched off in an adiabatic manner such that the signal-carrying photon will be absorbed. In this process, the quantum properties of the signal photon, which are encoded in its spatial structure, are coherently transferred to the atomic medium. When the strong control field is switched on again, the process is reversed. Hence, the stored photon will be emitted, ideally in its original structure, i.e. without losing the high-dimensional quantum information. See Fig. 8.9 on the right side for a sketch of the scheme. In experiments, storage times of hundreds of nanoseconds as well as state fidelities of more than 80% have been achieved. First experiments demonstrated the quantum storage of a two-dimensional state [63] that was encoded in the azimuthal DOF of light. Even though this demonstration was done using a weak laser, various follow-up works also showed the storage of high-dimensional quantum states of single photons and a memory storing high-dimensionally entanglement in two different atom clouds. Furthermore, other memory realizations such as a solid rare-earth-ion-doped crystal employing an atomic frequency comb protocol have also been put forward. The storage of quantum information encoded in additional degrees of freedom such as polarization and path as well as storing hybrid- and hyper-entangled states have been demonstrated. Challenges to be tackled in the future will include increasing the dimensionality of the stored state as well as the overall efficiency of the quantum storage [62].

8.4.4 Quantum interfaces

In a larger network, having the possibility of changing the frequency of the structured photon is another crucial requirement. At various stages of the distribution of photons, a quantum interface could match the frequencies for best suiting to the quantum channel, the memory and the detection, but also to match the frequencies of the photons coming from different sources. It is of special interest for structured photons that these frequency conversions not only have the highest efficiency, but also that it works irrespective of the spatial mode of the photons.

The most common techniques used for converting photons from one frequency into another use parametric processes in nonlinear optical media, such as sum frequency generation (SFG) [64] and difference frequency generation (DFG) [65]. The former is used to obtain higher frequencies, i.e. upconvert the photon to a shorter wavelength, and the latter leads to a lower frequency and, thus, longer wavelength. In both schemes, the signal photon and the strong pump field is fed into a nonlinear crystal, which is often (but not necessarily) put into a cavity to enhance the conversion efficiency (see Fig. 8.10 on the left). For SFG, the converted photon's frequency ω_c is the sum of the signal photon's frequency ω_s and the pump ω_p, i.e. $\omega_c = \omega_s + \omega_p$. For DFG, the converted photon's frequency is the difference between the two frequencies, i.e. $\omega_c = \omega_s - \omega_p$. The major benefit of these nonlinear processes is the great flexibility and adaptability when tuning the resulting frequency. Especially, phase matching obtained through periodically poling can be used to make a broad range of frequencies available. Unfortunately, the conversion efficiencies are only on the order of a few percent when pumped by a strong laser field. The efficiency can be further reduced for higher-order modes such that measures to counteract, like using different pump beam profile, are being developed.

8.4.5 Quantum router

The high-dimensional nature of entanglement of structured photons allows another exclusive application, namely to subdivide the large Hilbert space into smaller subspaces, in a so-called

232   Chapter 8

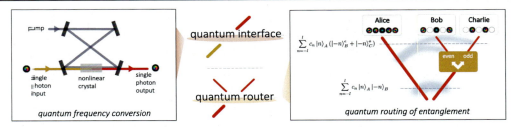

Figure 8.10: Sketch depicting a quantum interface for structured photons (left) and a high-dimensional quantum router (right). Left: In a quantum interface, structured photons are converted to another wavelength via a nonlinear optical process, where the efficiency can be increased by using a bow-tie or ring cavity, as shown here. Right: The high-dimensional quantum router is a device that redistributes OAM photons depending on the parity of the OAM value. The router can be used to distribute high-dimensionally entangled quantum states between more users, e.g. between Alice and Bob as well as Alice and Charlie.

routing scheme. Starting with a high-dimensional entangled state $|\psi\rangle$, it is possible to split the state into two lower-dimensionally entangled states using a quantum router [66]. While in general, it is possible to divide the full state space into arbitrary many subspaces and parties, where the minimal states space per party is of dimension two, the experimental implementation requires a unitary operation acting on the spatial mode and the beam path, i.e. coupling both DOFs together. One implementation of such a unitary operation is known as the OAM-parity-sorter, which redirects structured photons into different paths on the basis of the parity of the OAM value [22]. When such a sorter is placed into the beam path of one photon of a high-dimensionally entangled state, let us say Bob's photon, the state is split depending on the parity. As a result, the state can be shared between Alice, Bob, and a third party named Charlie as in Fig. 8.10 on the right. Mathematically, this process can be described by

$$|\psi\rangle = \sum_{n=-\ell}^{\ell} c_n |n\rangle_A |-n\rangle_B \quad \xrightarrow{\hat{R}} \quad \sum_{n=-\ell}^{\ell} c_n |n\rangle_A \left(|-n\rangle_B^e + |-n\rangle_C^o\right), \qquad (8.5)$$

where $\hat{R}$ stands for the action of the high-dimensional quantum router, the indices label the different parties, and the superscripts e and o stand for even and odd parity, respectively.
In an experiment, a state with at least 10-dimensional entanglement was split into five- and six-dimensionally entangled states, shared between three parties [66]. Moreover, if the router is controllable, the states can be actively re-distributed, e.g. by switching the parity for Bob and Charlie. If more than two photons are high-dimensionally entangled, more possibilities of routing can be imagined.

8.5 Conclusion

Quantum communication offers a secure, reliable and efficient technique for transmitting encoded information. The structured photons would further enlarge the possibilities in a quantum network. They provide the realization of high-dimensional quantum states in quantum communication applications. Not only due to enhanced information capacity and noise resilience of qudits, which are their best-known benefits, but also because of the advanced protocols, structured photons will help in pushing future quantum technologies to live up to their potential. Of course, the manifold benefits are accompanied by the need for an advanced experimental toolbox, especially if applied to a global quantum network. Along with the experimental efforts, theoretical studies are also needed to uncover conceivable additional features based on the increased complexity and diversity of the quantum state space.

The chapter has given an overview of several quantum protocols, which benefit or have been developed with inspiration from structured photons as qudit realizations. Moreover, the experimental tools and technologies that have been outlined will be required when implementing such protocols on a global scale. As most of the required techniques and methods have been demonstrated to work in proof-of-principle experiments, a bright future for quantum communication with structured photon can be anticipated.

Acknowledgments

The authors thank Markus Hiekkamäki and Lea Kopf for helpful discussions. Robert Fickler and Shashi Prabhakar acknowledge the support from Academy of Finland through the Competitive Funding to Strengthen University Research Profiles (decision 301820) and the Photonics Research and Innovation Flagship (PREIN - decision 320165).

References

[1] L. Allen, M.W. Beijersbergen, R. Spreeuw, J. Woerdman, Orbital angular momentum of light and the transformation of Laguerre-Gaussian laser modes, Phys. Rev. A 45 (11) (1992) 8185, https://doi.org/10.1103/PhysRevA.45.8185.
[2] A. Mair, A. Vaziri, G. Weihs, A. Zeilinger, Entanglement of the orbital angular momentum states of photons, Nature 412 (2001) 313–316, https://doi.org/10.1038/35085529.
[3] X.-M. Hu, Y. Guo, B.-H. Liu, Y.-F. Huang, C.-F. Li, G.-C. Guo, Beating the channel capacity limit for superdense coding with entangled ququarts, Sci. Adv. 4 (2018) eaat9304, https://doi.org/10.1126/sciadv.aat9304.
[4] C.H. Bennett, G. Brassard, Quantum cryptography: public key distribution and coin tossing Int, 1984.
[5] G. Molina-Terriza, A. Vaziri, R. Ursin, A. Zeilinger, Experimental quantum coin tossing, Phys. Rev. Lett. 94 (2005) 040501, https://doi.org/10.1103/PhysRevLett.94.040501.
[6] M. Hillery, V. Bužek, A. Berthiaume, Quantum secret sharing, Phys. Rev. A 59 (1999) 1829, https://doi.org/10.1103/PhysRevA.59.1829.
[7] A. Tavakoli, I. Herbauts, M. Żukowski, M. Bourennane, Secret sharing with a single d-level quantum system, Phys. Rev. A 92 (2015) 030302, https://doi.org/10.1103/PhysRevA.92.030302.

[8] J. Pinnell, I. Nape, M. De Oliveira, N. Tabebordbar, A. Forbes, Experimental demonstration of 11-dimensional 10-party quantum secret sharing, Laser Photonics Rev. 14 (2020) 2000012, https://doi.org/10.1002/lpor.202000012.

[9] M. Malik, M. Erhard, M. Huber, M. Krenn, R. Fickler, A. Zeilinger, Multi-photon entanglement in high dimensions, Nat. Photonics 10 (2016) 248, https://doi.org/10.1038/nphoton.2016.12.

[10] X.-M. Hu, W.-B. Xing, C. Zhang, B.-H. Liu, M. Pivoluska, M. Huber, Y.-F. Huang, C.-F. Li, G.-C. Guo, npj Quantum Inf. 6 (2020) 1, https://doi.org/10.1038/s41534-020-00318-6.

[11] M. Pivoluska, M. Huber, M. Malik, Layered quantum key distribution, Phys. Rev. A 97 (2018) 032312, https://doi.org/10.1103/PhysRevA.97.032312.

[12] E. Bolduc, N. Bent, E. Santamato, E. Karimi, R.W. Boyd, Exact solution to simultaneous intensity and phase encryption with a single phase-only hologram, Opt. Lett. 38 (2013) 3546–3549, https://doi.org/10.1364/OL.38.003546.

[13] L. Marrucci, C. Manzo, D. Paparo, Optical spin-to-orbital angular momentum conversion in inhomogeneous anisotropic media, Phys. Rev. Lett. 96 (2006) 163905, https://doi.org/10.1103/PhysRevLett.96.163905.

[14] E. Karimi, S.A. Schulz, I. De Leon, H. Qassim, J. Upham, R.W. Boyd, Generating optical orbital angular momentum at visible wavelengths using a plasmonic metasurface, Light Sci. Appl. 3 (2014) e167, https://doi.org/10.1038/lsa.2014.48.

[15] R.C. Devlin, A. Ambrosio, N.A. Rubin, J.B. Mueller, F. Capasso, Arbitrary spin-to-orbital angular momentum conversion of light, Science 358 (2017) 896–901, https://doi.org/10.1126/science.aao5392.

[16] F. Bouchard, N.H. Valencia, F. Brandt, R. Fickler, M. Huber, M. Malik, Measuring azimuthal and radial modes of photons, Opt. Express 26 (2018) 31925–31941, https://doi.org/10.1364/OE.26.031925.

[17] M. Hiekkamäki, S. Prabhakar, R. Fickler, Near-perfect measuring of full-field transverse-spatial modes of light, Opt. Express 27 (2019) 31456–31464, https://doi.org/10.1364/OE.27.031456.

[18] G.C. Berkhout, M.P. Lavery, J. Courtial, M.W. Beijersbergen, M.J. Padgett, Efficient sorting of orbital angular momentum states of light, Phys. Rev. Lett. 105 (2010) 153601, https://doi.org/10.1103/PhysRevLett.105.153601.

[19] N.K. Fontaine, R. Ryf, H. Chen, D.T. Neilson, K. Kim, J. Carpenter, Laguerre-Gaussian mode sorter, Nat. Commun. 10 (2019) 1–7, https://doi.org/10.1038/s41467-019-09840-4.

[20] R. Fickler, F. Bouchard, E. Giese, V. Grillo, G. Leuchs, E. Karimi, Full-field mode sorter using two optimized phase transformations for high-dimensional quantum cryptography, J. Opt. 22 (2020) 024001, https://doi.org/10.1088/2040-8986/ab6303.

[21] R. Fickler, M. Ginoya, R.W. Boyd, Custom-tailored spatial mode sorting by controlled random scattering, Phys. Rev. B 95 (2017) 161108, https://doi.org/10.1103/PhysRevB.95.161108.

[22] J. Leach, M.J. Padgett, S.M. Barnett, S. Franke-Arnold, J. Courtial, Measuring the orbital angular momentum of a single photon, Phys. Rev. Lett. 88 (2002) 257901, https://doi.org/10.1103/PhysRevLett.88.257901.

[23] Y. Zhou, M. Mirhosseini, D. Fu, J. Zhao, S.M.H. Rafsanjani, A.E. Willner, R.W. Boyd, Sorting photons by radial quantum number, Phys. Rev. Lett. 119 (26) (2017) 263602, https://doi.org/10.1103/PhysRevLett.119.263602.

[24] X. Gu, M. Krenn, M. Erhard, A. Zeilinger, Gouy phase radial mode sorter for light: concepts and experiments, Phys. Rev. Lett. 120 (10) (2018) 103601, https://doi.org/10.1103/PhysRevLett.120.103601.

[25] R. Fickler, M. Krenn, R. Lapkiewicz, S. Ramelow, A. Zeilinger, Real-time imaging of quantum entanglement, Sci. Rep. 3 (2013) 1914, https://doi.org/10.1038/srep01914.

[26] X. Gao, M. Krenn, J. Kysela, A. Zeilinger, Arbitrary d-dimensional Pauli X gates of a flying qudit, Phys. Rev. A 99 (2019) 023825, https://doi.org/10.1103/PhysRevA.99.023825.

[27] A. Babazadeh, M. Erhard, F. Wang, M. Malik, R. Nouroozi, M. Krenn, A. Zeilinger, High-dimensional single-photon quantum gates: concepts and experiments, Phys. Rev. Lett. 119 (2017) 180510, https://doi.org/10.1103/PhysRevLett.119.180510.

[28] J.-F. Morizur, L. Nicholls, P. Jian, S. Armstrong, N. Treps, B. Hage, M. Hsu, W. Bowen, J. Janousek, H.-A. Bachor, Programmable unitary spatial mode manipulation, J. Opt. Soc. Am. A 27 (2010) 2524–2531, https://doi.org/10.1364/JOSAA.27.002524.

[29] F. Brandt, M. Hiekkamäki, F. Bouchard, M. Huber, R. Fickler, High-dimensional quantum gates using full-field spatial modes of photons, Optica 7 (2020) 98–107, https://doi.org/10.1364/OPTICA.375875.

[30] M. McLaren, T. Mhlanga, M.J. Padgett, F.S. Roux, A. Forbes, Self-healing of quantum entanglement after an obstruction, Nat. Commun. 5 (2014) 3248, https://doi.org/10.1038/ncomms4248.

[31] S. Walborn, S. Pádua, C. Monken, Conservation and entanglement of Hermite-Gaussian modes in parametric down-conversion, Phys. Rev. A 71 (2005) 053812, https://doi.org/10.1103/PhysRevA.71.053812.

[32] M. Krenn, R. Fickler, M. Huber, R. Lapkiewicz, W. Plick, S. Ramelow, A. Zeilinger, Entangled singularity patterns of photons in Ince–Gauss modes, Phys. Rev. A 87 (2013) 012326, https://doi.org/10.1103/PhysRevA.87.012326.

[33] J. Romero, D. Giovannini, S. Franke-Arnold, S. Barnett, M. Padgett, Increasing the dimension in high-dimensional two-photon orbital angular momentum entanglement, Phys. Rev. A 86 (2012) 012334, https://doi.org/10.1103/PhysRevA.86.012334.

[34] E. Kovlakov, S. Straupe, S. Kulik, Quantum state engineering with twisted photons via adaptive shaping of the pump beam, Phys. Rev. A 98 (2018) 060301, https://doi.org/10.1103/PhysRevA.98.060301.

[35] J. Kysela, M. Erhard, A. Hochrainer, M. Krenn, A. Zeilinger, Experimental high-dimensional entanglement by path identity, arXiv:1904.07851, https://arxiv.org/abs/1904.07851.

[36] J. Svozilík, J. Peřina Jr, J.P. Torres, High spatial entanglement via chirped quasi-phase-matched optical parametric down-conversion, Phys. Rev. A 86 (2012) 052318, https://doi.org/10.1103/PhysRevA.86.052318.

[37] S. Liu, Z. Zhou, S. Liu, Y. Li, Y. Li, C. Yang, Z. Xu, Z. Liu, G. Guo, B. Shi, Coherent manipulation of a three-dimensional maximally entangled state, Phys. Rev. A 98 (2018) 062316, https://doi.org/10.1103/PhysRevA.98.062316.

[38] Y.L. Lim, A. Beige, Multiphoton entanglement through a Bell-multiport beam splitter, Phys. Rev. A 71 (2005) 062311, https://doi.org/10.1103/PhysRevA.71.062311.

[39] D. Cozzolino, D. Bacco, B. Da Lio, K. Ingerslev, Y. Ding, K. Dalgaard, P. Kristensen, M. Galili, K. Rottwitt, S. Ramachandran, et al., Orbital angular momentum states enabling fiber-based high-dimensional quantum communication, Phys. Rev. Appl. 11 (2019) 064058, https://doi.org/10.1103/PhysRevApplied.11.064058.

[40] N. Bozinovic, Y. Yue, Y. Ren, M. Tur, P. Kristensen, H. Huang, A.E. Willner, S. Ramachandran, Terabit-scale orbital angular momentum mode division multiplexing in fibers, Science 340 (2013) 1545–1548, https://doi.org/10.1126/science.1237861.

[41] H. Cao, S.-C. Gao, C. Zhang, J. Wang, D.-Y. He, B.-H. Liu, Z.-W. Zhou, Y.-J. Chen, Z.-H. Li, S.-Y. Yu, et al., Distribution of high-dimensional orbital angular momentum entanglement over a 1 km few-mode fiber, Optica 7 (2020) 232–237, https://doi.org/10.1364/OPTICA.381403.

[42] A. Sit, R. Fickler, F. Alsaiari, F. Bouchard, H. Larocque, P. Gregg, L. Yan, R.W. Boyd, S. Ramachandran, E. Karimi, Quantum cryptography with structured photons through a vortex fiber, Opt. Lett. 43 (2018) 4108–4111, https://doi.org/10.1364/OL.43.004108.

[43] F. Bouchard, A. Sit, F. Hufnagel, A. Abbas, Y. Zhang, K. Heshami, R. Fickler, C. Marquardt, G. Leuchs, E. Karimi, et al., Quantum cryptography with twisted photons through an outdoor underwater channel, Opt. Express 26 (2018) 22563–22573, https://doi.org/10.1364/OE.26.022563.

[44] N.H. Valencia, S. Goel, W. McCutcheon, H. Defienne, M. Malik, Unscrambling entanglement through a complex medium, Nat. Phys. 16 (2020) 1112–1116, https://doi.org/10.1038/s41567-020-0970-1.

[45] M. Krenn, J. Handsteiner, M. Fink, R. Fickler, R. Ursin, M. Malik, A. Zeilinger, Twisted light transmission over 143 km, Proc. Natl. Acad. Sci. 113 (2016) 13648–13653, https://doi.org/10.1073/pnas.1612023113.

[46] M.P. Lavery, Vortex instability in turbulent free-space propagation, New J. Phys. 20 (2018) 043023, https://doi.org/10.1088/1367-2630/aaae9e.

[47] Y. Zhang, S. Prabhakar, F.S. Roux, A. Forbes, T. Konrad, et al., Experimentally observed decay of high-dimensional entanglement through turbulence, Phys. Rev. A 94 (2016) 032310, https://doi.org/10.1103/PhysRevA.94.032310.

[48] S. Prabhakar, C. Mabena, T. Konrad, F.S. Roux, Turbulence and the Hong-Ou-Mandel effect, Phys. Rev. A 97 (2018) 013835, https://doi.org/10.1103/PhysRevA.97.013835.

[49] X. Gu, L. Chen, M. Krenn, Phenomenology of complex structured light in turbulent air, Opt. Express 28 (2020) 11033–11050, https://doi.org/10.1364/OE.386962.

[50] M. Krenn, J. Handsteiner, M. Fink, R. Fickler, A. Zeilinger, Twisted photon entanglement through turbulent air across Vienna, Proc. Natl. Acad. Sci. 112 (2015) 14197–14201, https://doi.org/10.1073/pnas.1517574112.

[51] A. Sit, F. Bouchard, R. Fickler, J. Gagnon-Bischoff, H. Larocque, K. Heshami, D. Elser, C. Peuntinger, K. Günthner, B. Heim, et al., High-dimensional intracity quantum cryptography with structured photons, Optica 4 (2017) 1006–1010, https://doi.org/10.1364/OPTICA.4.001006.

[52] S. Ecker, F. Bouchard, L. Bulla, F. Brandt, O. Kohout, F. Steinlechner, R. Fickler, M. Malik, Y. Guryanova, R. Ursin, et al., Overcoming noise in entanglement distribution, Phys. Rev. X 9 (2019) 041042, https://doi.org/10.1103/PhysRevX.9.041042.

[53] F. Hufnagel, A. Sit, F. Bouchard, Y. Zhang, D. England, K. Heshami, B.J. Sussman, E. Karimi, Investigation of underwater quantum channels in a 30 meter flume tank using structured photons, New J. Phys. 22 (2020) 093074, https://doi.org/10.1088/1367-2630/abb688.

[54] F. Bouchard, F. Hufnagel, D. Koutnỳ, A. Abbas, A. Sit, K. Heshami, R. Fickler, E. Karimi, Quantum process tomography of a high-dimensional quantum communication channel, Quantum 3 (2019) 138, https://doi.org/10.22331/q-2019-05-06-138.

[55] B. Ndagano, B. Perez-Garcia, F.S. Roux, M. McLaren, C. Rosales-Guzman, Y. Zhang, O. Mouane, R.I. Hernandez-Aranda, T. Konrad, A. Forbes, Characterizing quantum channels with non-separable states of classical light, Nat. Phys. 13 (2017) 397–402, https://doi.org/10.1038/nphys4003.

[56] C.M. Mabena, F.S. Roux, High-dimensional quantum channel estimation using classical light, Phys. Rev. A 96 (2017) 053860, https://doi.org/10.1103/PhysRevA.96.053860.

[57] S. Muralidharan, L. Li, J. Kim, N. Lütkenhaus, M.D. Lukin, L. Jiang, Optimal architectures for long distance quantum communication, Sci. Rep. 6 (2016) 20463, https://doi.org/10.1038/srep20463.

[58] J. Calsamiglia, Generalized measurements by linear elements, Phys. Rev. A 65 (2002) 030301, https://doi.org/10.1103/PhysRevA.65.030301.

[59] Y.-H. Luo, H.-S. Zhong, M. Erhard, X.-L. Wang, L.-C. Peng, M. Krenn, X. Jiang, L. Li, N.-L. Liu, C.-Y. Lu, et al., Quantum teleportation in high dimensions, Phys. Rev. Lett. 123 (2019) 070505, https://doi.org/10.1103/PhysRevLett.123.070505.

[60] X.-M Hu, C. Zhang, B.-H. Liu, Y. Cai, X.-J. Ye, Y. Guo, W.-B. Xing, C.-X. Huang, Y.-F. Huang, C.-F. Li, et al., Experimental high-dimensional quantum teleportation, Phys. Rev. Lett. 125 (2020) 230501, https://doi.org/10.1103/PhysRevLett.125.230501.

[61] Y. Zhang, M. Agnew, T. Roger, F.S. Roux, T. Konrad, D. Faccio, J. Leach, A. Forbes, Simultaneous entanglement swapping of multiple orbital angular momentum states of light, Nat. Commun. 8 (2017) 1–7, https://doi.org/10.1038/s41467-017-00706-1.

[62] B.-S. Shi, D.-S. Ding, W. Zhang, Quantum storage of orbital angular momentum entanglement in cold atomic ensembles, J. Phys. B 51 (2018) 032004, https://doi.org/10.1088/1361-6455/aa9b95.

[63] A. Nicolas, L. Veissier, L. Giner, E. Giacobino, D. Maxein, J. Laurat, A quantum memory for orbital angular momentum photonic qubits, Nat. Photonics 8 (2014) 234, https://doi.org/10.1038/nphoton.2013.355.

[64] Z.-Y. Zhou, Y. Li, D.-S. Ding, W. Zhang, S. Shi, B.-S. Shi, G.-C. Guo, Orbital angular momentum photonic quantum interface, Light Sci. Appl. 5 (2016) e16019, https://doi.org/10.1038/lsa.2016.19.

[65] S.-L. Liu, S.-K. Liu, Y.-H. Li, S. Shi, Z.-Y. Zhou, B.-S. Shi, Coherent frequency bridge between visible and telecommunications band for vortex light, Opt. Express 25 (2017) 24290–24298, https://doi.org/10.1364/OE.25.024290.

[66] M. Erhard, M. Malik, A. Zeilinger, A quantum router for high-dimensional entanglement, Quantum Sci. Technol. 2 (2017) 014001, https://doi.org/10.1088/2058-9565/aa5917.

CHAPTER 9

Optical angular momentum interaction with turbulent and scattering media

Mingjian Chen[a,b] and Martin Lavery[a]

[a]University of Glasgow, Glasgow, United Kingdom [b]Xidian University, Xi'an, Shaanxi, China

Contents

9.1 Atmospheric turbulence variations in real environments 238
9.2 Turbulence-induced phase variations 241
9.3 Turbulence's effect on structured beams 244
9.4 Degradation of beams that carry OAM 246
9.5 Scattering dynamics of beams that carry OAM 254
9.6 Conclusions 257
References 257

Optical beams that are structured in both phase and intensity have drawn considerable attention as a promising technology to increase the transmission capacity of communication links. A particular area of application of such beams is over free-space optical (FSO) links, allowing fiber optical speeds but without the cables. As mentioned earlier in this book, a range of different spatial mode sets exist and are used for a range of different applications. A determining factor in one's choice of spatial mode group for a specific challenge is the exact media that the beam is propagating through and the supported modes, or eigenmodes, of that particular channel. In the case of guided channels, such as optical fibers, the refractive index profile of that fiber determines the mode group that most efficiently propagates along the channel. Conversely, in FSO links the choice of mode-group is not so clearly defined as many different spatial modes can freely propagate over that channel. Therefore, a more nuanced decision is required that can depend on the physical optical components used with an experimental system or the natural permutations that occur in FSO links arising from variations in environmental conditions. See Fig. 9.1.

The temperature and pressure inhomogeneities in the atmosphere lead to variations of the refractive index along the transmission path, both spatially and temporally, that give rise to a

Figure 9.1: Free-space optical (FSO) systems transmit information between a sender and receiver over an unguided channel in the air. The air over this channel is perturbed by temperature and pressure variations that distort the optical field known as atmospheric turbulence [1].

range of distortions to the optical field, which include irradiance fluctuations (scintillation), beam wander, loss of spatial coherence, and the cross-coupling between spatial modes arising from phase fluctuations. Known as channel crosstalk, the distributed phase perturbations along the channel result in the cross-coupling of optical energy between free-space propagating optical modes. This corrupts the information transmitted within any mode-division or spatially division multiplexed communication system. In this chapter, we focus on optical modes that carry Orbital Angular Momentum (OAM) and how they interact with both atmospheric turbulence and scattering media that uniquely distort these specific types of structured optical modes.

9.1 Atmospheric turbulence variations in real environments

Turbulence is associated with the fluctuations in the optical density of atmosphere, where the air can be closely described as a moving viscous fluid. Similar to other liquids the atmosphere has two distinct states of motion: (1) laminar flow and (2) turbulent flow. These two states have distinct characteristic flow dynamics, where: (1) Laminar flow is a smooth continuous flowing motion in which no fluid mixing occurs. This can be readily understood by the flow of liquid in a pipe, where velocity can vary across the diameter, but the flow moves as a whole. (2) Turbulent flow is more chaotic, similar to that of white water rapids, where considerable mixing occurs resulting in eddying currents and fluid vortices of many different sizes.

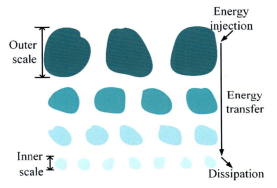

Figure 9.2: Turbulence energy cascade theory. Energy is injected into a fluid dynamic system by an external source such as sunlight. These large areas of high energy break down in a cascaded fashion into smaller eddies until they reach the inner scale size, where the energy is then completely dissipated.

Flow conditions can readily change from laminar to turbulent depending on the physical properties of the channel. An important non-dimensional number that characterizes the transition from laminar to turbulent flow is called the Reynolds number. The Reynolds number, Re, is involved in a numerical relationship between viscous and inertial forces, where low Re signifies dominance of viscous forces that lead to constant flow and high Re signifies that inertial forces dominate, leading to more chaotic behavior. This can be expressed as $Re = VL/v$, where v is the kinematic viscosity, V is velocity of flow and L the characteristic linear dimension of the fluid, respectively. Laminar flow is readily computable, however, turbulent flow is a nonlinear process governed by the Navier-Stokes equations and can often be very difficult to simulate at appropriate length scales for optical systems. Because of difficulties in solving the Navier-Stokes equations for fully developed turbulence, the classical statistical theory of turbulence was developed in Kolmogorov's seminal work in 1941 [2]. See Fig. 9.2.

The central principles of the Kolmogorov formulism of atmospheric turbulence is the energy cascade theory originally proposed by Lewis Fry Richardson in 1922 [3]. Atmospheric turbulence can be explained more generally as a cascade of random turbulent eddies, with various scale sizes, extending from an inner scale l_0 to an outer scale L_0. Large areas of air are heated, by the sun for example, forming large-scale turbulent eddies, L_0, which subsequently break down further into smaller eddies under the influence of inertial forces. This breakdown continues until the size of the turbulent eddy reaches the inner scale when the energy within that eddy dissipates. Kolmogorov postulated that during this energy cascade, geometrical and directional information is lost. This leads to the assumption that the small scale statistics have a universal character and are the same for all turbulent flows when the Reynolds number is sufficiently high. Kolmogorov then extended these assumptions for very high Reynolds number and postulated that the statistics is determined predominantly by kinematic viscosity, v, and energy dissipation, ε.

For the majority of naturally occurring turbulent environments, we only consider the propagation of optical beams through a restricted range of Reynolds numbers known as the inertial range. Under these conditions the Reynolds number is linked to the ratio of inner scale to outer scale, by the relationship $Re = (l_0/L_0)^{-4/3}$.

As these processes are inherently random, it is customary to consider a power spectrum or structure function to describe a turbulent environment. The critical parameter that affects optical fields is the extent of refractive index fluctuations that results from the eddies formed. The turbulence power spectrum models can give the statistical averages of the random variation process of atmosphere. Kolmogorov derived such a power spectrum for the inertial subrange as a $-11/3$ power law, which is the most commonly used approach to mathematically modeling turbulent environments for optical systems [2]. This power law is routed in the well-known Kolmogorov spatial frequency spectrum, formulated as

$$\Phi_n(\kappa) = 0.033 C_n^2 \kappa^{-11/3}, \quad 1/L_0 \ll \kappa \ll 1/l_0 \tag{9.1}$$

where κ is the angular spatial frequency, and C_n^2 is the refractive-index structure constant, which is a measure of the strength of the fluctuations in the refractive index.

The Kolmogorov turbulence theory is the basis for many subsequent theories of turbulence; although the Kolmogorov spectrum model is useful to relate temperature fluctuations to refractive index fluctuation, it has limited validity as the inner and outer scales are not explicitly included in this formulism. Tatarskii, von Kármán, and Hill have all developed modified variants of Kolmogorov's power spectrum that consider the effects of both inner and outer scales. Among these, the von Kármán spectrum model is the simplest and most frequently used one, given by

$$\Phi_n(\kappa) = 0.033 C_n^2 \frac{\exp\left(-\kappa^2/\kappa_m^2\right)}{\left(\kappa^2 + \kappa_0^2\right)^{11/6}} \tag{9.2}$$

where $\kappa_m = 5.92/l_0$ and $\kappa_0 = 1/L_0$.

For many practical applications, one wishes to consider the direct influence of the spatial spectral variations on a specific optical system. In 1965, Fried considered the turbulent effects on astronomical telescopes. Within imaging systems, the resolution is generally governed by the diffraction limit for a given aperture. However, the functional resolution will also be adversely affected by turbulence and lower the recorded image quality. Fried introduced a parameter, known widely as Fried's coherence length, r_0, that has a unit of length to allow for direct comparison with the diameter of an optical aperture, D. The coherence length is defined as the diameter of a circular area over which the route mean square (RMS) wavefront aberration is equal to 1 radian [4]. As r_0 is an accumulative turbulence effect, it can be calculated from the structure constant C_n^2 and has the form

$$r_0 = \left(0.423 k^2 \int_0^L C_n^2(z)\, dz\right)^{-3/5} \tag{9.3}$$

One can then characterize the effect of turbulence on an optical system by considering the ratio D/r_0. This ratio sets two limiting cases, first when $D/r_0 < 1$ the resolution of the system is limited by its aperture, and second in the case of $D/r_0 > 1$ the atmosphere limits the system's ability to resolve an object due to the introduction of phase variations from random fluctuations in refractive index.

9.2 Turbulence-induced phase variations

Fluctuations in refractive index, much like in bulk materials, lead to spatially varying phase delays that can sustainably alter the intensity and phase profile of optical images. In imaging systems such as telescopes, the phase distortions can increase the diameter of a particular point spread function and subsequently limit the ability to adequately resolve an image. For beams with both phase and intensity structuring this effect can be much more pronounced.

The Fried parameter formulism of turbulence can be used to represent the stochastic phase fluctuations in the turbulent atmosphere channel. The individual realization of the random turbulent process can be effectively described by the random phase screens, which have the same statistical variation. One can derive an extension of Kolmogorov theory introducing the phase structure function, defined by

$$D_n(\mathbf{r}_1, \mathbf{r}_2) = \left\langle [\phi(\mathbf{r}_1) - \phi(\mathbf{r}_2)]^2 \right\rangle = 6.88 \left|\frac{\mathbf{r}_1 - \mathbf{r}_2}{r_0}\right|^{5/3} \tag{9.4}$$

where $\phi(\mathbf{r}_1)$ and $\phi(\mathbf{r}_2)$ are two random phases [4].

Although this phase structure functions defines the relationship between any two random phase screens, one needs a specific method to simulate these phase variations. There are two widely used approaches to generate random phase screens of atmospheric turbulence, which are mode group expansion methods and power spectrum inversion methods. Mode group expansion methods consider a particular mode set, such as Zernike polynomials or orbital angular momentum, using a predefined weighted superposition of these modes to generate an approximation to a turbulent phase screen. Such approaches are commonly used in adaptive optical systems, where one selects a set of modes suitable for the particular correction system one is using. These approaches generally, however, only model the low-order spatial components of the optical aberrations. In the power-spectrum inversion methods, one uses a Fourier transform of a particular spatial power spectrum to generate more physically representative phase screens than those achieved through mode group expansion. However, care is required to mitigate numerical issues arising from the Fourier transform of regular arrays to faithfully reproduce all of the spatial frequencies in turbulent phase screens.

Let us first consider mode group expansion as a way to decompose the random phase aberrations induced by turbulence as a discrete set of polynomials. Zernike polynomials are a widely used mode group for the representation of optical aberrations in cylindrical coordinates. For the Zernike polynomial expansion method, the lower-order orthogonal Zernike polynomials with coefficients are adopted to represent the wavefront distortions $\phi(r, \theta)$ induced by the turbulence. The Zernike polynomials, Z_j, are classified by a mode ordering number j that are a function of mode indices n and m such that $j = \frac{n(n+2)+m}{2}$. The values of n is a positive integer known as radial order and m is an integer known as the angular frequency. These integers satisfy the relations as follows: $m \leq n$ and $n - |m|$ is always even. The polynomials are defined here as

$$Z_j(r,\theta) = \begin{cases} \sqrt{2(n+1)} R_n^m(r) \sqrt{2} \cos m\theta, & m \neq 0 \text{ and even } j \\ \sqrt{2(n+1)} R_n^m(r) \sqrt{2} \sin m\theta, & m \neq 0 \text{ and odd } j \\ \sqrt{2(n+1)} R_n^0(r), & m = 0 \end{cases} \quad (9.5)$$

where

$$R_n^m(r) = \sum_{s=0}^{(n-m)/2} \frac{(-1)^s (n-s)!}{s![(n+m/2-s)]![(n-m)/2-s]!} r^{n-2s}. \quad (9.6)$$

Optical phase aberrations arising from turbulence can be expressed as a weighted sum of these Zernike polynomials,

$$\psi(r,\theta) = \sum_{j=0}^{J} a_j Z_j(r,\theta) \quad (9.7)$$

where J is the maximum mode number resolvable by the optical system and a_j is a Gaussian random coefficient with zero mean. The weighting of particular Zernike polynomials is determined by the covariance between two separate Zernike polynomials a_j and $a_{j'}$ such that

$$\langle a_j a_{j'} \rangle = c_{jj'} \left(\frac{D}{r_0} \right)^{5/3} \quad (9.8)$$

Table 9.1: Noll Matrix defining the weighting parameter $c_{jj'}$.

j/j'	1	2	3	4	5	6	7
1		0	0	0	0	0	0
2	0	0.449	0	0	0	0	0
3	0	0	0.449	0	0	0	0.0142
4	0	0	0	0.0232	0	0	0
5	0	0	0	0	0.0232	0	0
6	0	0	0	0	0	0.0232	0
7	0	0	0.0142	0	0	0	0.00619

where $c_{jj'}$ are the matrix elements from the Noll matrix, Table 9.1, D is the system aperture diameter, and r_0 is the fried parameter as defined previously. We generally only consider values along the diagonal when simulating or decomposing a turbulent phase-front into Zernike polynomials [5]. You will notice there is no value included in the first cell of the Noll matrix, corresponding to c_{11}, this is because that value is the first order Zernike polynomial that represents a global phase offset over the entire aperture of the system, known as pilar shift, which does not generally contribute to any noticeable optical aberrations [6].

This method is an efficient way to describe atmospheric turbulence when considering a small range of spatial frequencies and is widely used to decompose the distorted wavefronts for real time correction using adaptive optics due to the low number of actuators these devices commonly have. In many astronomical imaging applications, one is limited by the aperture size of the telescopes and therefore only needs to consider a small number of spatial frequencies. As free-space optical systems have been evolving and are being used for a wider variety of applications including sensing and communication systems, there is a requirement to consider other computational approaches that more readily replicate the high order spatial frequencies that are generated by atmospheric turbulence. This can be achieved by computing the inverse Fourier transform of the expected random distribution of spatial frequencies as determined by a particular power spectral density function. The spectral density function that is generally considered the most reliable for atmospheric turbulence is that proposed by von Kármán [7]. The von Kármán spectrum, using coefficients $c_{n,m}$, has the form

$$c_{n,m} = h(n,m) \frac{1}{L_x L_y} \sqrt{0.023} r_{0,j}^{-5/6} \exp\left(-\frac{f_{x_n}^2 + f_{y_m}^2}{f_m^2}\right)^{1/2} \left(f_{x_n}^2 + f_{y_m}^2 + f_0^2\right)^{-11/12} \quad (9.9)$$

where $h(n,m)$ is the random array that obeys complex circular Gaussian statistics with zero-mean and unit-variance; L_x and L_y are the sizes of the phase screen. $f_0 = 1/L_0$, and $f_m = 5.92/(2\pi l_0)$; $f_{xn} = n/L$ and $f_{ym} = m/L$ are the spatial frequencies.

244   Chapter 9

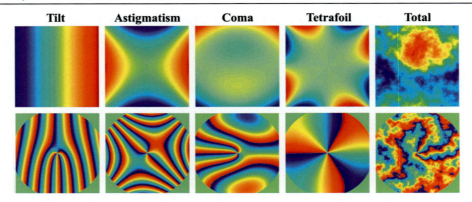

Figure 9.3: Zernike polynomial mode group can be used to simulate a range of base deformation commonly arising in cylindrical coordinates including: Tilt, Astigmatism, Coma, and Tetrafoil aberrations. Specifically weighted superpositions of these aberrations can be used to generate turbulent phase screens that represent atmospheric turbulence. These phase aberrations can distort structured beam, such as $\ell = 3$ OAM phase profiled depicted in lower the half of this figure.

Turbulence-induced phase $\phi(x, y)$ can be represented as a Fourier series:

$$\phi(x, y) = \sum_{n=-\infty}^{\infty} \sum_{m=-\infty}^{\infty} c_{n,m} \exp\left[i 2\pi \left(f_{x_n} x + f_{y_m} y\right)\right] \qquad (9.10)$$

where f_{x_n} and f_{x_m} are the discrete x- and y-directed spatial frequencies, and the $c_{n,m}$ are the Fourier series coefficients. Unfortunately, the above equation cannot sample faithfully the spatial frequencies low enough to accurately represent modes like tilt aberration. The subharmonic method described by Lane et al. [8] has been successfully employed to compensate for this shortcoming:

$$\phi_{LF}(x, y) = \sum_{p=1}^{N_p} \sum_{n=-1}^{1} \sum_{m=-1}^{1} c_{n,m} 3^{-p} \exp\left[i 2\pi \left(f_{x_n} x + f_{y_m} y\right)\right] \qquad (9.11)$$

The common approach is to use a 3×3 ($p \times N_p$) grid of 2D arrays of spatial frequencies. The frequency grid spacing for each value of p is $1/(3^p L)$. See Fig. 9.3.

9.3 Turbulence's effect on structured beams

As structured light propagates through atmospheric turbulence, the wavefronts are distorted by the fluctuations in refractive index. For a range of applications, it is important for one to have robust methods for modeling the physical interaction of structured light with turbulent channels. For any optical system the starting point is Maxwell's equations, where the electric field $U(\mathbf{r})$ of the light beam in the atmospheric turbulence must satisfy the scalar Helmholtz equation:

$$\left[\nabla^2 + k^2 n^2(\mathbf{r})\right] U(\mathbf{r}) = 0 \tag{9.12}$$

where $\mathbf{r} = (r, \theta, z)$ is the position vector, $k = (2\pi)/\lambda$ is the wave number and n is the refractive index. Generally, the scalar Helmholtz equation in the atmospheric turbulence can be solved in the form of the extended Huygens-Fresnel integral [9]. However, the analytical solution of the extended Huygens-Fresnel integral has proven difficult to solve, due to the complex electric field of structured light. The best-known classical approaches to solving Eq. (9.12) are the Rytov approximation and the Born approximation methods.

Under the assumption that the fluctuations in atmospheric turbulence are very mild over the optical channel being modeled, the square of the index of refraction term can be written as $n^2(\mathbf{r}) = 1 + 2n_1(\mathbf{r})$, then the wave equation (9.12) becomes

$$\left\{\nabla^2 + k^2 [1 + 2n_1(\mathbf{r})]\right\} U(\mathbf{r}) = 0 \tag{9.13}$$

By considering the Rytov approximation method, the electric field $U(\mathbf{r})$ at the z plane can be expressed as a sum of terms of the form

$$U(\mathbf{r}) = U_0(\mathbf{r}) \exp[\psi(\mathbf{r})] \tag{9.14}$$

where $U_0(\mathbf{r})$ is the electric field of light beam in the vacuum and $\psi(\mathbf{r})$ is a complex phase perturbation due to the atmospheric turbulence that takes the form

$$\psi(\mathbf{r}) = \psi_1(\mathbf{r}) + \psi_2(\mathbf{r}) + \cdots \tag{9.15}$$

where ψ_1 and ψ_2 as the first-order and second-order complex phase perturbations, respectively.

In such a form, the statistical variation can be somewhat abstract from the physical behavior that is giving rise to these specific superpositions of structured modes. These perturbations redistribute the energy among the spatial modes used within a given system. This concept can be characterized through the use of the superposition theory of the spiral harmonics [10] electric field $U(\mathbf{r})$, which can be written as

$$U(\mathbf{r}) = \frac{1}{\sqrt{2\pi}} \sum_{n=-\infty}^{\infty} a_n(\mathbf{r}) \exp(inz) \tag{9.16}$$

where $a_n = 1/\sqrt{2\pi} \int_0^{2\pi} U(\mathbf{r}) \exp(-inz) \, d\varphi$. These superpositions, and therefore their resultant spatial spectra, can contain a considerable amount of information about the environmental channel that the beam passes through that can be useful for both the correction of the optical fields that are transmitted over these channels and determination of the systems one needs to develop for different application scenarios.

9.4 Degradation of beams that carry OAM

Optical modes are degraded by four main forms of distortion in free-space communication links: (1) scintillation, random time varying changes in the intensity of a received optical field that arise from changes in optical absorption and some scattering interactions; (2) over long range propagation we have the divergence of an optical beam as it propagates; (3) misalignment of an optical beam with the receiver aperture, arising from mechanical and environmental factors between a sender and receiver; and (4) phase aberrations that arise from random changes in the refractive index along an FSO channel. See Fig. 9.3.

In this section, we focus on the latter two types of distortions that have led to distinct forms of channel crosstalk. Firstly, we cover the specific effect of misalignment on beams that carry OAM. A pure OAM mode is defined with respect to a specific axis [11]. This axis is the z-axis of the cylindrical polar-coordinate system where the complex amplitude cross-section is a transverse plane ($z = $ const.) for a pure mode with a specific ℓ value written in the form $\exp(i\ell\phi)$ This is commonly referred to as the beam axis. When described with respect to a different axis, the measurement axis, a pure OAM state becomes a well-understood superposition of a number of these states [12]. To derive this superposition of OAM modes we can consider specifically the Laguerre-Gaussian (LG) laser modes already mentioned earlier in this chapter and in other chapters throughout this book. An LG mode is described by two indices, ℓ and p. In the waist plane of a LG mode the complex amplitude is given by

$$\psi_{\ell,p}(x, y) = C_{\ell p} \left(\frac{2(x^2 + y^2)}{w_0^2}\right)^{\frac{|\ell|}{2}} L_p^{|\ell|} \left(\frac{2(x^2 + y^2)}{w_0^2}\right)$$

$$\times \exp\left[\frac{-(x^2 + y^2)}{w_0^2}\right] \exp[i\ell \operatorname{atan2}(y, x)], \quad (9.17)$$

where w_0 is the waist size in the paraxial regime, $L_p^{|\ell|}(x)$ is the Laguerre polynomial for the mode indices ℓ, and p, and we have the amplitude normalization term [13,14], given by

$$C_{\ell p} = \begin{cases} \frac{1}{w_0}\sqrt{2p!/[\pi(p+\ell)!]} & \text{for } \ell \geq 0, \\ \frac{1}{w_0}\sqrt{2p!/[\pi(p-\ell)!]} & \text{for } \ell \leq 0. \end{cases} \quad (9.18)$$

Note that we describe the beam in Cartesian coordinates, as this is convenient to deal with the misalignments we consider here. The term $\exp(i\ell\phi)$ then takes the form $\exp[i\ell \text{atan2}(y, x)]$, where $\text{atan2}(y, x)$ is used in place of $\arctan(y/x)$ as arctan does not distinguish between angles that differ by π. As the LG modes form an orthonormal basis, any beam cross-section $\psi(x, y)$ can be written, using bra-ket notation $\Psi_{\ell,p}(x, y) = |lp\rangle$, $|\psi\rangle = \psi(x, y)$, in the form

$$|\psi\rangle = \sum_{\ell=-\infty}^{\infty} \sum_{p=0}^{\infty} |lp\rangle\langle lp|\psi\rangle. \tag{9.19}$$

The power in the component $|lp\rangle$ is then given by the modulus squared of the coefficient for that component, namely:

$$P_{\ell,p} = |\langle \ell p|\psi\rangle|^2 \tag{9.20}$$

The power in all components with the same value of ℓ is given by

$$P_\ell = \sum_{p=0}^{\infty} P_{\ell,p} = \sum_{p=0}^{\infty} |\langle lp|\psi\rangle|^2. \tag{9.21}$$

The set of these powers P_ℓ is the OAM spectrum we expect to measure. It should be noted that this particular formalism is valid for all distortions of beams carrying OAM that lead to a distribution of power between modal components.

Focusing specifically on misaligned LG modes, the misalignment of the beam axis can be defined with respect to the measurement axis described by the parameters Δx, Δy, α and β; see Fig. 9.4. The complex amplitude of such a misaligned LG beam is

$$\psi(x, y) = \Psi_{\ell,p}(x - \Delta x, y - \Delta y) \exp\left[i\frac{2\pi}{\lambda}(x \sin\alpha + y \sin\beta)\right], \tag{9.22}$$

where $\Psi_{\ell,p}$ is the complex amplitude of an LG mode whose beam axis coincides with the measurement axis (Fig. 9.4), and $k = 2\pi/\lambda$, where λ is the wavelength of the light. The corresponding OAM spectrum can then be calculated using Eq. (9.21) above; see Fig. 9.4. As we can write a closed form equation for the expected crosstalk, the effect of particular variations in the system can be readily simulated and corrected for in FSO communication systems. Mechanical stabilization is critical for many long-distance channels, as the beam wander generated by this can lead to substantial channel crosstalk. At the transmitter, these stability issues will lead to a change in beam axes of propagation (α, β) and beam wander $(\Delta x, \Delta y)$ variations with respect to the pre-determined receiver location. A further source of this kind of distortion are large-scale turbulent eddies that cause a beam to be randomly deflected along

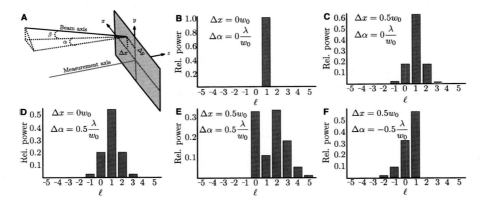

Figure 9.4: (A) Mechanical variations and atmospheric turbulence can result in an optical beam misaligned with respect to the receiver aperture. (B-F) This misalignment has a specific effect on the modal spectrum that represents the misaligned optical field [13].

its path, also resulting in beam wander and axis-of-propagation fluctuations at the receiver aperture [13].

Higher-order aberrations arising from turbulence can have a more complex effect on the optical field. Patterson considered this superposition approach to determine the normalized energy carried by the corresponding OAM modes and represented this as $s_n = P_n / \sum_{\Delta=-\infty}^{\infty} P_\Delta$ with $P_n = \int_0^\infty |a_n(r,z)|^2 r dr$ [14]. To represent this superposition arising from turbulence, the phase structure function can be introduced, where the resulting integral simplifies dramatically and becomes

$$s_\Delta = \frac{1}{\pi} \int_0^1 d\rho \rho \int_0^{2\pi} \exp\left[-3.44 \left(\frac{D}{r_0}\right)^{5/3} \left(\frac{\rho \sin(\varphi)}{2}\right)^{5/3} \cos(\Delta \varphi)\right] d\varphi \qquad (9.23)$$

where $\rho = 2r/D$, and D is the aperture of the system, Δ is the relative step in mode index from a chosen value of ℓ. This theory assumes the distribution of energy into neighboring OAM modes is invariant to the specific mode propagated over the turbulent channel.

Similar to the consideration one makes for astronomical imaging systems, the aperture size makes a difference to the magnitude of the distortion that occurs from atmospheric turbulence for beams that carry OAM, see Fig. 9.5. For a small receiver aperture $D/r_0 \to 0$, the exponent of Eq. (9.23) can be expanded in a power series in D/r_0 and one may retain only the first two terms, yielding

$$s_\Delta = \begin{cases} 1 - 1.01 (D/r_0)^{5/3} & \text{for } \Delta = 0 \\ 0.142 \frac{\Gamma(\Delta - 5/6)}{\Gamma(\Delta + 11/6)} \left(\frac{D}{r_0}\right)^{5/3} & \text{otherwise} \end{cases} \qquad (9.24)$$

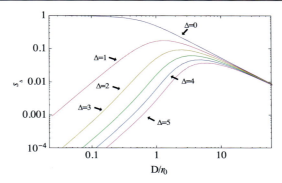

Figure 9.5: Patterson formulated a relationship for the crosstalk between neighboring optical modes. This theory predicted that under certain conditions that on average, the optical power for a given OAM mode will be distributed evenly between neighboring modes as determined by Eq. (9.23).

where $\Gamma(x)$ is the usual gamma function. For a very large receiver aperture with $D/r_0 \gg 1$, the integral can also be evaluated as

$$s_\Delta = \frac{12\Gamma(3/5)}{5\pi (3.44)^{3/5}} \left(\frac{D}{r_0}\right)^{-1} = 0.542 \left(\frac{D}{r_0}\right)^{-1}. \tag{9.25}$$

The results are valid for certain realizations of turbulent environments where the transmission channel can be considered as a thin phase (<400 m) turbulence, but are not fully valid for longer distance (>400 m) link lengths [1]. The thin-phase turbulence regime is where the accumulative effect of the turbulence can be reduced to a single phase-screen that perturbs the optical field similar to the aberrations that occur in an optical lens. This turbulence condition is commonly considered in astronomy, where light from astronomical objects propagates largely unhindered through the vacuum of space and only experiences non-negligible aberrations when entering the earth's atmosphere. Adaptive optical systems have become a central part of large aperture telescope systems, which record the phase aberrations of those that use wavefront sensing and display the reciprocal phase distortion to an adaptive mirror that corrects the effect of turbulence. Although this technique is very powerful, a study of the appropriateness of its application is required for optical fields with complexity beyond that of Gaussian optical beams. These models indicated that one would expect the received OAM expectation value to be the same as the transmitted mode order [15], [16]. Further theories were extended to incorporate thick atmospheric turbulence effects and found a similar result to those for the thin phase regime just with a broader range of modal constituents [17].

For longer propagation paths, the single plane distortion approach is not valid due to the repeated interactions with varying turbulent layers, followed by subsequent propagation of the

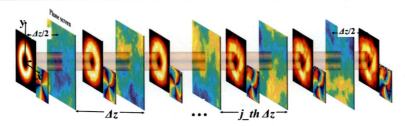

Figure 9.6: To faithfully represent atmospheric turbulence over realistic FSO links one should use multiplane numerical simulation.

optical field. Long distance propagation studies have shown that single plane approximation theories do not hold for optical fields with $|\ell| > 1$, where the measured OAM expectation value at the receiver does not match that of the transmitted mode [18]. Instead, one should consider the interaction with multiple planes of phase distortion to more accurately represent this form of optical turbulence [16]. However, for such systems it is rather challenging to develop closed form solutions due to the cascaded and independent nature of the multiple, randomly varying phase screens, therefore numerical simulations are largely used as an efficient tool to quantitatively investigate the propagation effects of light beams in the atmospheric turbulence. A commonly used modeling technique is to numerically simulate propagation through cascaded phase screens that, both individually and collectively, are obedient to the statistical relationship originally postulated by Kolmogorov turbulence [19]. This is achieved through a split-step beam propagation method [20].

As free space propagation and the phase modulation induced by turbulence phase screen are regarded as two processes that are independent and simultaneously complete, the turbulent atmosphere can be simulated by a series of thin random phase screens that are equally spaced over a link being simulated, as illustrated in Fig. 9.6. The initial optical field after propagation over a particular length, Δz, is phase-modulated by a phase only turbulent screen, simulated with a von Kármán spectrum model as discussed above. After a further propagation length of Δz, both phase and intensity are distorted due to diffraction of the perturbed optical field. At this point, a further phase only turbulence screen is superimposed onto the optical field to provide the cascaded aberrations that would occur in a real channel. This step-by-step approach can be extended to provide numerically accurate simulations of optical propagation [14]. In such a link, the accumulative turbulence range are decomposed sufficiently into a series of j weak thin turbulence screens $(D/r_0)_{total}$, where each phase screen has $(D/r_0)_{step} = (D/r_0)_{total}/j$.

As an example, Fig. 9.7 presents a simulation for a particular long distance FSO link; the channel can be broken up into 10 m cells corresponding to the micro-scale atmospheric circulation, corresponding roughly to a size scale of 10% of a normal the outer-scale turbulence

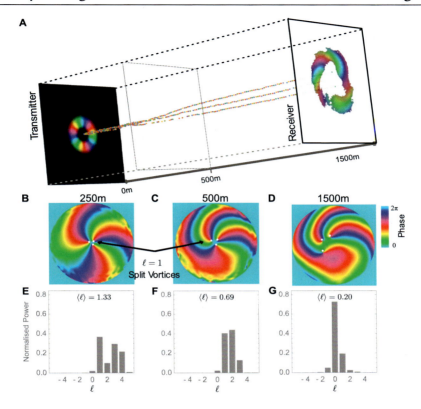

Figure 9.7: (A) Using a split-step numerical simulation, the location of vortices can be tracked over a FSO optical link. (B-D) This results in progressively increasing splitting of the optical vortices within the beam. (E-G) This leads distinctive change to the OAM spectrum and the expectation value [14].

value, L. The propagation within each cell can be simulated through plane-wave decomposition. Within each of these cells, a turbulent phase screen that follows Kolmogorov turbulence theory is added to the propagating optical field [4]. As shown in Fig. 9.6, the turbulence in each cell will be compounded over the length of the channel to yield the turbulence over the full length of the link.

Momentum conservation in an optical field leads to preservation of the total number of vortices under a perturbation. However, in the presence of mild turbulence or any weak non-cylindrically symmetric aberration, any high-order vortex of index ℓ will break up to give ℓ individual vortices of index 1 upon propagation [21–24]. Hence, it is not unexpected for vortices to split during the propagation over an FSO link and this can be problematic for single plane correction schemes that are widely used in astronomy. In Fig. 9.7, it can be clearly seen

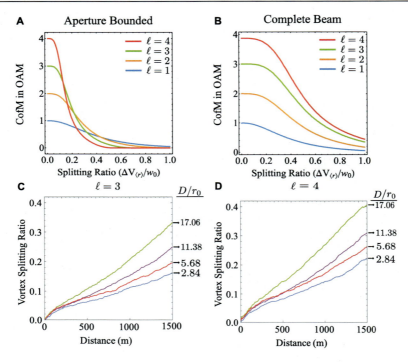

Figure 9.8: (A-B) The center of mass of the OAM spectrum is heavily influenced by the splitting ratio V and the size receiver aperture. (C-D) This splitting ratio can be used to determine the strength of the optical turbulence [14,18].

that, over the free-space channel, an $\ell = 3$ vortex breaks up into three $\ell = 1$ vortices that gradually split over the free-space channel. Weak aberrations early in the channel lead to a vortex to be split, with subsequent aberrations over the propagation length then further amplifying this modal breakup; see Fig. 9.7. Therefore, at the receiver the optical field is more naturally represented by ℓ spatially distributed $\ell = 1$ vortices, rather than a single high-order optical field. Spatially distributed $\ell = 1$ vortices can lead to an OAM distribution at the receiver no longer centered on the expected OAM value, $\langle \ell \rangle$, as is shown in Fig. 9.7.

This vortex-splitting can be characterized by a ratio defined as $V = \frac{\Delta V_{\langle r \rangle}}{w_0}$, where $\Delta V_{\langle r \rangle}$ is the average radial distance from the beam origin for the individual vortices and w_0 is the beam waist of the transmitted mode. This splitting will change in average OAM; however, the effect is amplified in the presence of a bounding aperture that is smaller than the annulus of the optical mode. In Fig. 9.8, the expected average OAM is shown for two cases: first, where no bounding aperture is impeding the propagation beam and second, where the 150 mm aperture of our experimental system is imposed [18].

By considering the same simulation, at each 10 m step of the simulated propagation the vortex locations can be determined and the average radial distance from the beam origin is calculated to determine the corresponding V value. As the turbulence is randomly varying, the vortex splitting slightly varies. The results are the average of 30 random phase aberrations, similar to the effect of measuring experimentally the vortex locations with a camera with a shutter time of 0.1 s. The error in V was determined by the standard error of the averaged measurements $\sqrt{\sigma}$, where σ is the standard deviation. As the vortex position search is limited to a single pixel in the modeled field, a small offset is applied to V. The range of D/r_0 for both $\ell = 3$ and $\ell = 4$ was determined, Fig. 9.8. As the computational time required to generate a turbulent phase screen is considerable, pre-determined turbulent phase screens were used. The effective Fresnel number of the simulated link was scaled to yield the presented D/r_0 values. The link is designed to model a 15 mm beam collected by a 150 mm aperture after propagation over a 1500 m link. These results also indicate that higher-order modes are more sensitive to the effects of weak non-cylindrically symmetric aberration than lower-order modes. This vortex splitting could arise from system back-reflections, static optical aberrations, or scattered light in the preparation of the optical mode. In addition, atmospheric turbulence near the transmitter that perturbs the mode under-propagation could result in vortex splitting, hence changing the measured average OAM.

Krenn et al. presented a study showing that the intensity profiles of superpositions of OAM modes seem largely unaffected by propagation through turbulent environment of 3 km over the city of Vienna. Modal superpositions were generated and numerically propagated over the same simulated 1.5 km channel. It can be seen in Fig. 9.9 that after long distance propagation the intensity structure of modal superpositions are fairly well preserved as expected from experimental studies [25]. As the intensity profile of these modal superpositions is encoded at generation, under propagation through mild turbulence the intensity structure is relatively unaffected and the effect on these modes is similar to the aberrations that occur when imaging a distant object.

A segmented approach to modeling turbulence is critically important for the characterization of real channels with variance in turbulence conditions along the length of the link. This limits the appropriateness of many current optical corrections to the system that are largely based on both the less effective mode group turbulence modeling approaches and single plane correction systems. Further, it is important that for near-perfect correction one would need to collect the entire optical field at the receiver or to have a more advanced correction technique. This experimental evidence inspired the modeling method presented in this chapter, which will allow for the development and testing of future methods to mitigate atmospheric turbulence.

254 Chapter 9

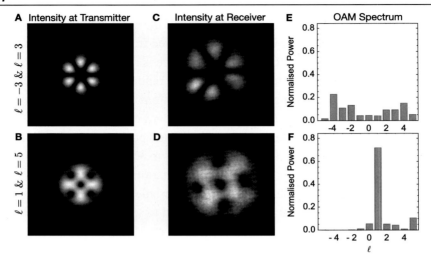

Figure 9.9: Superpositions of OAM modes lead to shaping of the intensity of the optical field. These unique intensity patterns are maintained in even fairly strong turbulence; however the phase is still considerably distorted due to the phase aberrations that are accumulated over the FSO link [14].

9.5 Scattering dynamics of beams that carry OAM

OAM multiplexing has become the SDM technique that many have chosen to explore, not only for FSO applications but also for underwater communications [26]. In such environmental channels localized heating, turbidity, salinity, and mechanical mixing, such as flowing water, all result in degradation of the optical field [27]. Scattering from suspended particulates and optical aberrations from density variations are the main sources of channel degradation in any environment [9,27]. Suspended particulates further result in the attenuation of a propagating optical beam and local deflection of light incident on these particles leads to multi-path interference. The density of particles strongly influences water attenuation measurements. Thermal variations in water due to current flow and mixing between stratified layers in the water column can further degrade an optical field in a similar manner to atmospheric turbulence.

In coastal, river, and marine environments particulates have broad distributions of particle size that will collectively affect the propagation dynamics of an optical beam. For both communications and sensing applications underwater, it is important to accurately simulate the distortion of OAM modes due to their interaction with particulates in that turbid channel. To fully model the optical interaction with a channel comprising particles distributed evenly over a particular volume of water (or air), the channel is represented as a cascaded set of successive

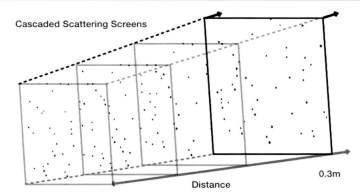

Figure 9.10: Similar to the methods used for simulating atmospheric turbulence, one can similarly numerically simulate a dense scattering environment with multiple planes of randomly distributed absorbing spheres [28].

scattering screens with randomly displaced circular absorbing regions with radius, r_p, corresponding to the size of simulated floating particles; see Fig. 9.10. Such an approach allows for the simulation of a large number of distributed particles ($>10^5$ particles for sizes of 11 μm), which would be computationally challenging in ray tracing approaches. For a given input optical beam we can model the propagation between the cascaded scattering screens, separated by 10 cm, computationally by plane-wave decomposition to calculate the spatial intensity and phase profiles of a propagating field. This split-step channel simulation approach is repeated over a number of steps, N_s, to simulate the accumulative effects of a long-distance scattering channel. To compute each scattering screen, we first determine the number of absorbing particles required in each scattering screen by considering the total channel loss A_T, where the expected loss for each scattering screen is the ratio $A_s = A_T/N_s$. By considering the area illuminated by the simulated input beam and the total particulate area coverage of absorbers that would result in a loss equivalent to A_s the number of simulated particles in each screen is calculated to be

$$N_p = \frac{A_T w_0^2}{N_s r_p^2}, \qquad (9.26)$$

where w_0 is the beam size of the propagating optical field. In each scattering screen, N_p circles are randomly placed over the aperture. Using a random number generator distributes the particles across the entire phase screen, with an even random distribution. At each propagation step an absorbing boundary of 5% of the total simulation area is added to prevent numerical re-introduction of wavefront components that leave the simulation area. In this way, the propagation model neglects numerical errors that could result in non-physical results arising from rays that would normally have left the optical aperture to be reintroduced.

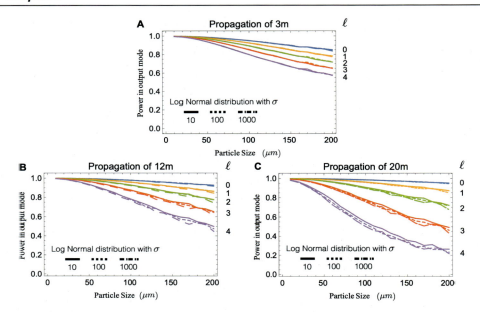

Figure 9.11: Increased particle size results in increased distortion of a propagating OAM beam that leads to optical power being spread among neighboring orders and subsequently reduces the power in the expected received mode. (A-C) In each case the total attenuation is equivalent, however increased propagation between transmitter and receiver results in increased optical distortion due to the cascaded perturbations arising from particulates over the optical channel [28].

To explore the connection between channel length and measured crosstalk from distributed particles, let us consider a channel with a fixed total attenuation of $A_T = 20$ dB, the same as for the 3 m underwater link. Fig. 9.11 shows the expected results for a propagation length of 3 m, 12 m and 20 m, respectively [28]. It can be seen in Fig. 9.11 that increasing both distance and ℓ dramatically increases the inter-channel crosstalk. These results indicate that OAM mode crosstalk has a strong dependence on particle size and increased crosstalk at longer link lengths. In any real-world environment scattering channel particulates will have a random distribution in size. To simulate the relatively narrow size distributions used in real world channels, one can consider a log normal distribution with the form

$$R(r) = \frac{1}{\sigma\sqrt{2\pi}r} \exp\left(\frac{-(\ln r - r_p)^2}{2\sigma^2}\right) \quad (9.27)$$

where r is the particle radius from 0 to 20 mm, r_p is the average particle radius and σ is the standard deviation of the distribution.

The physical reason for the presented optical behavior is that the propagation distance after a small perturbation is critical in determining the resulting phase front distortion. Each absorbing particle creates diffraction at its edges, generating additional spatial modes in the optical field from deflection of the wavefront similar in form to an Airy disk. During propagation these phase fluctuations will diverge at a rate different from the input mode, resulting in an increased modal degradation. Successive interactions with absorbing particles over long distances or being greater in diameter will result in compounded diffractive effects, as seen in Fig. 9.11. These distinctive crosstalk profiles are influenced by the diameter of randomly distributed particles in the optical channels.

9.6 Conclusions

FSO communications has been a fantastic driver for exploring the fundamental interaction of structured optical fields with the environment. The intended use of structured beams for point-to-point optical links has illuminated a range of new interesting challenges, where novel interactions with turbulent and scattering environment have highlighted the need for new technology to overcome the distorting effects of atmospheric turbulence. Although this chapter only focuses on a handful of recent studies to highlight some of the fundamental interactions, one should observe, when developing FSO systems, there is a global community of researchers pushing the boundaries of the field and exciting developments are continually being published. One such development, inspired by these advances in communications, is the use of these unique optical interactions of OAM with the environment that has instigated novel sensor technologies and could be a new frontier for structured optical fields.

References

[1] M.P.J. Lavery, Vortex instability in turbulent free-space propagation, New J. Phys. 20 (4) (Apr. 2018) 043023.
[2] A.N. Kolmogorov, The local structure of turbulence in incompressible viscous fluid for very large Reynolds numbers, Proc. R. Soc. 434 (1991) 9–13.
[3] L.F. Richardson, Weather Prediction by Numerical Process, Mon. Weather Rev., Cambridge University Press, 1922.
[4] D.L. Fried, Statistics of a geometric representation of wavefront distortion, JOSA 55 (11) (Nov. 1965) 1427–1431.
[5] R.J. Noll, Zernike polynomials and atmospheric turbulence, J. Opt. Soc. Am. 66 (3) (1976) 207–211.
[6] Robert K. Tyson, Principles of Adaptive Optics, Academic Press, Boston, MA, 1998.
[7] J.D. Schmidt, Numerical Simulation of Optical Wave Propagation with Examples in MATLAB, SPIE Press Monograph Series, SPIE, Bellingham, 2010.
[8] R.G. Lane, A. Glindemann, J.C. Dainty, Simulation of a Kolmogorov phase screen, Waves Random Media 2 (3) (1992) 209–224.
[9] V.I. Tatarski, Wave Propagation in a Turbulent Medium, McGraw-Hill, New York, 1961.
[10] L. Torner, J.P. Torres, S. Carrasco, Digital spiral imaging, Opt. Express 13 (3) (2005) 873–881.

[11] L. Allen, M.W. Beijersbergen, R.J.C. Spreeuw, J.P. Woerdman, Orbital angular-momentum of light and the transformation of Laguerre-Gaussian laser modes, Phys. Rev. A 45 (11) (1992) 8185–8189.
[12] M.V. Vasnetsov, V.A. Pas'ko, M.S. Soskin, Analysis of orbital angular momentum of a misaligned optical beam, New J. Phys. 7 (2005) 46.
[13] M.P.J. Lavery, G.C.G. Berkhout, J. Courtial, M.J. Padgett, Measurement of the light orbital angular momentum spectrum using an optical geometric transformation, J. Opt. 13 (6) (Jun. 2011) 64006.
[14] C. Paterson, Atmospheric turbulence and orbital angular momentum of single photons for optical communication, Phys. Rev. Lett. 94 (2005) 153901.
[15] B. Rodenburg, et al., Influence of atmospheric turbulence on states of light carrying orbital angular momentum, Opt. Lett. 37 (17) (May 2012) 3736–3737.
[16] G.A. Tyler, R.W. Boyd, Influence of atmospheric turbulence on the propagation of quantum states of light carrying orbital angular momentum, Opt. Lett. 34 (2) (Jan. 2009) 142–144.
[17] B. Rodenburg, et al., Simulating thick atmospheric turbulence in the lab with application to orbital angular momentum communication, New J. Phys. 16 (3) (Mar. 2014) 33020.
[18] M.P.J. Lavery, et al., Free-space propagation of high-dimensional structured optical fields in an urban environment, Sci. Adv. 3 (10) (2017).
[19] L.C. Andrews, R.L. Phillips, Laser Beam Propagation Through Random Media, second edition, 2005.
[20] A.J. Lambert, D. Fraser, Linear systems approach to simulation of optical diffraction, Appl. Opt. 37 (34) (1998) 7933–7939.
[21] J.F. Nye, M.V. Berry, Dislocations in wave trains, R. Soc. Lond. A 336 (1974) 165–190.
[22] M.S. Soskin, V.N. Gorshkov, M.V. Vasnetsov, J.T. Malos, N.R. Heckenberg, Topological charge and angular momentum of light beams carrying optical vortices, Phys. Rev. A 56 (5) (1997) 4064–4075.
[23] I.V. Basistiy, V.Y. Bazhenov, M.S. Soskin, M.V. Vasnetsov, Optics of light beams with screw dislocations, Opt. Commun. 103 (5–6) (Dec. 1993) 422–428.
[24] N. Matsumoto, T. Inoue, T. Ando, Y. Ohtake, Y. Takiguchi, Structure of optical singularities in coaxial superpositions of Laguerre–Gaussian modes, JOSA A 27 (12) (Dec. 2010) 2602–2612.
[25] M. Krenn, et al., Twisted light transmission over 143 km, Proc. Natl. Acad. Sci. 113 (48) (Nov. 2016) 13648–13653.
[26] A.E. Willner, et al., Optical communications using orbital angular momentum beams, Adv. Opt. Photonics 7 (1) (Mar. 2015) 66–106.
[27] O. Korotkova, Light propagation in a turbulent ocean, in: Progress in Optics, 2019.
[28] S. Vicla, et al., Degradation of light carrying orbital angular momentum by ballistic scattering, Phys. Rev. Res. 2 (3) (2020) 033093.

CHAPTER 10

Causes and mitigation of modal crosstalk in OAM multiplexed optical communication links

Alan E. Willner[a], Haoqian Song[a], Cong Liu[a], Runzhou Zhang[a], Kai Pang[a], Huibin Zhou[a], Nanzhe Hu[a], Hao Song[a], Xinzhou Su[a], Zhe Zhao[a], Moshe Tur[b], Hao Huang[a], Guodong Xie[a], and Yongxiong Ren[a]

[a]*University of Southern California, Los Angeles, CA, United States* [b]*Tel Aviv University, Ramat Aviv, Israel*

Contents

10.1 Introduction and overview 260
10.2 Causes for channel crosstalk in an OAM multiplexed link 263
    10.2.1 Atmospheric turbulence 263
    10.2.2 Misalignment 264
    10.2.3 Obstruction 265
    Summary 266
10.3 Adaptive optics (AO) for crosstalk (XT) mitigation 266
    10.3.1 AO using wavefront sensor (WFS) and Gaussian probe beam 266
    10.3.2 AO using WFS and Gaussian probe beam in a quantum communication link 269
    10.3.3 AO using camera for beam intensity measurement 271
    10.3.4 Simultaneous demultiplexing and XT mitigation by using multi-plane light converter (MPLC) 273
    Summary 274
10.4 Spatial modes manipulation for crosstalk mitigation 275
    10.4.1 Turbulence pre-compensation by OAM mode combination 275
    10.4.2 Simultaneous orthogonalizing and shaping of multiple LG beams 277
    10.4.3 Utilizing Bessel-Gaussian (BG) beams with non-zero OAM order 279
    Summary 280
10.5 Digital signal processing for crosstalk mitigation 280
    10.5.1 MIMO equalization for crosstalk mitigation in laboratory 281
    10.5.2 MIMO equalization for crosstalk mitigation in the link through a flying UAV 282
    Summary 284

10.6 Summary 284
Acknowledgment 284
References 284

10.1 Introduction and overview

Spatial modes are specially structured light fields that have unique spatial amplitude and phase profiles. Different modes from the same orthogonal modal set are mutually orthogonal [1]. One example of such spatial modal set is the orbital angular momentum (OAM) modes, which is a subset of the Laguerre-Gaussian (LG) modes [2]. The beams carrying OAM modes, i.e., OAM beams, have unique phase profiles in the transverse plane that could be described as $\exp(i\ell\theta)$, where θ is the azimuthal coordinate, and ℓ (OAM order) is the number of 2π phase shifts occurring in the azimuthal direction [3]. An OAM beam with a non-zero OAM order has a "twisting" wavefront along the propagation axis and a ring-shaped intensity profile, as shown in Fig. 10.1. OAM beams with different orders are mutually orthogonal [4,5].

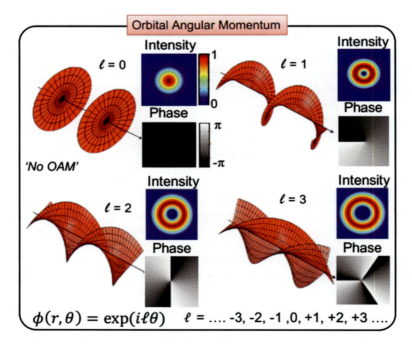

Figure 10.1: The wavefronts (left), intensity profiles (right-top), and phase profiles (right-bottom) of OAM beams. The OAM beam with a non-zero order has a donut shape intensity profile and helical phasefront. (©2011 Optical Society of America. Reprinted from [3].)

Causes and mitigation of modal crosstalk in OAM multiplexed optical communication links

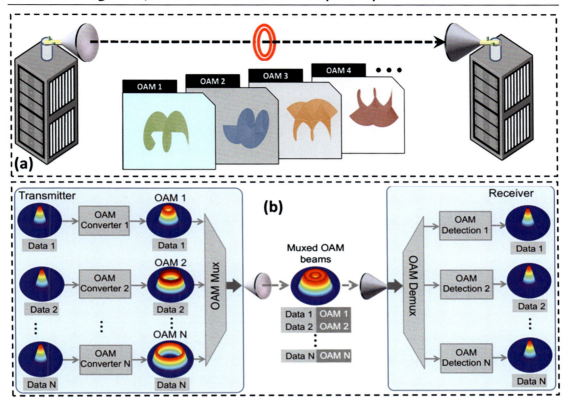

Figure 10.2: The concept of OAM multiplexed FSO links. (a) Multiple OAM beams are co-axially transmitted through free space. (b) Each orthogonal OAM beam carries an independent data stream. (©2013 Nature Research. Reprinted from Nat. Commun. (5) (1) (2014) 1-9.)

Recently, there has been an increased interest in the use of OAM beams for mode division multiplexing (MDM) [6–12], which itself is a subset of space-division multiplexing (SDM) [13,14]. More specifically, multiple independent data streams, carried by beams with different OAM orders, are spatially multiplexed at the transmitter, co-axially transmitted and demultiplexed at the receiver, and therefore bring about the potential to increase the system capacity [15], as shown in Fig. 10.2. Previous reports have shown the OAM multiplexed links in the laboratory environment with a data rate of up to ∼1 Pbit/s [16].

However, the MDM links (e.g., OAM multiplexed links) are sensitive to various degradations [17–19]. Many potential problems could arise in an MDM link, causing power coupling among the neighboring modes and thereby reducing orthogonality and increasing system crosstalk among data channels. Specifically, problems causing crosstalk include: (a) atmo-

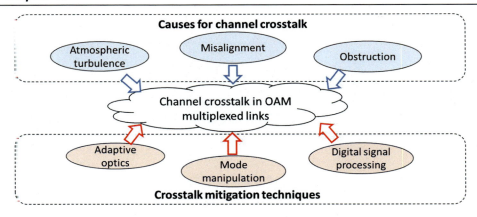

Figure 10.3: The causes and mitigation techniques for crosstalk in an OAM multiplex link.

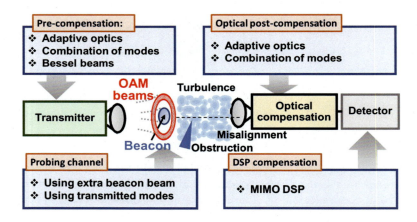

Figure 10.4: The crosstalk compensation approaches in OAM multiplexed links.

spheric turbulence [17], (b) misalignment between the transmitter and receiver apertures [18], and (c) obstructions [19]. See Fig. 10.3.

The causes and mitigation techniques for modal crosstalk in OAM multiplexed optical communication links will be discussed in this chapter, including but not limited to the approaches shown in Fig. 10.4. Specifically, Section 10.2 will describe the potential causes for crosstalk in an OAM multiplexed link; Sections 10.3, 10.4, and 10.5 will introduce the techniques for crosstalk mitigation including adaptive optics, manipulating transmitted spatial modes, and digital signal processing, respectively.

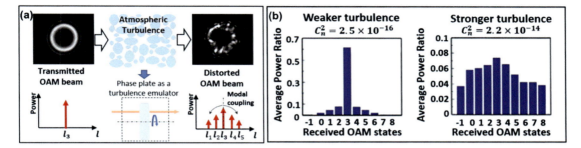

Figure 10.5: (a) An ideal OAM beam emerges as a distorted OAM after passing through the turbulence (emulated using a phase plate). (b) The measured power distribution of an OAM +3 beam after passing through the turbulence emulator. (©2013 Optical Society of America. Reprinted from [17].)

10.2 Causes for channel crosstalk in an OAM multiplexed link

OAM multiplexing can help increase the capacity and spectral efficiency of a communication system. However, various potential challenges exist when designing such an OAM-based multiplexing system. Several important issues need to be carefully considered: (i) In a practical FSO communication system, atmospheric turbulence induces a critical challenge by distorting the phase front of the light beam. This could be more important for OAM-based communications since the efficient demultiplexing of OAM beams relies on the helical phase front [17]. (ii) Aperture misalignment might induce power loss and crosstalk in OAM multiplexed systems [18]. (iii) Moreover, obstructions may partially block the beam path, which would also affect the profiles of the transmitted beams and induce crosstalk [19].

10.2.1 Atmospheric turbulence

Atmospheric turbulence is one important issue that needs to be considered for an OAM multiplexed communication system [20–25]. It was shown that the inhomogeneities in the temperature and pressure of the atmosphere lead to variations of the refractive index along the transmission path [26–31]. In an OAM multiplex link, there are multiple copropagating OAM beams and their orthogonality depends on the helical phase front. The turbulence-induced refractive index inhomogeneities can distort the phasefront of OAM beams, thereby causing intermodal crosstalk between different data channels with different OAM orders, as shown in Fig. 10.5(a) [17].

Reference [17] experimentally investigated the effects of atmospheric turbulence on the performance of OAM multiplexed systems. The concept of using a rotating phase plate to emulate the turbulence effects is shown in Fig. 10.5(a). The phase screen plate is mounted on a

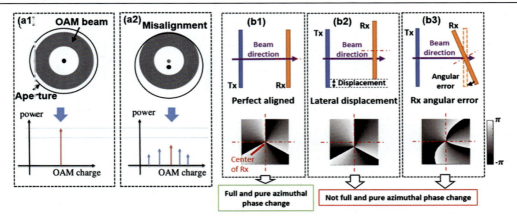

Figure 10.6: (a) Two different cases of alignment between the transmitter and receiver: (a1) a perfectly aligned system, (a2) a system with misalignment. (b) Two types of misalignments: (b2) lateral displacement and (b3) receiver angular error. (b1) is the perfectly aligned system. Tx is for the transmitter; Rx for the receiver. (©2015 Optical Society of America. Reprinted from [18].)

rotation stage and placed in the optical path. On the rotating plate, the pseudorandom phase distribution obeys Kolmogorov-spectrum statistics. The strength of the emulated turbulence is generally dependent on the Fried coherence length r_0, and the beam size that is incident on the plate. To evaluate the turbulence effects, the modal crosstalk is characterized by measuring the power of the distorted beam in each OAM mode. Fig. 10.5(b) presents the normalized power distribution among the neighboring OAM modes under weak or strong turbulence when an OAM +3 beam is transmitted. It is shown that under the weak turbulence, the majority of the power is still in the transmitted OAM mode (i.e., OAM +3), and only a small part of power is coupled into neighboring OAM modes. However, as the turbulence strength increases, the power coupling into other OAM modes would become stronger, which could induce more severe signal fading and crosstalk.

10.2.2 Misalignment

The efficient multiplexing and de-multiplexing of OAM beams are enabled by the orthogonality between beams with different OAM orders. This orthogonality is ensured when all OAM beams are co-axially propagated and the receiver is perfectly aligned on-axis with the beams, as shown in Fig. 10.6(a1). However, due to the jitter and vibration of the transmitter/receiver platform, there might be some misalignments between the transmitter and the receiver, including lateral displacement and angular error, as shown in Figs. 10.6 (b2) and (b3), respectively. As a result, these misalignments would degrade the orthogonality of OAM

Causes and mitigation of modal crosstalk in OAM multiplexed optical communication links 265

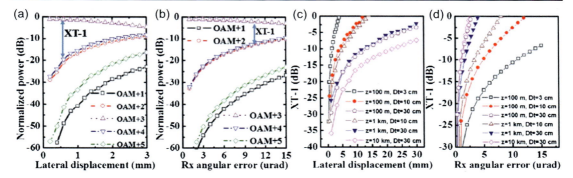

Figure 10.7: (a-b) Simulated power distribution among different OAM modes as a function of lateral displacement/receiver angular error over a 100-m link for which only the OAM 3 is transmitted. (c-d) XT-1 as functions of lateral displacement/receiver angular error for different transmission distances with different transmitted beam sizes. XT-1 is for the ratio of the received power on OAM +4 to the received power on OAM +3; Dt for the transmitted beam size; z for the transmission distance. (©2015 Optical Society of America. Reprinted from [18].)

beams due to a phase mismatch and cause power to be coupled into other modes, i.e., severe intermodal crosstalk [18,32], as shown in Fig. 10.6(a2).

Reference [18] showed that, as the lateral displacement or the angular error increases, the power coupling into other modes increases, whereas the power on OAM +3 decreases. This is because a larger displacement causes a larger mismatch between the received OAM beams and the receiver. The power coupled into OAM +2 and OAM +4 is greater than that of OAM +1 and OAM +5 due to their smaller mode spacing with respect to OAM +3. This indicates that a system with larger mode spacing is more tolerant to the lateral displacement and the angular error. One of the important concerns is the ratio of the power coupled to the nearest mode (here, OAM +4 is used) to the power on the desired mode (i.e. OAM +3), namely, relative crosstalk XT-1. Figs. 10.7(c) and (d) show the relative crosstalk XT-1 as a function of the lateral displacement or the angular error with various transmission distances and transmitted beams sizes. The results show that a larger beam size at the receiver will result in two opposing effects: (i) a smaller lateral displacement-induced crosstalk because the differential phase change per unit area is smaller, and (ii) a larger tilt phase error-induced crosstalk because the phase error scales with a larger optical path delay.

10.2.3 Obstruction

It is known that free-space optical communication links rely on the line-of-sight (LoS) operation. For all the LoS communication links, another potential critical issue might be obstructions in the beam path [33–35]. Such a problem could be particularly important for OAM

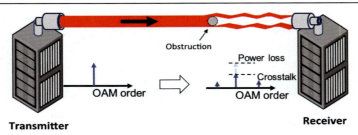

Figure 10.8: Concept of obstruction in an OAM communication link.

communications since completely or partially blocking the beam path will produce distortions to each beam. This would reduce the orthogonality between different OAM beams and cause power coupling from the desired OAM mode to other neighboring OAM modes, thereby leading to signal fading and channel crosstalk [19], as shown in Fig. 10.8.

Summary

In summary, various degradation effects could induce channel crosstalk in an OAM multiplexed link. In this section, we mainly focus on the channel crosstalk caused by: (i) atmospheric turbulence, (ii) misalignment, and (iii) obstruction. Several reports that have investigated those degradation effects are discussed in this section to show the potential channel crosstalk induced by each of these effects.

10.3 Adaptive optics (AO) for crosstalk (XT) mitigation

The atmospheric turbulence can introduce phase distortion to the helical phasefront of OAM beams, thereby increasing modal coupling and inter-modal XT, as discussed in Section 10.2. AO is one of the turbulence mitigation approaches [36–38], and its concept is shown in Fig. 10.9. In a typical AO system, the distorted beam profile would be measured, and an error correction pattern could be derived and then sent to the beam correction module through a feed-back loop. By compensating the distorted OAM beams, the AO approach can help to improve the performance of OAM-multiplexed FSO communication links. Several approaches for AO are discussed in the following subsections, including AO using: (a) wavefront sensor, (b) camera and (c) multi-plane light converter (MPLC).

10.3.1 AO using wavefront sensor (WFS) and Gaussian probe beam

A well-known method of AO turbulence mitigation is using WFS to detect the wavefront distortion of a beam. However, due to the phase structure of an OAM beam, it would be chal-

Causes and mitigation of modal crosstalk in OAM multiplexed optical communication links 267

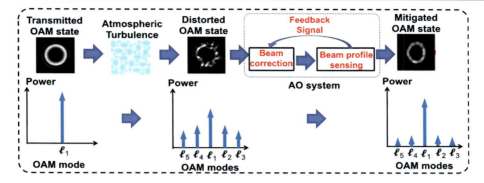

Figure 10.9: Concept of AO approach for turbulence-induced XT mitigation for OAM beams. (©2014 Optical Society of America. Reprinted from [39].)

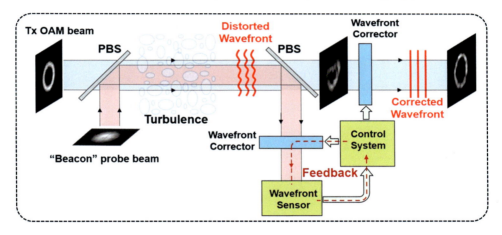

Figure 10.10: Concept of AO approach for turbulence mitigation using a Gaussian beam as a probe to detect wavefront distortion by using WFS.

lenging to directly measure its phasefront. This is because there is a singular point at the center of the OAM beam, which could lead to an inaccurate measurement when measuring the phasefront using typical Shack-Hartmann WFSs. One approach could be using a Gaussian probe beam on a different polarization or wavelength that can be measured by a commercially available WFS and be easily separated. The concept of this approach is shown in Fig. 10.10.

This approach was demonstrated in Ref. [39]. Fig. 10.11 presents the intensity profiles of the Gaussian probe and OAM beams for a particular turbulence realization (a1-a6) without and (b1-b6) with AO compensation. It was shown that by using the correction pattern obtained from the Gaussian probe beam in the AO system, the distorted OAM beams up to OAM $\ell = 9$ are efficiently compensated.

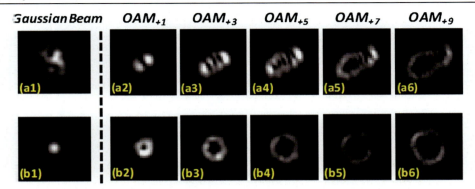

Figure 10.11: Far-field intensity images of the Gaussian probe beam and OAM beams ($\ell = 1, 3, 5, 7,$ and 9) before [upper, (a1)–(a6)] and after [lower, (b1)–(b6)] AO compensation. (©2014 Optical Society of America. Reprinted from [39].)

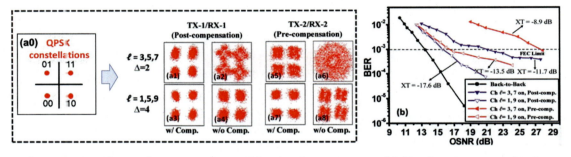

Figure 10.12: The performance of AO mitigation in an OAM-multiplexed FSO communication link. (a0) Concept of the constellation for QPSK signal. (a1)–(a8) Recovered QPSK constellations of OAM channel with $\ell = 5$, for forward (TX-1/RX-1) and backward (TX-2/RX-2) links; (b) bit error rate (BER) for OAM channel $\ell = 5$, when transmitting three multiplexed channels ($\ell = 3, 5,$ and 7 with $\Delta = 2$, or $\ell = 1, 5,$ and 9, with $\Delta = 4$) under a random turbulence realization, with and without post- and pre-compensation. (©2014 Optical Society of America. Reprinted from [40].)

By compensating the distorted OAM beams, the AO approach can help to reduce the turbulence-induced XT and improve the performance of OAM-multiplexed FSO communication links. This single-end AO module was also used for simultaneously pre- and post-turbulence compensation in a bi-directional OAM link [40]. Fig. 10.10 shows the concept of post-compensation, where forward propagating OAM beams are first distorted by atmospheric turbulence, then compensated by the AO system. For the pre-compensation of turbulence, the backward propagating OAM first experiences compensation by the AO system, and then propagates through the same turbulence with the forward channel. The AO mitigation for bi-directional OAM-multiplexed link was demonstrated in Ref. [40] and the results are shown in Fig. 10.12. In Ref. [40], 3 OAM-multiplexed channels each carrying 100 Gbit/s

Causes and mitigation of modal crosstalk in OAM multiplexed optical communication links   269

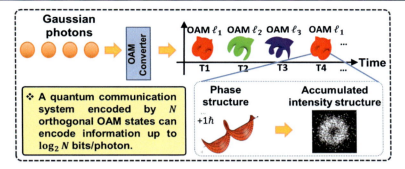

Figure 10.13: Concept of OAM-based quantum data encoding. Within each symbol period, the coming Gaussian photon is converted to OAM photon occupying one of the N OAM states, results in information encoding of up to $\log_2 N$ bits/photon.

quadrature-phase-shift-keying (QPSK) signal propagated through emulated weak-to-moderate atmospheric turbulence. The QPSK modulation encodes 2 bits/symbol of information with four phase points on the constellation as shown in Fig. 10.12(a0). The constellations of the received channels with and without compensations at an optical signal-to-noise ratio (OSNR) of 23.5 dB are shown in Fig. 10.12(a1-a8) for the post- and pre-compensations (forward and backward links). Link performance under different OAM mode spacing (Δ) was compared for $\Delta = 2$ and $\Delta = 4$. The result shows that smaller Δ results in higher crosstalk before compensation and the link under both scenarios can be compensated by the AO approach. The BER curves for the data channels are presented as functions of OSNR, as shown in Fig. 10.12(b). It shows that the BER curves for both forward and backward links are below the forward error correction (FEC) limit of 1×10^{-3} at OSNR values of 21 and 27.5 dB (for $\Delta = 2$), and 15.5 and 16.5 dB (for $\Delta = 4$), respectively. We note that the FEC limit is a BER upper limit under which the FEC coding method can be performed to mitigate the errors over unreliable or noisy communication channels.

10.3.2 AO using WFS and Gaussian probe beam in a quantum communication link

OAM modes can be used for encoding in quantum communication systems, providing potential advantages in system capacity and security [41–44]. Fig. 10.13 shows the concept of OAM-based quantum encoding. A single photon can carry one of the many orthogonal OAM modes. By using an OAM mode converter, within each symbol period, the coming single Gaussian photon is converted to the designed OAM photon that occupies one of the N OAM states, resulting in information encoding of up to $\log_2 N$ bits/photon [45–49]. The system performance degradation due to the atmospheric turbulence is also a key limitation of OAM-based quantum FSO links. Turbulence could affect the wavefront of a photon, such that

270   Chapter 10

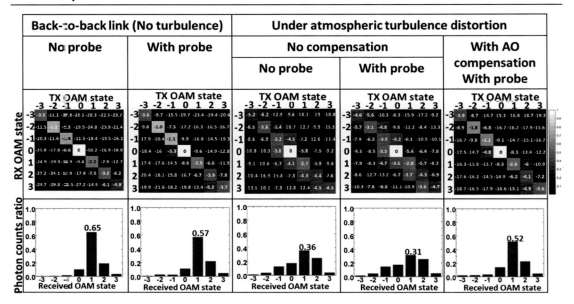

Figure 10.14: The performance of AO mitigation in OAM-based quantum link. (Upper row) Channel transfer matrices when sending OAM modes $\{\ell = -3, \ldots, +3\}$, respectively, for different cases. The numbers are measured in the quantum domain as the ratio of the measured photon counts to the maximum photon counts in this matrix in a unit of dB. (Lower row) The photon counts ratio on received OAM modes $\{\ell = -3, \ldots, +3\}$ when sending only OAM $\ell = +1$ photons for different cases. TX is for the transmitter; RX for the receiver. (©2019 American Association for the Advancement of Science. Reprinted from [51].)

the unique spatial phase profile that defines its OAM state would be distorted [50]. Therefore, turbulence can decrease the OAM modal purity and thereby increase intermodal crosstalk in quantum communication systems.

The AO approach could be applied to OAM-encoded quantum communication links by using a classical Gaussian (OAM $\ell = 0$) probe beam. To separate the probe beam from the quantum channel, one could transmit the probe beam with different polarization, wavelength, and OAM orders with the quantum channel. The single-end AO approach was demonstrated for emulated turbulence compensation in an OAM-encoded, bi-directional FSO quantum communication link at 10 Mbit/s per channel [51]. Fig. 10.14 shows the channel transfer matrices (upper row), as well as the photon counts ratio (lower row) on the received OAM modes when sending OAM modes $\{\ell = -3, \ldots, +3\}$ in different cases: (from left to right column) the back-to-back link (without the turbulence emulator in the optical path) with the probe being turned (i) off and (ii) on, under a random turbulence realization with the probe (iii) off and (iv) on without AO mitigation, and (v) with AO mitigation while the probe is on. The performance of the AO approach in quantum communications was also analyzed in Ref. [51]

Causes and mitigation of modal crosstalk in OAM multiplexed optical communication links 271

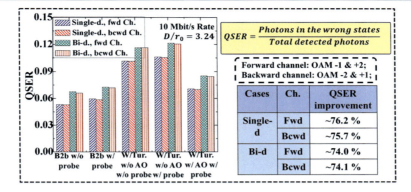

Figure 10.15: The performance of AO mitigation in an OAM-encoded, bidirectional quantum link: the QSER of an OAM-encoded quantum communication link for the single-directional and bi-directional cases under emulated turbulence effects. OAM $\ell = -1$ and $+2$ are used for forward channel encoding and OAM $\ell = -2$ and $+1$ are used for backward channel encoding. The turbulence with a strength of $D/r_0 = 3.24$ is applied. The improvements of the AO mitigation are calculated and shown in the inset table; b2b: back-to-back. (©2019 American Association for the Advancement of Science. Reprinted from [51].)

by measuring the quantum-symbol-error-rate (QSER) which is the ratio of the photons in the wrong OAM-states over the total detected photons in an encoding sequence. Quantum information was encoded with OAM values of $\{\ell = -1, +2\}$ and $\{\ell = -2, +1\}$ for the forward and backward quantum channels, respectively. The AO system reduced the turbulence-effects-increased QSER by ~76% and ~74%, for both channels in the uni-directional and bi-directional links, respectively, as shown in Fig. 10.15.

10.3.3 AO using camera for beam intensity measurement

In an AO turbulence compensation system, one can also measure the incoming beam intensity profile by using a camera and derive the phase correction pattern by applying a stochastic-parallel-gradient-descent (SPGD) algorithm [52–54]. This approach does not need the additional Gaussian probe beam in an OAM-based link. The concept of this approach is shown in Fig. 10.16. At the receiver, the intensity profile is measured by an infrared (IR) camera to provide the feedback signal to the SPGD algorithm. The algorithm will generate a phase-correction pattern to load on the wavefront corrector for compensation. The information collected on one OAM order can be used to correct beams with other OAM orders as the beams pass through the same distorting transmission medium.

This approach was demonstrated for an OAM-multiplexed communication link in Ref. [54]. OAM $\ell = +3$ was firstly transmitted through the emulated turbulence and the SPGD algo-

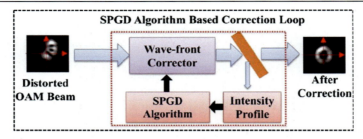

Figure 10.16: Concept of phase correction for a distorted OAM beam using a camera for intensity profile detection and stochastic parallel gradient descent (SPGD) algorithm. (©2015 Optical Society of America. Reprinted from [54].)

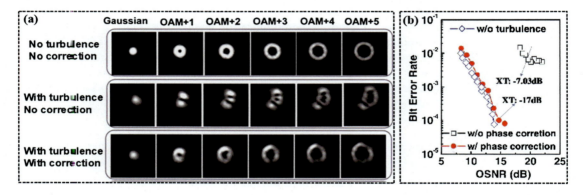

Figure 10.17: The measured beam profiles of the AO mitigation by using a camera: (a) far-field intensity profiles for various OAM beams (OAM $\ell = 0, \ldots, +5$) before and after phase correction under a specific turbulence realization with $r_0 = 1$ mm and (b) measured BERs as functions of OSNR when the phase pattern derived from OAM $\ell = +3$ is used to correct the phases of three simultaneously transmitted OAM beams (OAM $\ell = +1, +3$ and $+5$). (©2015 Optical Society of America. Reprinted from [54].)

rithm was used to generate the phase-correction pattern. The turbulence emulation follows a Kolmogorov-spectrum and the Fried parameter was held fixed to be $r_0 = 1$ mm. This correction pattern was then used to correct the wavefront distortions of a Gaussian beam, as well as those of OAM $\ell = +1$ to OAM $\ell = +5$ beams. The far-field intensity profiles of the received OAM beams are shown in Fig. 10.17(a). The emulated turbulence distorted the intensity profile of the OAM beams and after phase correction the intensity profiles are compensated. A 50-Gbaud QPSK signal was encoded onto three multiplexed channels with OAM $\ell = +1$, $+3$, and $+5$. The OAM-multiplexed channels propagated co-axially through emulated turbulence effects. The correction pattern derived from OAM $\ell = +3$ was applied to a multiplexed beam, comprising three channels (OAM $\ell = +1, +3$ and $+5$), propagating through the same turbulence. Fig. 10.17(b) shows the BERs of the OAM $\ell = +3$ channel before and after phase

Causes and mitigation of modal crosstalk in OAM multiplexed optical communication links 273

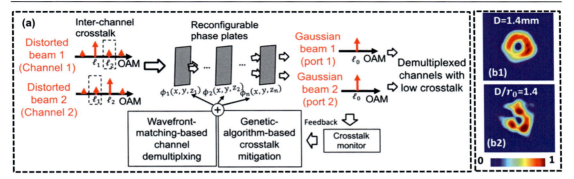

Figure 10.18: (a) Concept diagram of simultaneously mitigating turbulence-induced crosstalk and demultiplexing OAM modes using one MPLC. (b) Normalized intensity profile of OAM $\ell = +1$ (b1) without and (b2) with turbulence distortion ($D/r_0 = 1.4$, beam diameter $D = 1.4$ mm, fried parameter $r_0 = 1$ mm). SMF: single mode fiber. (©2020 Optical Society of America. Reprinted from [59].)

correction for one turbulence realization. Without phase correction, the BER could barely reach the FEC limit of 3.8×10^{-3}, because of the large amount of crosstalk from the other two channels. With the phase correction, the BER could achieve the FEC limit.

10.3.4 Simultaneous demultiplexing and XT mitigation by using multi-plane light converter (MPLC)

For OAM-based MDM links in turbulent atmosphere, there are two potential challenges: (a) the compensation of power coupling among modes caused by atmospheric turbulence and (b) efficient demultiplexing for various OAM-carrying data channels. The MPLC, which is composed of multiple phase plates, has been shown to demultiplex multiple modes from HG or LG modal sets [55–58]. Moreover, it could also be applied in such an OAM-based MDM link through turbulence to simultaneously demultiplex modes and reduce turbulence-induced crosstalk. Without using wavefront sensor or MIMO DSP processing, one MPLC module consisting of several-plane phase can: (i) demultiplex two OAM modes and (ii) mitigate turbulence-induced crosstalk by using wavefront matching method and genetic algorithm [59].

The concept of this approach is shown in Fig. 10.18(a). Atmospheric turbulence introduces distortion to the wavefront of OAM beams, which will induce power coupling between neighboring modes and crosstalk in the OAM-multiplexed link. After propagation, the beam with distorted wavefront evolves into a beam with a distorted intensity profile, as shown in Fig. 10.18(b). At the receiver, the wavefront of distorted beams will be spatially manipulated

274   Chapter 10

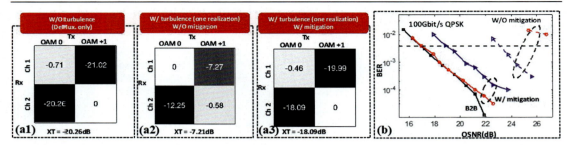

Figure 10.19: The performance of the MPLC-based simultaneous demultiplexing and XT mitigation. (a1-a3) Crosstalk matrix of OAM mode ($\ell = 0$ and $+1$) demultiplexing and with/without crosstalk mitigation under turbulence. (b) Measured BERs as functions of OSNR for two OAM-($\ell = 0$ and $+1$) multiplexed channels with and without MPLC crosstalk mitigation. B2B: back-to-back. The red and blue curves are for channels 1 and 2. (©2020 Optical Society of America. Reprinted from [59].)

when propagating through the several consecutive phase patterns in the MPLC. The generation of such phase patterns consists of two steps: (a) applying the wavefront matching method to generate initial phase patterns which could demultiplex undistorted OAM beams, and (b) iteratively updating phase patterns using genetic algorithm to reduce the turbulence-induced modal crosstalk. The genetic algorithm could iteratively search for the optimized phase patterns utilizing the measured crosstalk from each OAM channel. At the output side, the data channels carried by each distorted beam would be demultiplexed and the channel crosstalk would also be compensated.

Reference [59] demonstrated this approach for a 2-mode, 200-Gbit/s OAM-multiplexed link. The crosstalk matrix and BERs were measured. For the case with turbulence and without using turbulence-induced crosstalk mitigation (Fig. 10.19(a2)), the crosstalk is 13.25 dB worse than that for the case without turbulence (Fig. 10.19(a1)). After applying turbulence-induced crosstalk mitigation, the crosstalk for channel 1 ($\ell = 0$) and channel 2 ($\ell = +1$) is improved by 11.26 dB and 6.42 dB as shown in Fig. 10.19(a3). As shown in Fig. 10.19(b), without using turbulence-induced crosstalk mitigation, channel 1 ($\ell = 0$) cannot achieve below 3.8e-3 FEC limit and channel 2 ($\ell = +1$) has ~6 dB power penalty at the 3.8e-3 FEC limit in BER performance. Using turbulence-induced crosstalk mitigation, there are ~0.2 dB and ~2 dB power penalty for channel 1 and channel 2, respectively.

Summary

In summary, AO, as an optical method, can be used in both classical and quantum OAM-multiplexed free-space optical communication systems for turbulence-induced crosstalk mitigation. The WFS can be used with a Gaussian probe beam to measure phase distortion,

Causes and mitigation of modal crosstalk in OAM multiplexed optical communication links

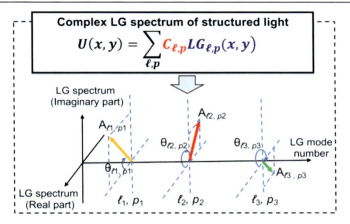

Figure 10.20: Concept of the LG mode decomposition and the complex OAM spectrum. The wavefront of a structured beam can be decomposed into a set of LG modes carrying OAM. (©2021 AIP Publishing LLC. Reprinted from APL Photonics (6) (3) (2021) 030901.)

while the camera can also be used to measure the intensity profile of the distorted OAM beams to derive the phase correction pattern. In addition, one can consider using the MPLC components for simultaneously demultiplexing the OAM modes and mitigating the inter-model crosstalk.

10.4 Spatial modes manipulation for crosstalk mitigation

It would be valuable to develop an alternative optical approach for XT mitigation in OAM-multiplexed FSO links without sensing the beam profile. One potential approach would be to select and design the transmitted mode or the combination of modes, namely, spatial mode manipulation. This method is achievable based on that the wavefront of a structured beam can be decomposed into a set of LG modes carrying OAM. The coefficient of each LG mode in the decomposition can be complex, containing both amplitude and phase information [60–64], as shown in Fig. 10.20. On the other hand, one can also generate a structured beam by coherently combining multiple LG modes. Approaches using spatial modes manipulations for OAM beam crosstalk mitigation will be discussed in this section, including: (a) transmitting OAM modes combination, (b) simultaneously orthogonalizing and shaping multiple LG beams, and (c) utilizing Bessel beams with non-zero OAM order.

10.4.1 Turbulence pre-compensation by OAM mode combination

The concept of turbulence compensation by mode combination is shown in Fig. 10.21(a). At the transmitter side, the inverse transmission matrix is applied using a compensation phase

276 Chapter 10

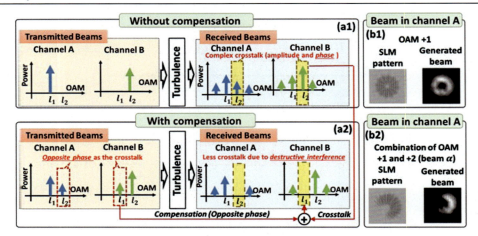

Figure 10.21: (a) The concept of the turbulence compensation utilizing inverse transmission matrix method: (a1) transmitting pure OAM modes, no compensation approach applied; and (a2) transmitting the combination of OAM modes to compensation the turbulence effects. (b) An example of transmitted beams, in (b1) no compensation and in (b2) with compensation cases; the SLM patterns and the generated beams intensity profiles are shown. (©2020 IEEE. Reprinted from [65].)

patterns at the transmitter for each channel, so that the signal from each transmitted channel is carried by the combination of multiple (e.g., 2) OAM modes with designed complex weights. The weights are calculated based on the inverse of the complex transmission matrix under the corresponding turbulence realization. After combining the beams from the two channels at the output port of the transmitter, the data-carrying beams from both channels become coaxial, and each of the two OAM modes in the combined beams carries the combination of signals from both channels. When the beams are transmitted through the turbulence, the signals on the two transmitted modes will couple to their neighboring modes and experience interference on those modes. Thus, the channels could have little power on the designated modes due to destructive interference and relatively high power on the others. By receiving the mode on which the undesired channel has little power, the desired channel could be recovered with little inter-channel crosstalk. The same concept can be applied to recover the second channel when receiving another mode. As an example, the compensation phase patterns used for beam generation in channel A and the intensity profiles of the generated beams are shown in Fig. 10.21(b).

This approach was demonstrated in Ref. [65], where two OAM-multiplexed channels each carrying a 100-Gbit/s QPSK signal were transmitted. Channels A and B transmitted OAM $\ell = +1$ and $\ell = +2$, respectively, in the link without the compensation. To apply the compensation approach, channels A and B transmitted beams α and β, which were combinations

Causes and mitigation of modal crosstalk in OAM multiplexed optical communication links   277

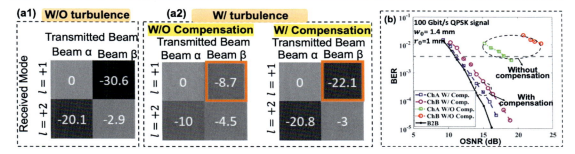

Figure 10.22: The performance of the XT mitigation utilizing inverse transmission matrix method: (a1-a2) The normalized transmission intensity matrices and (b) BER performance for the transmitted channels without and with compensation. In the demonstration, channel A receives OAM $\ell = +1$, and channel B receives OAM $\ell = +2$. In the link without compensation, channels A and B transmit OAM $\ell = +1$, $\ell = +2$, respectively. Beams α and β are the combinations of OAM +1 and +2 transmitted by channels A and B, respectively, when the compensations are applied. (©2020 IEEE. Reprinted from [65].)

of OAM $\ell = +1$ and $\ell = +2$. The receivers of channels A and B always receive and demultiplex OAM $\ell = +1$ and $\ell = +2$. The results of this demonstration are shown in Fig. 10.22. Fig. 10.22(a1) shows the back-to-back crosstalk. As can be seen from Fig. 10.22(a2), with the turbulence effect, the inter-channel crosstalk increases to -8.7 dB and -5.5 dB for channels A and B in the absence of compensation, respectively. This crosstalk decreases to -22.1 dB when receiving OAM $+1$, and -17.8 dB when receiving OAM $+2$ with compensation. The BER performance for the channels is shown in Fig. 10.22(b). Through applying precompensation phase patterns in the forward link, the BER performance could be improved for both forward channels.

10.4.2 Simultaneous orthogonalizing and shaping of multiple LG beams

In a perfectly aligned LG mode multiplexed FSO link, the receiver with a full aperture could receive the whole transmitted beam with little power loss and crosstalk. However, in a real FSO link, the Rx aperture size might be limited and there might be a misalignment between the Tx and the Rx; these issues might induce power loss and crosstalk [18,66–68], as discussed in Section 10.2.

An approach that can mitigate both of the above issues would be always desired. One can consider to re-orthogonalize the combinations of multiple LG modes to mitigate power loss and crosstalk simultaneously. Fig. 10.23(a) presents the concept of transmitting each data channel on a designed combination of multiple LG modes to mitigate the effect of the limited-size aperture or misalignment in an MDM link. Generally, the mode coupling between a set

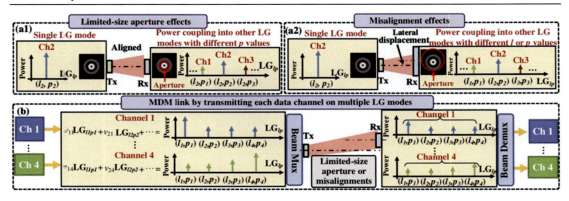

Figure 10.23: (a) Concept of (a1) limited-size aperture and (a2) misalignment effects on an FSO link using LG modes. (b) Concept diagram of transmitting each data channel on a designed beam that is a combination of multiple LG modes to mitigate the effects of the limited-size aperture or misalignments in an MDM link. Ch: Channel; Tx: Transmitter; Rx: Receiver. (©2020 Optical Society of America. Reprinted from [70].)

of LG modes in a given link can be described as a complex transmission matrix H. H can be factorized by SVD and written as $\mathbf{H} = \mathbf{U} \cdot \sum \cdot \mathbf{V}^*$ [69]. At the Tx side, the orthogonal beams are generated by complex combinations of multiple LG modes, of which the complex weights (amplitude and phase) are given by the column vectors of the V. After passing through the limited-size aperture or being affected by the misalignment, the transmitted beams on different channels are still orthogonal to each other. As a result, these independent beams could be demultiplexed with little crosstalk based on the row vectors of the inverse of the U matrix. When the beams are orthogonalized, the intensity profiles of the transmitted beams would also be shaped, which might simultaneously reduce the power loss. As an example, the intensity profiles of pure LG_{10}, pure LG_{11} beams, and the generated orthogonal beam are shown in Fig. 10.23(b).

Reference [70] demonstrated the above approach to a four-channel multiplexed link, each carrying a 100-Gbit/s QPSK signal. The results are shown in Fig. 10.24. The left panel of Fig. 10.24 presents that when transmitting data channels on pure LG modes with different p values (LG_{10}, LG_{11}), the crosstalk would become larger with the horizontal displacement. However, when using the orthogonal beams (beam 1 and 2) which are generated by designed complex combinations of LG_{10}, LG_{11} modes, the crosstalk for both of the two generated beams could stay at a relatively low level (<-27 dB) with the displacement. The right panel of Fig. 10.24 shows the effects of displacement on LG modes with different l values (LG_{10}, LG_{-10}). The crosstalk of LG_{10}, LG_{-10} modes increases with the displacement and could stay at a relatively low level (<-17 dB) with the displacement for both channels.

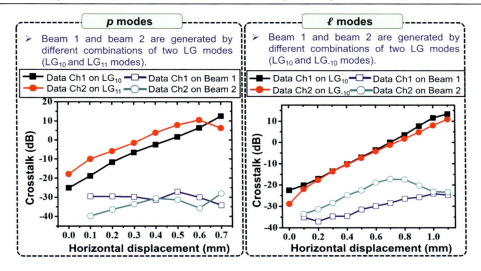

Figure 10.24: The performance of simultaneous orthogonalizing and shaping of multiple LG beams method for XT mitigation: Displacement-induced channel crosstalk under various horizontal displacements when transmitting data channels on (left panel) pure LG modes with different p values (LG_{10}, LG_{11}) or designed orthogonal beams and (right panel) pure LG modes with different l values (LG_{10}, LG_{-10}) or designed orthogonal beams. (©2020 Optical Society of America. Reprinted from [70].)

10.4.3 Utilizing Bessel-Gaussian (BG) beams with non-zero OAM order

If the MDM channels are transmitted over short ranges (e.g., tens of meters) with obstruction in the path, a key advance in multiple-channel free-space links would be if the basic phase and amplitude profiles of orthogonal beams remain intact. One potential approach could be to use BG beams for transmission over short ranges. BG beams are propagation invariant over a length determined by the generation method and can extend to a few tens of meters [71–74]. BG beams have the unique property to reconstruct or 'self-heal' the transverse intensity and phase profiles after experiencing an obstruction. The self-healing property of the BG beams may have important applications in short-range free-space communication links. The concept diagram is shown in Fig. 10.25(a). A free-space link employs spatially multiplexed data-carrying beams. N distinct input OAM beams, each carrying a distinct data channel, are spatially multiplexed and transmitted through an axicon to be transformed into BG beams. Within the 'Bessel-region', beams are propagation invariant and, therefore, can sustain partial obstructions. At the end of the Bessel-region, an exit axicon having opposite cone angles is placed to remove conical phases. Finally, an OAM mode demultiplexer separates each OAM beam. The ability of BG beams to sustain the adverse effects of an object inadvertently blocking the beam path within the Bessel region could be beneficial in recovering data channels.

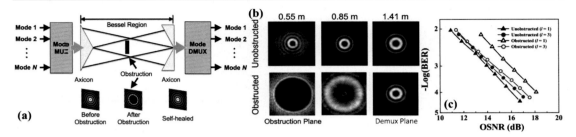

Figure 10.25: (a) Conceptual diagram of an MDM link using multiplexed BG beams. 'Bessel-region' is the distance over which BG beams are propagation invariant and retain their profile. Insets depict transverse intensity profiles of a BG beam before and after an opaque disk and in the receiver plane. (b) Transverse intensity profiles of obstructed and unobstructed BG beam $\ell = +3$ after an obstruction of radius 1.5 mm ($\zeta = 0.71$) at different locations along the propagation direction. (c) Measured BER for the multiplexed BG beams obstructed by obstructions of radii $r_{Obs} = 1$ mm. (©2016 Nature Research. Reprinted from [19].)

Reference [19] demonstrated this approach. Two OAM-multiplexed channels (OAM $\ell = +1$ and $\ell = -3$) each carrying uncorrelated 50-Gbaud QPSK channels were transmitted through a path with obstructions. As an example, the transverse intensity profiles of the obstructed and unobstructed BG beam with $\ell = +3$ are presented in Fig. 10.25(b). An obstruction of radius 1.5 mm ($\zeta = 0.71$) was placed at the beam center and images of the transverse intensity profiles were taken at various locations along the propagation direction. A comparison of obstructed and unobstructed beams in the plane of the demultiplexer revealed reconstruction of the BG beam. Fig. 10.25(c) shows the BER measurement for the unobstructed and obstructed BG beams. Each channel achieves a raw BER of 3.8×10^{-3}. At a BER of 3.8×10^{-3}, the OSNR penalty for the 1 mm obstructions is <2.3 dB. The large penalty for the channel on the $\ell = +1$ beam could be explained by noting that the $\ell = +1$ beam has a smaller spot size and, therefore, encounters a relatively large obstruction as compared with the $\ell = +3$ beam.

Summary

In summary, the spatial modes manipulation approaches could be considered for crosstalk mitigation in OAM-multiplexed FSO links, which do not require beam profile sensing at the receiver side, when compared with the AO approach. The approaches discussed in this section could help to mitigate the intermodal crosstalk caused by atmospheric turbulence, limited size receiver aperture, transmitter/receiver misalignment, and obstructions in the optical link.

10.5 *Digital signal processing for crosstalk mitigation*

Sections 10.3 and 10.4 address the mitigation of channel crosstalk using optical approaches. An alternative option for crosstalk mitigation in OAM multiplexed links would be to shift the

Causes and mitigation of modal crosstalk in OAM multiplexed optical communication links

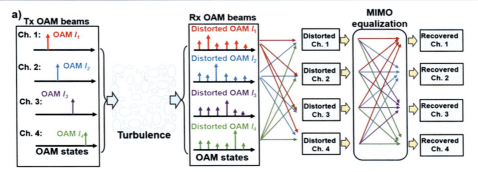

Figure 10.26: The concept of using MIMO DSP equalization for crosstalk mitigation.

complexity of the optical setup to the digital domain, i.e., using digital signal processing. This section will introduce digital approaches for the crosstalk mitigation in OAM multiplexed FSO links by: (1) using multiple-input-multiple-output (MIMO) digital signal processing (DSP) equalization in the laboratory environment, and (2) using MIMO DSP equalization in a link through a flying UAV.

10.5.1 MIMO equalization for crosstalk mitigation in laboratory

The wavefront of the OAM beams could be affected by various degradation effects, as described in Section 10.2. Therefore, after demultiplexing, the received signal carried on a particular OAM mode may include the signal leaked from other channels. One potential crosstalk mitigation approach is to utilize MIMO DSP-based equalization [75,76], as indicated in Fig. 10.26.

Reference [77] demonstrated the implementation of a 4 × 4 adaptive MIMO equalizer in a four-channel OAM multiplexed link using heterodyne detection. Four OAM modes, each carrying 20 Gbit/s QPSK data, were collinearly propagated through a turbulence emulator to introduce crosstalk and are demultiplexed at the receiver. In Ref. [77], the turbulence was emulated using a rotatable phase screen that obeys Kolmogorov-spectrum statistics with a Fried parameter of ∼5 mm.

Figs. 10.27(a) and (b) illustrate the recovered constellations of all four 20 Gbit/s QPSK signals (with and without MIMO equalization) in each channel for a single turbulence realization. In the experiment, channels 1-4 corresponded to OAM modes with $l = -1, -3, +1, +3$, respectively. With the assistance of MIMO DSP equalization, the Error vector magnitude (EVM) of four channels was improved from 0.24, 0.46, 0.33, and 0.46 to 0.14, 0.14, 0.15, and 0.21, respectively. The EVM is a parameter that could quantify the performance of a communication system, and a lower EVM value indicates a better signal quality. In the report,

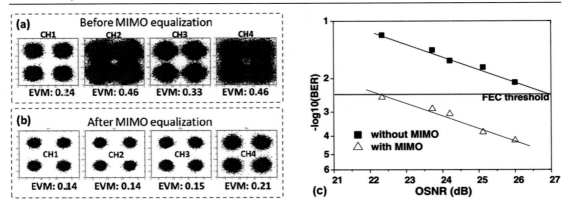

Figure 10.27: Recovered constellations of 20-Gbit/s QPSK signal in each of the four channels (a) without and (b) with MIMO equalization. The four channels in (a) and (b) are OAM modes with $l = -1, -3, +1, +3$. (c) Measured BER as a function of OSNR for the 20 Gbit/s QPSK signal on channel 4. The four OAM modes in (c) are $l = +2, +4, +6$, and $+8$. (©2014 Optical Society of America. Reprinted from [77].)

the MIMO equalization using another four OAM modes ($l = +2, +4, +6$, and $+8$) was also demonstrated. Fig. 10.27(c) shows the calculated BER as a function of OSNR for the channel carried on $l = +8$ (the channel that experienced the most crosstalk). Without MIMO equalization, the measured BER for this channel is above 3.8×10^{-3}, i.e., the 7% FEC limit. The curve crosses the FEC threshold at the OSNR of ~26.8 dB. After MIMO equalization, the estimated required OSNR at the BER of 3.8×10^{-3} is ~22.3 dB.

10.5.2 MIMO equalization for crosstalk mitigation in the link through a flying UAV

The communication capacity needs of unmanned aerial vehicle (UAV) have been increasing dramatically over the past several years, thereby driving the need for higher-capacity links between UAVs and their ground stations [78–80]. A recent report demonstrated a 100-m round-trip OAM-multiplexed link between a ground station and a retro-reflecting flying UAV [81]. Fig. 10.28 shows the concept of such an OAM-multiplexed FSO link. Multiple independent data-carrying OAM beams are multiplexed and transmitted from the ground station to the UAV and retro-reflected back to the same ground station. However, due to atmospheric turbulence, the OAM beams may be distorted during their free-space propagation, such that signal power on each particular OAM mode may be coupled to its neighboring modes and induce channel crosstalk. MIMO equalization can be used to reduce crosstalk among channels by applying the inverse channel matrix to the received signals, thus mitigating the performance degradation in this scenario.

Figure 10.28: Concept of OAM-multiplexed FSO communication link between a ground station and a retro-reflecting UAV through atmospheric turbulence. (©2019 Optical Society of America. Reprinted from [81].)

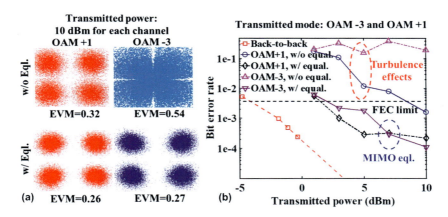

Figure 10.29: (a) Recovered 20-Gbit/s QPSK constellations for two OAM multiplexed channels (OAM +1 and +3) with and without MIMO equalization; (b) measured BERs as functions of transmitted power for both channels with and without MIMO equalization. Each channel carries a 20-Gbit/s QPSK signal. (©2019 Optical Society of America. Reprinted from [82].)

Reference [82] showed the mitigation of emulated turbulence in a 40-Gbit/s retro-reflected FSO link multiplexing two OAM modes by using MIMO equalization. In this experiment, two OAM channels, i.e., OAM +1 and −3, each carrying a 20 Gbit/s QPSK signal, were transmitted between transmitter and receiver. Fig. 10.29(a) shows the received QPSK constellation diagrams and corresponding EVMs for both channels when the UAV is hovering ∼50 m away with the phase plate fixed at a random angle. The transmitted power for each channel is 10 dBm. The EVM for the two channels improves from 0.32 and 0.54 to 0.26 and 0.27, respectively, after MIMO equalization. Fig. 10.29(b) shows BERs for both channels as functions of transmitted power. The result shows that the BERs dramatically decrease to below the 7% overhead FEC limit of 3.8×10^{-3} for all channels after MIMO equalization.

Summary

In summary, digital signal processing approaches can be used for crosstalk mitigation in OAM-multiplexed FSO links without using additional optical components. The reports discussed in this section show the potential for turbulence-induced crosstalk mitigation using MIMO equalization both in a laboratory environment and in a link through a flying UAV.

10.6 Summary

This chapter discusses the causes and mitigation techniques for crosstalk in OAM multiplexed FSO communication links. Several related works for crosstalk mitigation are discussed, including the approaches of adaptive optics, spatial modes manipulation, and digital signal processing.

To further explore and develop the crosstalk mitigation techniques for OAM multiplexed links, one may need to consider several issues, including but not limited to (1) scalability: for scaling the compensation scheme to the system that multiplexes a larger number of modes; (2) tuning speed: for real-time mitigation of the atmospheric turbulence effects; and (3) complexity: for enabling the deployment of a compensation scheme in an actual OAM multiplexed link.

Acknowledgment

We thank the support from Defense Advanced Research Projects Agency (DARPA) under the InPho (Information in a Photon) program; Vannevar Bush Faculty Fellowship sponsored by the Basic Research Office of the Assistant Secretary of Defense (ASD) for Research and Engineering and funded by the Office of Naval Research (ONR) N00014-16-1-2813; Office of Naval Research through a MURI award N00014-20-1-2558; National Science Foundation (NSF) ECCS-1509965 and IIP-1622777; Intel Labs Research Office; NSF MRI program; Air Force Office of Scientific Research (AFOSR) FA-9550–15–C-0024 and FA9550-16-C-0008; NxGen Partners; Air Force Research Laboratory (AFRL) FA8650-18-P-1699 and FA8650-20-C-1105; Office of Naval Research (ONR) N00014-15-1-2635; Office of Naval Research (ONR) N000141812352; Defense Security Cooperation Agency (DSCA) DSCA-4440646262 and DSCA-4441006051; Naval Air Warfare Center Aircraft Division N68335-18-C-0588; Nippon Telegraph and Telephone (NTT); Airbus and Korean Air; Qualcomm Innovation Fellowship (QIF).

References

[1] M.J. Padgett, Orbital angular momentum 25 years on, Opt. Express 25 (10) (2017) 11265–11274.

[2] L. Allen, M.W. Beijersbergen, R.J.C. Spreeuw, J.P. Woerdman, Orbital angular momentum of light and the transformation of Laguerre-Gaussian laser modes, Phys. Rev. 45 (11) (1992) 8185.

[3] A.M. Yao, M.J. Padgett, Orbital angular momentum: origins, behavior and applications, Adv. Opt. Photonics 3 (2) (2011) 161–204.

[4] N. Bozinovic, Y. Yue, Y. Ren, M. Tur, P. Kristensen, H. Huang, A.E. Willner, S. Ramachandran, Terabit-scale orbital angular momentum mode division multiplexing in fibers, Science 340 (6140) (2013) 1545–1548.

[5] J. Wang, Advances in communications using optical vortices, Photon. Res. 4 (5) (2016) B14–B28.

[6] G. Gibson, J. Courtial, M.J. Padgett, M. Vasnetsov, V. Pas'ko, S.M. Barnett, S. Franke-Arnold, Free-space information transfer using light beams carrying orbital angular momentum, Opt. Express 12 (22) (2004) 5448–5456.

[7] J. Wang, J. Yang, I.M. Fazal, N. Ahmed, Y. Yan, H. Huang, Y. Ren, Y. Yue, S. Dolinar, M. Tur, A.E. Willner, Terabit free-space data transmission employing orbital angular momentum multiplexing, Nat. Photonics 6 (7) (2012) 488–496.

[8] G. Milione, M.P.J. Lavery, H. Huang, Y. Ren, G. Xie, T.A. Nguyen, E. Karimi, L. Marrucci, D.A. Nolan, R.R. Alfano, A.E. Willner, 4 × 20 Gbit/s mode division multiplexing over free space using vector modes and a q-plate mode (de) multiplexer, Opt. Lett. 40 (9) (2015) 1980–1983.

[9] K. Pang, H. Song, Z. Zhao, R. Zhang, H. Song, G. Xie, L. Li, C. Liu, J. Du, A.F. Molisch, M. Tur, A.E. Willner, 400-Gbit/s QPSK free-space optical communication link based on four-fold multiplexing of Hermite-Gaussian or Laguerre-Gaussian modes by varying both modal indices, Opt. Lett. 43 (16) (2018) 3889–3892.

[10] Y. Ren, Z. Wang, P. Liao, L. Li, G. Xie, H. Huang, Z. Zhao, Y. Yan, N. Ahmed, A. Willner, M.P.J. Lavery, N. Ashrafi, S. Ashrafi, R. Bock, M. Tur, I.B. Djordjevic, M.A. Neifeld, A.E. Willner, Experimental characterization of a 400 Gbit/s orbital angular momentum multiplexed free-space optical link over 120 m, Opt. Lett. 41 (3) (2016) 622–625.

[11] Y. Zhao, J. Liu, J. Du, S. Li, Y. Luo, A. Wang, L. Zhu, J. Wang, Experimental demonstration of 260-meter security free-space optical data transmission using 16-QAM carrying orbital angular momentum (OAM) beams multiplexing, in: 2016 Optical Fiber Communication Conference, OSA, 2016, p. Th1H-3.

[12] M. Krenn, J. Handsteiner, M. Fink, R. Fickler, R. Ursin, M. Malik, A. Zeilinger, Twisted light transmission over 143 km, Proc. Natl. Acad. Sci. 113 (48) (2016) 13648–13653.

[13] D.J. Richardson, J.M. Fini, L.E. Nelson, Space-division multiplexing in optical fibres, Nat. Photonics 7 (5) (2013) 354–362.

[14] P.J. Winzer, Making spatial multiplexing a reality, Nat. Photonics 8 (5) (2014) 345–348.

[15] A.E. Willner, H. Huang, Y. Yan, Y. Ren, N. Ahmed, G. Xie, C. Bao, L. Li, Y. Cao, Z. Zhao, J. Wang, M.P.J. Lavery, M. Tur, S. Ramachandran, A.F. Molisch, N. Ashrafi, S. Ashrafi, Optical communications using orbital angular momentum beams, Adv. Opt. Photonics 7 (1) (2015) 66–106.

[16] J. Wang, S. Li, M. Luo, J. Liu, L. Zhu, C. Li, D. Xie, Q. Yang, S. Yu, J. Sun, X. Zhang, W. Shieh, A.E. Willner, N-dimensional multiplexing link with 1.036-Pbit/s transmission capacity and 112.6-bit/s/Hz spectral efficiency using OFDM-8QAM signals over 368 WDM pol-muxed 26 OAM modes, in: 2014 the European Conference on Optical Communication (ECOC), IEEE, 2014, pp. 1–3.

[17] Y. Ren, H. Huang, G. Xie, N. Ahmed, Y. Yan, B.I. Erkmen, N. Chandrasekaran, M.P.J. Lavery, N.K. Steinhoff, M. Tur, S. Dolinar, M. Neifeld, M.J. Padgett, R.W. Boyd, J.H. Shapiro, A.E. Willner, Atmospheric turbulence effects on the performance of a free space optical link employing orbital angular momentum multiplexing, Opt. Lett. 38 (20) (2013) 4062–4065.

[18] G. Xie, L. Li, Y. Ren, H. Huang, Y. Yan, N. Ahmed, Z. Zhao, M.P.J. Lavery, N. Ashrafi, S. Ashrafi, R. Bock, M. Tur, A.F. Molisch, A.E. Willner, Performance metrics and design considerations for a free-space optical orbital-angular-momentum–multiplexed communication link, Optica 2 (4) (2015) 357–365.

[19] N. Ahmed, Z. Zhao, L. Li, H. Huang, M.P.J. Lavery, P. Liao, Y. Yan, Z. Wang, G. Xie, Y. Ren, A. Almaiman, A.J. Willner, S. Ashrafi, A.F. Molisch, M. Tur, A.E. Willner, Mode-division-multiplexing of multiple Bessel-Gaussian beams carrying orbital-angular-momentum for obstruction-tolerant free-space optical and millimetre-wave communication links, Sci. Rep. 6 (2016) 22082.

[20] S. Fu, C. Gao, Influences of atmospheric turbulence effects on the orbital angular momentum spectra of vortex beams, Photon. Res. 4 (5) (2016) B1–B4.
[21] J.A. Anguita, M.A. Neifeld, B.V. Vasic, Turbulence-induced channel crosstalk in an orbital angular momentum-multiplexed free-space optical link, Appl. Opt. 47 (13) (2008) 2414–2429.
[22] N. Chandrasekaran, J.H. Shapiro, Photon information efficient communication through atmospheric turbulence–Part I: channel model and propagation statistics, J. Lightwave Technol. 32 (6) (2014) 1075–1087.
[23] C. Paterson, Atmospheric turbulence and orbital angular momentum of single photons for optical communication, Phys. Rev. Lett. 94 (15) (2005) 153901.
[24] M.A. Cox, N. Mphuthi, I. Nape, N. Mashaba, L. Cheng, A. Forbes, Structured light in turbulence, IEEE J. Sel. Top. Quantum Electron. 27 (2) (2020) 1–21.
[25] Malik, M. O'Sullivan, B. Rodenburg, M. Mirhosseini, J. Leach, M.P.J. Lavery, M.J. Padgett, R.W. Boyd, Influence of atmospheric turbulence on optical communications using orbital angular momentum for encoding, Opt. Express 20 (12) (2012) 13195–13200.
[26] Larry C. Andrews, Ronald L. Phillips, Laser beam propagation through random media, in: SPIE, 2005.
[27] S. Clifford, G.R. Ochs, R.S. Lawrence, Saturation of optical scintillation by strong turbulence, JOSA 64 (2) (1974) 148–154.
[28] D.P. Greenwood, Bandwidth specification for adaptive optics systems, JOSA 67 (3) (1977) 390–393.
[29] M.R. Bhatnagar, Z. Ghassemlooy, Performance analysis of gamma– gamma FSO MIMO links with pointing errors, J. Lightwave Technol. 34 (9) (2016) 2158–2169.
[30] H. Samimi, Distribution of the sum of k-distributed random variables and applications in free-space optical communications, IET optoelectron. 6 (1) (2012) 1–6.
[31] D.L. Fried, Optical resolution through a randomly inhomogeneous medium for very long and very short exposures, JOSA 56 (10) (1966) 1372–1379.
[32] M. Vasnetsov, V. Pas' Ko, M. Soskin, Analysis of orbital angular momentum of a misaligned optical beam, New J. Phys. 7 (1) (2005) 46.
[33] G. Zhu, Y. Wen, X. Wu, Y. Chen, J. Liu, S. Yu, Obstacle evasion in free-space optical communications utilizing airy beams, Opt. Lett. 43 (6) (2018) 1203–1206.
[34] L. Zhu, Z. Yang, S. Fu, Z. Cao, Y. Wang, Y. Qin, A. Koonen, Airy beam for free-space photonic interconnection: generation strategy and trajectory manipulation, IEEE/OSA J. Lightwave Technol. 38 (23) (2020) 6474–5480.
[35] L. Zhu, A. Wang, J. Wang, Free-space data-carrying bendable light communications, Sci. Rep. 9 (1) (2019) 1–8.
[36] G.A. Tyler, Adaptive optics compensation for propagation through deep turbulence: a study of some interesting approaches, Opt. Eng. 52 (2) (2012) 021011.
[37] S. Chen, S. Li, Y. Zhao, J. Liu, L. Zhu, A. Wang, J. Du, L. Shen, J. Wang, Demonstration of 20-Gbit/s high-speed Bessel beam encoding/decoding link with adaptive turbulence compensation, Opt. Lett. 41 (20) (2016) 4680–4683.
[38] T. Weyrauch, Mikhail A. Vorontsov, Free-space laser communications with adaptive optics: atmospheric compensation experiments, in: Free-Space Laser Communications, Springer, 2004, pp. 247–271.
[39] Y. Ren, G. Xie, H. Huang, C. Bao, Y. Yan, N. Ahmed, M.P.J. Lavery, B.I. Erkmen, S. Dolinar, M. Tur, M.A. Neifeld, M.J. Padgett, R.W. Boyd, J.H. Shapiro, A.E. Willner, Adaptive optics compensation of multiple orbital angular momentum beams propagating through emulated atmospheric turbulence, Opt. Lett. 39 (10) (2014) 2845–2848.
[40] Y. Ren, G. Xie, H. Huang, N. Ahmed, Y. Yan, L. Li, C. Bao, M.P.J. Lavery, M. Tur, M.A. Neifeld, R.W. Boyd, J.H. Shapiro, A.E. Willner, Adaptive-optics-based simultaneous pre- and post-turbulence compensation of multiple orbital-angular-momentum beams in a bidirectional free-space optical link, Optica 1 (6) (2014) 376–382.
[41] M. Mirhosseini, O.S. Magaña-Loaiza, M.N. O'Sullivan, B. Rodenburg, M. Malik, M.P.J. Lavery, M.J. Padgett, D.J. Gauthier, R.W. Boyd, High-dimensional quantum cryptography with twisted light, New J. Phys. 17 (3) (2015) 033033.

[42] M. Erhard, R. Fickler, M. Krenn, A. Zeilinger, Twisted photons: new quantum perspectives in high dimensions, Light, Sci. Appl. 7 (3) (2018) 17146.
[43] M. Mafu, A. Dudley, S. Goyal, D. Giovannini, M. McLaren, M.J. Padgett, T. Konrad, F. Petruccione, N. Lütkenhaus, A. Forbes, Higher-dimensional orbital-angular-momentum-based quantum key distribution with mutually unbiased bases, Phys. Rev. A 88 (3) (2013) 032305.
[44] J. Liu, I. Nape, Q. Wang, A. Vallés, J. Wang, A. Forbes, Multidimensional entanglement transport through single-mode fiber, Sci. Adv. 6 (4) (2020) eaay0837.
[45] G. Vallone, V. D'Ambrosio, A. Sponselli, S. Slussarenko, L. Marrucci, F. Sciarrino, P. Villoresi, Free-space quantum key distribution by rotation-invariant twisted photons, Phys. Rev. Lett. 113 (6) (2014) 060503.
[46] F. Bouchard, A. Sit, F. Hufnagel, A. Abbas, Y. Zhang, K. Heshami, R. Fickler, C. Marquardt, G. Leuchs, R.w. Boyd, E. Karimi, Quantum cryptography with twisted photons through an outdoor underwater channel, Opt. Express 26 (17) (2018) 22563–22573.
[47] D. Cozzolino, E. Polino, M. Valeri, G. Carvacho, D. Bacco, N. Spagnolo, L.K. Oxenløwe, F. Sciarrino, Air-core fiber distribution of hybrid vector vortex-polarization entangled states, Adv. Photonics 1 (4) (2019) 046005.
[48] H. Cao, S.C. Gao, C. Zhang, J. Wang, D.Y. He, B.H. Liu, Z.W. Zhou, Y.J. Chen, Z.H. Li, S.Y. Yu, J. Romero, Y.F. Huang, C.-F. Li, G.-C. Guo, Distribution of high-dimensional orbital angular momentum entanglement over a 1 km few-mode fiber, Optica 7 (3) (2020) 232–237.
[49] A. Sit, F. Bouchard, R. Fickler, J. Gagnon-Bischoff, H. Larocque, K. Heshami, D. Elser, C. Peuntinger, K. Günthner, B. Heim, C. Marquardt, G. Leuchs, R.W. Boyd, E. Karimi, High-dimensional intracity quantum cryptography with structured photons, Optica 4 (9) (2017) 1006–1010.
[50] J. Zhao, Y. Zhou, B. Braverman, C. Liu, K. Pang, N.K. Steinhoff, G.A. Tyler, A.E. Willner, R.W. Boyd, Performance of real-time adaptive optics compensation in a turbulent channel with high-dimensional spatial mode encoding, Opt. Express 28 (10) (2020) 15376–15391.
[51] C. Liu, K. Pang, Z. Zhao, P. Liao, R. Zhang, H. Song, Y. Cao, J. Du, L.L. Li, H. Song, Y. Ren, G. Xie, Y.-F. Zhao, J. Zhao, S.M.H. Rafsanjani, A.N. Willner, J.H. Shapiro, R.W. Boyd, M. Tur, A.E. Willner, Single-end adaptive optics compensation for emulated turbulence in a bi-directional 10-mbit/s per channel free-space quantum communication link using orbital-angular-momentum encoding, Research (2019) 8326701.
[52] L. Liu, M.A. Vorontsov, Phase-locking of tiled fiber array using SPGD feedback controller, in: Proc. SPIE 5895, 2005, p. 58950P.
[53] S. Zommer, E.N. Ribak, S.G. Lipson, J. Adler, Simulated annealing in ocular adaptive optics, Opt. Lett. 31 (7) (2006) 939–941.
[54] G. Xie, Y. Ren, H. Huang, M.P.J. Lavery, N. Ahmed, Y. Yan, C. Bao, L. Li, Z. Zhao, Y. Cao, M. Willner, M. Tur, S.J. Dolinar, R.W. Boyd, J.H. Shapiro, A.E. Willner, Phase correction for a distorted orbital angular momentum beam using a Zernike polynomials-based stochastic-parallel-gradient-descent algorithm, Opt. Lett. 40 (7) (2015) 1197–1200.
[55] G. Labroille, B. Denolle, P. Jian, P. Genevaux, N. Treps, J. Morizur, Efficient and mode selective spatial mode multiplexer based on multi-plane light conversion, Opt. Express 22 (13) (2014) 15599–15607.
[56] J.-F. Morizur, L. Nicholls, P. Jian, S. Armstrong, N. Treps, B. Hage, M. Hsu, W. Bowen, J. Janousek, H.-A. Bachor, Programmable unitary spatial mode manipulation, J. Opt. Soc. Am. A 27 (11) (2010) 2524–2531.
[57] N.K. Fontaine, R. Ryf, H. Chen, D.T. Neilson, K. Kim, J. Carpenter, Laguerre-Gaussian mode sorter, Nat. Commun. 10 (1) (2019) 1–7.
[58] S. Bade Satyanarayana, B. Denolle, G. Trunet, N. Riguet, P. Jian, O. Pinel, G. Labroille, Fabrication and characterization of a mode-selective 45-mode spatial multiplexer based on multi-plane light conversion, in: Optical Fiber Communications Conference and Exposition (OFC), OSA, 2018, pp. 1–3.
[59] H. Song, X. Su, H. Song, R. Zhang, Z. Zhao, C. Liu, K. Pang, N. Hu, A. Almaiman, S. Zach, N. Cohen, A. Molisch, R. Boyd, M. Tur, A.E. Willner, Simultaneous turbulence mitigation and mode demultiplexing using one MPLC in a two-mode 200-Gbit/s free-space OAM-multiplexed link, in: Optical Fiber Communication Conference, OSA, 2020, W1G–3.

[60] G. Xie, C. Liu, L. Li, Y. Ren, Z. Zhao, Y. Yan, N. Ahmed, Z. Wang, A.J. Willner, C. Bao, Y. Cao, P. Liao, M. Ziyadi, A. Almaiman, S. Ashrafi, M. Tur, A.E. Willner, Spatial light structuring using a combination of multiple orthogonal orbital angular momentum beams with complex coefficients, Opt. Lett. 42 (5) (2017) 991–994.

[61] A. D'Errico, R. D'Amelio, B. Piccirillo, F. Cardano, L. Marrucci, Measuring the complex orbital angular momentum spectrum and spatial mode decomposition of structured light beams, Optica 4 (11) (2017) 1350–1357.

[62] J. Du, R. Zhang, Z. Zhao, G. Xie, L. Li, H. Song, K. Pang, C. Liu, H. Song, A. Almaiman, B. Lynn, M. Tur, A.E. Willner, Single-pixel identification of 2-dimensional objects by using complex Laguerre–Gaussian spectrum containing both azimuthal and radial modal indices, Opt. Commun. 481 (2021) 126557.

[63] P. Zhao, S. Li, X. Feng, K. Cui, F. Liu, W. Zhang, Y. Huang, Measuring the complex orbital angular momentum spectrum of light with a mode-matching method, Opt. Lett. 42 (6) (2017) 1080–1083.

[64] F. Tang, X. Lu, L. Chen, The transmission of structured light fields in uniaxial crystals employing the Laguerre-Gaussian mode spectrum, Opt. Express 27 (20) (2019) 28204–28213.

[65] H. Song, H. Song, R. Zhang, K. Manukyan, L. Li, Z. Zhao, K. Pang, C. Liu, A. Almaiman, R. Bock, B. Lynn, M. Tur, A.E. Willner, Experimental mitigation of atmospheric turbulence effect using pre-signal combining for uni- and bi-directional free-space optical links with two 100-Gbit/s OAM-multiplexed channels, J. Lightwave Technol. 38 (1) (2020) 82–89.

[66] X. Zhong, Y. Zhao, G. Ren, S. He, Z. Wu, Influence of finite apertures on orthogonality and completeness of Laguerre-Gaussian beams, IEEE Access 6 (2018) 8742–8754.

[67] Z. Mei, D. Zhao, The generalized beam propagation factor of truncated standard and elegant Laguerre–Gaussian beams, J. Opt. A, Pure Appl. Opt. 6 (11) (2004) 1005.

[68] L. Li, G. Xie, Y. Yan, Y. Ren, P. Liao, Z. Zhao, N. Ahmed, Z. Wang, C. Bao, A.J. Willner, S. Ashrafi, M. Tur, A.E. Willner, Power loss mitigation of orbital-angular-momentum-multiplexed free-space optical links using nonzero radial index Laguerre–Gaussian beams, J. Opt. Soc. Am. B 34 (1) (2017) 1–6.

[69] G. Lebrun, J. Gao, M. Faulkner, MIMO transmission over a time varying channel using SVD, IEEE Trans. Wirel. Commun. 4 (2) (2005) 757–764.

[70] K. Pang, H. Song, X. Su, K. Zou, Z. Zhao, H. Song, A. Almaiman, R. Zhang, C. Liu, N. Hu, S. Zach, N. Cohen, B. Lynn, A.F. Molisch, R.W. Boyd, M. Tur, A.E. Willner, Experimental mitigation of the effects of the limited size aperture or misalignment by singular-value-decomposition-based beam orthogonalization in a free-space optical link using Laguerre–Gaussian modes, Opt. Lett. 45 (22) (2020) 6310–6313.

[71] D. McGloin, K. Dholakia, Bessel beams: diffraction in a new light, Contemp. Phys. 46 (1) (2005) 15–28.

[72] S. Li, J. Wang, Adaptive free-space optical communications through turbulence using self-healing Bessel beams, Sci. Rep. 7 (1) (2017) 1–8.

[73] N. Mphuthi, L. Gailele, I. Litvin, A. Dudley, R. Botha, A. Forbes, Free-space optical communication link with shape-invariant orbital angular momentum Bessel beams, Appl. Opt. 58 (16) (2019) 4258–4264.

[74] A. Dudley, M. Lavery, M. Padgett, A. Forbes, Unraveling Bessel beams, Opt. Photonics News 24 (6) (2013) 22–29.

[75] P.J. Winzer, G.J. Foschini, MIMO capacities and outage probabilities in spatially multiplexed optical transport systems, Opt. Express 19 (17) (2011) 16680–16696.

[76] B. Yousif, E.E. Elsayed, Performance enhancement of an orbital-angular-momentum-multiplexed free-space optical link under atmospheric turbulence effects using spatial-mode multiplexing and hybrid diversity based on adaptive MIMO equalization, IEEE Access 7 (2019) 84401–84412.

[77] H. Huang, Y. Cao, G. Xie, Y. Ren, Y. Yan, C. Bao, N. Ahmed, M.A. Neifeld, S.J. Dolinar, A.E. Willner, Crosstalk mitigation in a free-space orbital angular momentum multiplexed communication link using 4×4 MIMO equalization, Opt. Lett. 39 (15) (2014) 4360–4363.

[78] A.K. Majumdar, Free-space optical (FSO) platforms: unmanned aerial vehicle (UAV) and mobile, in: Advanced Free Space Optics, Springer, 2015, pp. 203–225.

[79] K.E. Zarganis, A. Hatziefremidis, Performance analysis of coherent optical OFDM applied to UAV mobile FSO systems, J. Opt. Photonics 3 (1) (2015) 5–12.

[80] A. Kaadan, H.H. Refai, P.G. LoPresti, Multielement fso transceivers alignment for inter-uav communications, J. Lightwave Technol. 32 (24) (2014) 4785–4795.
[81] L. Li, R. Zhang, Z. Zhao, G. Xie, P. Liao, K. Pang, H. Song, C. Liu, Y. Ren, G. Labroille, P. Jian, D. Starodubov, B. Lynn, R. Bock, M. Tur, A.E. Willner, High-capacity free-space optical communications between a ground transmitter and a ground receiver via a UAV using multiplexing of multiple orbital-angular-momentum beams, Sci. Rep. 7 (1) (2017) 1–12.
[82] L. Li, Runzhou Zhang, Peicheng Liao, Yinwen Cao, Haoqian Song, Yifan Zhao, Jing Du, Zhe Zhao, Cong Liu, Kai Pang, Hao Song, Ahmed Almaiman, Dmitry Starodubov, Brittany Lynn, Robert Bock, Moshe Tur, Andreas F. Molisch, Alan E. Willner, Mitigation for turbulence effects in a 40-Gbit/s orbital-angular-momentum-multiplexed free-space optical link between a ground station and a retro-reflecting UAV using MIMO equalization, Opt. Lett. 44 (21) (2019) 5181–5184.

Index

A

Accumulative turbulence
    effect, 241
    range, 250
Adaptive optics (AO)
    approach, 266, 268–270, 280
    system, 164
    turbulence compensation, 271
    turbulence mitigation, 266
Advanced
    detection methods, 217
    protocols, 233
Ancillary photon, 150, 151, 229, 230
Angle
    azimuthal, 8, 22, 64, 70, 128, 178
    relative, 165
Atmospheric turbulence, 95, 162, 165, 226, 238, 239, 241, 243–245, 262, 263, 266, 268
Authenticated classical channel, 152
Azimuthal
    angle, 8, 22, 64, 70, 128, 148, 178
    phase, 39, 44
    phase structure, 79
    phase variation, 206
    polarizer, 99, 103
    quantum number, 209

B

Beam
    amplitude, 58
    axis, 44, 82, 84, 87, 129, 165, 197, 198, 206, 246, 247
    combinations, 81
    distorted, 264, 274
    dynamics, 172
    focus, 81, 86
    frequency, 69
    gaussian, 42, 78, 96, 100, 101, 146, 272
    generation, 276
    intensity, 271
    parameters, 45, 125
    path, 232, 263, 265, 266, 279
    photons, 188
    pointing, 165
    polarization, 148
    profile, 41, 266, 275, 280
    propagation, 169, 252
    properties, 48
    radius, 46, 49, 55, 167
    size, 255, 264
    spread, 164
    structure, 227
    waist, 44, 45, 49, 51, 52, 61, 74, 85, 120, 170, 221, 222, 252
    wander, 164, 165, 238, 247, 248
    wandering, 166, 167, 170
    wavelength, 89
    width, 44
Beam splitter (BS), 108, 150
Bell states, 10, 22, 24–26, 229
Bessel beams, 39, 46, 78, 87, 96, 101, 102, 275
Bessel beams Poincaré modes, 100
Bessel vortex beams, 46
Bessel-Gaussian (BG) beams, 279, 280
Bosonic commutation rules, 110, 124, 125, 129, 131
Bosonic ladder operators, 133, 135

C

Channel
    attenuation, 168, 169
    capacity, 209, 210
    crosstalk, 238, 246, 247, 263, 266, 274, 280, 282
    degradation, 254
    depth, 169
    lengths, 168, 172, 256
    loss, 255
    matrix, 282
    quantum, 20–22, 27, 152, 158, 159, 210, 217, 219, 220, 223–227, 270, 271
    transmitted, 276
Chlorophyll concentration, 169, 170
Circular
    polarization, 5, 6, 15, 39, 42, 57, 61, 62, 78, 80, 140, 148, 165, 178, 182, 190
    handedness, 180, 183
    states, 80, 86, 148
    vectors, 68, 123
    stationary states, 126
    transverse polarization, 51
Circularly polarized
    beams, 81
    light, 5, 42, 72, 188
    photons, 80, 186
Collimated laser beams, 168

Index

Communication
    channel, 143, 195, 198, 269
    channel capacity, 142
    security, 142
    underwater, 254
Commutation rules, 111, 113, 114, 118, 124, 130–132
Compensation phase, 276
Computer generated holograms, 145, 215
Conical diffraction, 188
Conventional polarization, 78
Crosstalk
    channel, 238, 246, 247, 263, 256, 274, 280, 282
    matrix, 274
    mitigation, 262, 266, 275, 280, 231, 284
    mitigation techniques, 284
    modal, 262, 264
    OAM mode, 256
Curvature phase, 46, 52, 55
Cylindrical vector beams, 149

D

Data channels, 261, 263, 274, 277, 279
Degenerate down-conversion (DDC), 84
Degrees of freedom (DOF), 8, 11, 20, 38, 71, 73, 74, 77–79, 82, 108, 135, 142, 143, 155, 167, 171, 188, 206, 211, 221
Degrees photonic, 3, 142, 143, 162, 168
Detection
    algorithm, 101
    efficiency, 147
    method, 100, 215, 217
    photonic, 147
    processes, 81
    schemes, 160
    stage, 161
    system, 172, 217
    technique, 226
Difference frequency generation (DFG), 231

Diffraction
    effects, 96, 181
    efficiency, 147
    length, 120
    limit, 240
    pattern, 145
Digital Micromirror Device (DMD), 147
Digital signal processing (DSP), 281
Distorted
    beam, 264, 274
    beam wavefront, 273
    OAM beams, 266–268, 275
    wavefront, 164, 273
Doughnut
    beams, 51, 52, 54, 55, 58, 74
    modes, 51, 64

E

Eavesdropper, 20, 22, 140, 151, 152, 173, 209, 227
Eigenstates
    HO, 121
    OAM, 15, 178, 179, 188
    TAM, 179, 198
Electric field, 5, 6, 41, 48, 49, 52, 64, 82, 83, 181, 190, 245
Electric field vector, 5, 40, 41, 46, 47, 51, 57, 62, 65, 66, 68
Electric polarization, 117
Electromagnetic
    field, 5, 15, 80–82, 108, 113, 122, 134, 177, 178, 180, 190
    frequency, 100
Electron vortex beams, 46
Electronic states, 86, 88
Elliptical polarization, 6, 129, 130, 193
Emulated turbulence, 264, 270–272, 283
Encoded qubits, 206
Encoding
    information, 95
    polarization, 15, 29
    processes, 22
    quantum states, 216
    qubits, 206

qudits, 206
sequence, 271
states, 161
Entangled
    particles, 10
    photon, 10, 22, 25, 144, 222, 228
    photon pairs, 223, 228
    quantum states, 31
    qubit, 26, 27
    states, 9, 24, 29, 89, 108, 136, 211, 213, 221, 228, 229, 232
    structured photons, 226
Entanglement
    degree, 84
    distribution, 29, 30, 162, 167, 219, 227
    OAM, 226
    quantum, 5, 9, 10, 22, 23, 88, 89
    source, 22, 208, 219
    swapping, 219, 228–230
Error vector magnitude (EVM), 281, 283

F

Fiber
    axis, 190
    bending, 195
    channels, 173
    design, 193
    infrastructure, 224
    length, 224
    modes, 190, 191, 195
    network, 224
    optics, 190
    quantum channels, 224
    transmission, 224
Focal plane, 39, 41, 44, 46, 49, 54, 55, 57–59, 61, 62, 69, 120, 186
Forward channels, 268, 277
Forward error correction (FEC) limit, 269, 273, 274, 282
Free quantum particle, 115
Free space, 3, 27, 29, 47, 108, 110, 188
Free space propagation, 250

Index

Free-space optical (FSO)
    channel, 246
    communication, 257
    systems, 238
Frequency
    beam, 69
    conversion, 231
    difference, 61
    electromagnetic, 100
    rotation, 100, 101
    shift, 60
Fried parameter, 164, 166, 272, 281

G

Gaussian
    beam, 42, 78, 96, 100, 101, 146, 272
    beam profile, 215
    paraxial modes, 124
    probe beam, 266, 269, 271
Gouy phase, 44, 46, 50, 52, 55, 104, 120, 121
Greenberger-Horne-Zeilinger (GHZ) states, 212, 214
Ground state, 124

H

Half-wave plate (HWP), 144, 145
Harmonic oscillator (HO), 120
    eigenstates, 121
    states, 122, 125, 126
Helical
    phase, 178, 217
    phase front, 263
    wavefront, 141
Helicoidal wavefront, 79
Helmholtz equation, 40, 41, 46, 48, 49, 79, 190, 245
Heralded
    photon source, 167
    photons, 89
Hermite-Gauss (HG) modes, 108, 109, 120, 121, 124–126, 129, 141
Heterodyne detection, 281
Hologram
    phase, 145, 146, 155
    surface, 147

I

Idler photon, 167
Impart rotation, 100
Information encoding, 269
Information encoding quantum, 209
Interfering beams, 57, 70
Intermodal crosstalk, 263, 265, 270, 280

K

Kolmogorov turbulence, 240, 250, 251

L

Ladder operators, 124–126, 132, 133, 135
Laguerre-Gaussian (LG)
    beams, 39, 45, 46, 49, 54, 57, 58, 61, 70, 72, 74, 85, 275, 277
    beams superpositions focal plane, 128
    laser modes, 246
    light beams, 44, 48, 74
    modes, 44, 45, 50, 73, 79, 84, 108, 109, 123, 126–129, 143, 146, 246, 247, 275, 277, 278
    optical vortex, 74
    set, 43, 44
Laser
    beam, 38, 41, 42, 45, 104, 226, 228
    communications experiments, 29
Linear
    combinations, 116, 122, 126, 129, 130, 191, 192
    polarization, 5, 6, 51, 57, 62, 64, 66, 70, 72, 128, 165, 198
    polarization states, 149
    polarization vectors, 40, 128
    polarizer, 99, 100, 103, 104
    superpositions, 122
Linearly polarized (LP)
    beams, 45
    mode, 160, 161

M

Mathieu beams, 47
Maximally entangled
    photon qubits, 26
    qubit, 26
    state, 10, 21, 23, 24, 26, 89, 221, 222
Modal
    combinations, 104
    crosstalk, 262, 264
    superpositions, 253
Mode division multiplexing (MDM), 261
    channels, 279
    link, 261, 273, 277, 280
Momentum quantum operator, 123
Multi-mode fiber (MMF), 160
Multi-plane light conversion (MPLC), 266, 273, 274
    components, 275
    module, 273
    technique, 217
Multicore fibers, 193
Multimode fibers, 225, 226
Multiple qubits, 196
Multiple-input-multiple-output (MIMO), 281
    equalization, 281–284
    technology, 195
Mutually orthogonal polarizations, 70, 71
Mutually unbiased base (MUB), 152, 154, 157, 212

O

Orbital angular momentum (OAM)
    beams, 224, 260, 261, 263–268, 272, 279, 281, 282
    crosstalk mitigation, 275
    helical phasefront, 266
    phasefront, 263
    wavefront, 273
    channels, 274, 283
    detection, 83
    eigenstates, 15, 178, 179, 188
    encoding, 15
    entangled photon, 219
    entanglement, 226

Index

mode, 15, 86, 195, 206, 248, 254, 264, 266, 269, 275, 276, 279, 281
  combination, 275
  converter, 269
  crosstalk, 256
  spacing, 269
  superposition, 246, 253
orders, 261, 263, 264, 270, 271
photon, 87, 269
property, 38, 39, 71, 74
quantum states, 38
spectrum, 147, 221, 247
states, 47, 71, 148, 154, 155, 158, 165, 171, 185, 196, 223, 269, 270

P

Pancharatnam–Berry Optical Element (PBOE), 147, 148
Paraxial
  beams 51, 81, 123, 125
  light beams, 110, 123
  limit, 109, 111, 115–118, 120, 122, 123, 134, 135
  modes 108, 114–117, 119, 121, 122, 125, 126, 129, 131, 134, 135
  propagation, 136
  quantum fields, 114
  regime, 39, 40, 43, 49, 50, 74, 142
  wave equation, 114–116, 120, 121, 125, 135, 206
Path encoding, 15, 17, 18
Pattern rotation, 69
Phase
  aberrations, 246, 249
  azimuthal, 39, 44
  correction, 271, 273, 275
  delays, 241
  difference, 58, 60, 96, 97, 101
  distortion, 241, 250, 266, 274
  error, 255
  factions, 65
  factors, 178, 185, 198, 265
  flattening, 147
  fluctuations, 238, 257
  function, 40, 43, 44, 46, 50, 52, 55, 57, 58
  gradient, 217
  hologram, 145, 146, 155
  information, 146
  matching, 221, 231
  modulations, 212, 216–218, 250
  patterns, 274
  profiles, 164, 260, 279
  relations, 222
  relative, 6, 17, 98, 99, 155
  retardation, 181
  screen, 242, 243, 250, 253, 255, 263
  shift, 154, 260
  structure, 185, 241, 248, 266
  structure functions, 241
  variations, 102, 241
  vortex, 44, 172
Phasefront, 263, 267
Photon
  annihilation operator, 83
  creation, 84, 109, 134
  flux, 82
  level, 78
  measurements, 89
  number, 140, 142, 143
  number operator, 80
  number states, 108, 110
  OAM, 87, 269
  pairs, 22, 88, 136, 216, 221, 222, 230
  polarization, 3, 8, 170
  polarization states, 3, 7
  probability, 141
  SAM, 186
  sources, 228
  state, 21, 24, 30, 113, 141, 142, 171, 195, 198
  TAM, 186
  travel, 136
  wavefunction, 83
  wavelength, 219
Photonic
  degrees, 3, 142, 143, 162, 168
  detection, 147
  gear, 198, 199
  polarization states, 145
  quantum states, 143, 223
  qubit, 3, 31
  qudit, 143, 196
  setup, 198
  state, 15, 198
Plane wave propagation, 44, 50
Poincaré
  beams, 96, 102
  modes, 95, 96, 100, 103, 104
  sphere, 78, 79, 96, 128, 129, 144, 145, 148
  state, 99
Polarization
  arrangements, 39, 57, 70
  basis, 20, 190
  beam, 148
  beam splitters, 18
  channels, 170
  component, 104
  degree, 19, 142, 215, 220, 228
  direction, 110
  distribution, 69, 70
  DOF, 207
  ellipse, 96, 98
  ellipticity, 178
  encoded photons, 172, 223
  encoding, 15, 29
  gradients, 39, 68
  handedness, 182–186
  holograms, 183
  index, 72
  linear, 5, 6, 51, 57, 62, 64, 66, 70, 72, 128, 165, 198
  manipulation, 148
  optics, 142
  orientation, 102
  pattern, 71, 97, 193
  photon, 3, 8, 170
  Poincaré sphere, 135
  projections, 104
  quantum state, 145
  qubits, 15, 142, 145, 152
  radial, 101
  singularities, 183
  state, 8, 78, 80, 89, 96, 97, 129, 140, 141, 144, 145, 147, 148, 178, 196
  state entanglement
    distribution, 171

Index

state tomography, 145
structure, 79
superpositions, 95
transformation, 182
vector, 6, 39, 41, 57, 62, 63, 83, 110, 117, 128, 156, 190
Polarizer
    angles, 99, 100
    azimuthal, 99, 103
    linear, 99, 100, 103, 104
    set, 145
Polarizing beamsplitter (PBS), 144
Probe beam, 267, 270, 274
Probe beam gaussian, 266, 269, 271
Propagation
    beam, 169, 252
    beam axes, 247
    coordinate, 109, 116, 120, 133
    direction, 5, 78–80, 96, 108, 114, 117, 122, 123, 140–142, 180, 188, 280
    distance, 257
    dynamics, 104, 254
    effects, 250
    length, 250, 252, 256
    paraxial, 136
    path, 172
    step, 255
Protocols
    quantum, 162, 208, 233
    quantum communication, 5, 15, 31, 157, 206, 210, 223
Pump beam, 221, 222
    profile, 231
    shaping, 221
Pump photon, 221, 222

Q

Quantum
    aspects, 78, 90
    attack, 22
    bits, 7, 8
    channel, 20–22, 27, 152, 158, 159, 210, 217, 219, 220, 223–227, 270, 271
    cloning, 5, 13, 14, 143, 149–151, 158

cloning machine, 150, 151
communication
    channels, 38, 71, 151, 157
    for quantum computation, 14
    in free space, 29
    information, 15
    infrastructure, 164
    link, 223, 269, 270
    network, 14
    protocols, 5, 15, 19, 31, 157, 206, 210, 223
    satellite, 27
    schemes, 163, 165, 209, 216
    security, 158
    tasks, 3
    technology, 31
computation, 14, 17
contextuality, 152
correlations, 11, 84, 225
counterpart, 8
cryptographic
    schemes, 196
    tasks, 218
cryptography, 14, 20, 89, 149, 152, 208
cryptography protocols, 19
description, 108, 136
devices, 30
domain, 224
dot, 3
electrodynamics, 80
enhancement, 199
entanglement, 5, 9, 10, 22, 23, 88, 89
experiments, 28
fiber networks, 162
field, 109, 119
field operator, 109, 110, 117
field paraxial limit, 115
field theory, 77, 112
formulation, 82, 85
gate, 218
information, 3, 8, 14, 71, 84, 108, 150, 151, 157, 173, 196, 206, 207, 215, 271
information encoding, 209

information sciences, 149, 206
interface, 231
interference, 88, 230
issues, 83, 87
key, 2, 20, 152, 208, 209
language, 187
laws, 2
mechanical
    effects, 31
    structure, 86
    superposition, 79
    wavefunction, 83
mechanics, 7, 9, 10, 14, 23, 27, 90, 113, 122, 130, 135, 140, 149, 152, 155, 209
memories, 219, 228, 230
nature, 10, 29, 134
network, 206, 219, 223, 224, 230, 233
nonlocality, 10
numbers, 119, 122, 126, 130, 131, 133, 190, 209, 218
objects, 108
operations, 218
operators, 87, 108–113, 122–124, 134, 135
optics, 90, 132, 147, 206, 207, 219
optics community, 141
phenomena, 7, 10
physics, 2, 8, 9, 20, 108, 177, 208, 209
process, 156, 157
process tomography, 155–158, 227, 228
properties, 230
protocols, 162, 208, 233
regime, 162
router, 231, 232
schemes, 11
spin, 72
state space, 233
storage, 230, 231
system, 8–12, 14, 20, 119, 155, 156, 206, 212, 217
technology, 30, 233
teleportation, 10, 19, 22, 25, 26, 30, 228
teleportation protocol, 22–24

Index

theory, 77, 79, 108
transmission, 198
uncertainty, 2, 85, 87
world, 149
Quantum bit error rate (QBER), 152
Quantum key distribution (QKD), 2, 20, 152, 159, 161, 208–210, 212, 213
Quantum key distribution (QKD) protocols, 20, 22, 152, 157, 173, 209, 210
Quantum repeater (QR), 29, 208, 219, 223, 228, 230
Quantum secret sharing (QSS), 208, 211
Quantum state tomography (QST), 154–157, 217
Quarter-wave plate (QWP), 144
Qubits
  encoding, 206
  polarization, 15, 142, 145, 152
  representation, 9
  robustness, 198
  states, 9, 10, 13, 18, 19, 29, 142
Qutrit states, 211

R

Radial
  coordinate, 97, 108
  dependence, 99
  direction, 98
  distance, 61, 252, 253
  extent, 47
  index, 102, 108
  lines, 98, 99, 103
  mode, 126, 128, 129, 133, 135, 137
  number, 44–46, 73, 74
  polarization, 101
  polarizers, 104
  positions, 55, 63, 70
  quantum number, 209
  separation, 61
  spokes, 99
Random phase
  aberrations, 253
  screens, 241, 250

Receiver aperture, 246, 248, 249, 262, 280
Reciprocal phase distortion, 249
Relative
  angle, 165
  phase, 6, 17, 98, 99, 155
  rotation, 198
Rotatable phase screen, 281
Rotation
  dependence, 198
  frequency, 100, 101
  invariance, 190
  rate, 99, 100, 104
  relative, 198
Rotational
  sensitivity, 199
  symmetry, 186, 188, 191
Round robin differential phase shift (RRDPS) protocol, 154, 155
Route mean square (RMS), 241
Rytov-Vladimirskii-Berry (RVB)
  geometric phase, 188
  phase, 180
  transverse phase-gradient, 189

S

Secret key, 21, 30, 89, 140, 152–155, 159, 167
Secure quantum communication, 20, 168
Security
  analysis tools, 22
  for qubit states, 211
  in quantum channels, 159
  loopholes, 22
  quantum communication, 158
Shifted beam, 61
Sifted photon, 154, 167
Signal photon, 167, 230, 231
Space-division multiplexing (SDM), 159, 160
Spatial light modulator (SLM), 45, 104, 146, 147, 173, 194
Spin
  chirality, 54
  component, 81, 86, 180
  coupling, 87
  density, 72
  direction, 129

Hall effect, 187
  quantum, 72
  rotation matrices, 130
  states, 3
Spin angular momentum (SAM), 38–40, 71, 80, 178, 179
Spin-Hall effect of light (SHEL), 187–189
  phases, 191
  phenomena, 188
Spontaneous parametric down-conversion (SPDC), 147, 219
Spontaneous parametric down-conversion (SPDC) process, 219–223
State
  distinguishability, 172
  evolution, 148
  fidelity, 159, 230
  maximally entangled, 10, 21, 23, 24, 26, 221, 222
  photon, 21, 24, 30, 113, 141, 142, 171, 195, 198
  photonic, 15, 198
  Poincaré, 99
  polarization, 8, 78, 80, 89, 96, 97, 129, 140, 141, 144, 145, 147, 148, 178, 196
  space, 108, 129, 157, 206, 210, 232
  transmitted, 226
  vectors, 9
Stationary states, 121, 122, 124–126, 129
Sum frequency generation (SFG), 231
Summed orbital angular momentum, 88
Superposition
  approach, 248
  principle, 7, 9
  states, 14, 19, 142, 154, 172, 198, 210, 225, 227
Symmetric informationally complete positive operator-valued measure (SIC-POVM), 154

296

Index

T
Target beam, 167
Tomographic protocols, 154
Topological charge, 78–80, 84, 86, 87, 96, 97, 99, 101, 102, 148, 198
Topological charge superposition, 101
Total angular momentum (TAM)
 eigenstates, 179, 198
 photon, 186
 photonic states, 197
 states, 197
Transmitted
 beam, 263, 265, 276–278
 beam radius, 166
 channel, 276
 OAM mode, 264
 quantum signal, 227
 state, 226
Turbid channel, 254
Turbulence
 compensation, 275
 conditions, 164
 effects, 263, 264, 277
 emulation, 272
 emulator, 270, 281
 modeling approaches, 253
 phase screens, 250
 realization, 267, 270, 273, 276, 281
 screens, 250
 underwater, 173

Turbulent
 channel, 172
 phase screen, 241, 242, 251, 253
Twisted beam input, 89

U
Underwater
 channels, 143, 169–171, 173, 227
 communications, 254
 quantum channels, 227
 turbulence, 173
Unmanned aerial vehicle (UAV), 282, 283
Unobstructed
 beams, 280
 BG beam, 280

V
Vacuum state, 119
Vector
 beams, 95
 modes, 99, 160, 190, 191, 193
 polarization, 6, 39, 41, 57, 62, 63, 83, 110, 117, 128, 156, 190
 potential, 40, 53, 80, 110, 111, 114, 116, 117, 178
Vertical polarization, 5, 6, 20
Vortex
 beams, 38, 39, 42, 46, 47, 55, 57, 61, 67, 74, 212
 fibers, 160–162, 173

locations, 253
modes, 74, 97, 160, 172
phase, 44, 172
position, 253
splitting, 253
states, 167, 172
structured beams, 83

W
Wavefront
 aberration, 241
 components, 255
 corrector, 271
 distorted, 164, 273
 distortions, 165, 226, 242, 266, 272
 OAM beams, 273
 sensing, 249
 sensor, 266, 273
Wavelength-division multiplexing (WDM), 159
Waveplate angles, 144
Weak turbulence, 163, 226, 264
Weak-guidance approximation (WGA), 190
 fiber mode, 190
 propagation, 190
Winding number, 39, 44–46, 50, 52, 54, 57, 58, 62–64, 68, 70, 73, 74, 80

Z
Zernike polynomials, 164, 241, 242